D1695013

Springer Complexity

Springer Complexity is a publication program, cutting across all traditional disciplines of sciences as well as engineering, economics, medicine, psychology and computer sciences, which is aimed at researchers, students and practitioners working in the field of complex systems. Complex Systems are systems that comprise many interacting parts with the ability to generate a new quality of macroscopic collective behavior through self-organization, e.g., the spontaneous formation of temporal, spatial or functional structures. This recognition, that the collective behavior of the whole system cannot be simply inferred from the understanding of the behavior of the individual components, has led to various new concepts and sophisticated tools of complexity. The main concepts and tools – with sometimes overlapping contents and methodologies – are the theories of self-organization, complex systems, synergetics, dynamical systems, turbulence, catastrophes, instabilities, nonlinearity, stochastic processes, chaos, neural networks, cellular automata, adaptive systems, and genetic algorithms.

The topics treated within Springer Complexity are as diverse as lasers or fluids in physics, machine cutting phenomena of workpieces or electric circuits with feedback in engineering, growth of crystals or pattern formation in chemistry, morphogenesis in biology, brain function in neurology, behavior of stock exchange rates in economics, or the formation of public opinion in sociology. All these seemingly quite different kinds of structure formation have a number of important features and underlying structures in common. These deep structural similarities can be exploited to transfer analytical methods and understanding from one field to another. The Springer Complexity program therefore seeks to foster cross-fertilization between the disciplines and a dialogue between theoreticians and experimentalists for a deeper understanding of the general structure and behavior of complex systems.

The program consists of individual books, books series such as "Springer Series in Synergetics", "Institute of Nonlinear Science", "Physics of Neural Networks", and "Understanding Complex Systems", as well as various journals.

J. Portugali (Editor)

Complex Artificial Environments

Simulation, Cognition and VR in the Study and Planning of Cities

With 94 Figures

Prof. Juval Portugali, Ph.D.

Department of Geography and the Human Environment
Environmental Simulation Laboratory
Tel Aviv University
Tel Aviv 69978, Israel
e-mail: juval@post.tau.ca.il
http//:www.eslab.tau.ac.il

ISBN-10 3-540-25917-1 Springer Berlin Heidelberg New York
ISBN-13 978-3-540-25917-6 Springer Berlin Heidelberg New York

Library of Congress Control Number: 2005935701

Springer is a part of Springer Science+Business Media
springeronline.com

Printed in Germany

Print data prepared by LE-TeX GbR, Leipzig
Cover design: Erich Kirchner, Heidelberg
Printed on acid-free paper 54/3141/YL - 5 4 3 2 1 0

Contents

Part two: Specific experiences

Part three: Urban simulation models

Part four: Cognition and VR

Part five: Planning

Contributors

Theo A. Arentze
Associate Professor, Eindhoven University of Technology, Design & Decision Support Systems Research Programme (DDSS). PO Box 513, 5600 MB Eindhoven, The Netherlands. Email: t.a.arentze@bwk.tue.nl,

Michael Batty
Professor of Spatial Analysis and Planning, Director of CASA – Centre for Advanced Spatial Analysis, University College London, 1-19 Torrington Place, London, WC1E 6BT, UK. Email: mbatty@geog.ucl.ac.uk

Itzhak Benenson
Senior Lecturer, Department of Geography and the Human Environment, Deputy to the Director of ESLab – Environmental Simulation Laboratory, Tel Aviv University, Tel-Aviv 69978, Israel. Email: bennya@post.tau.ac.il

Slava Birfur
Ph.D student, Department of Geography and the Human Environment. Researcher at ESLab – Environmental Simulation Laboratory, Porter School of Environmental Studies, Tel Aviv University. Email: slavab@tvuna.co.il

Lars Bodum
Associate Professor, Division of Geomatics, Director of Centre for 3D GeoInformation, Aalborg University, Niels Jernes Vej 14, DK-9220 Aalborg, Denmark. Email: lbo@3Dgi.dk

Richard K. Brail
Professor, Urban Planning and Policy Development, Edward J. Bloustein School of Planning and Public Policy, Rutgers University, New Brunswick, New Jersey, USA. Email: rbrail@rci.rutgers.edu

Stephen C. Hirtle
Professor of Information Science and Telecommunications, School of Information Sciences, Pittsburgh, PA 15260, USA. Email: hirtle@pitt.edu

Ran Goldblatt
Graduate student, Department of Geography and the Human Environment. Researcher at ESLab – Environmental Simulation Laboratory, Porter School of Environmental Studies, Tel Aviv University, Tel-Aviv 69978, Israel. Email: ran@eslab.tau.ac.il

Hermann Haken
Professor and Director emeritus, Institute for Theoretical Physics,
Center of Synergetics, University of Stuttgart, 70550 Stuttgart, Germany.
Email: hermann.haken@itp1.uni-stuttgart.de

Bin Jiang
Senior lecturer in GIS and Cartography, Division of Geomatics, Department of Technology and Built Environment, University of Gävle, SE-801 76, Gävle, Sweden. Email: bin.jiang@hig.se

Vlad Kharbash
Ph.D student, Department of Geography and the Human Environment. Researcher at ESLab – Environmental Simulation Laboratory, Porter School of Environmental Studies, Tel Aviv University. Email: vlad@lipman.co.il

Michael Kwartler
Architect, planner and urban designer. President of ESC – Environmental Simulation Center, Ltd. (ESC), 631 West 35th Street, New York, NY 10001 USA.
Email: kwartler@simcenter.org

Sylvie Occelli
Architect. Senior Researcher at IRES - Istituto di Ricerche Economico Sociali del Piemonte, Via Nizza 18, 10125 Turin, Italy. Email: occelli@ires.piemonte.it

Itzhak Omer
Lecturer, Department of Geography and the Human Environment, Deputy to the Director of ESLab – Environmental Simulation Laboratory, Tel Aviv University, Tel-Aviv 69978, Israel. Email: omery@post.tau.ac.il

Juval Portugali
Professor of cognitive geography and urban dynamics. Department of Geography and the Human Environment, Head of ESLab – Environmental Simulation Laboratory, Tel Aviv University, Tel Aviv 69978, Israel. Email: juval@post.tau.ac.il

Denise Pumain
Professor, University Paris I Panthéon-Sorbonne, Institut Universitaire de France. Director, Research Group Libergeo (CNRS). University Paris I, UMR Géographie-cités, 13 rue du Four, 75006 Paris, France.
Email: pumain@parisgeo.cnrs.fr

Giovanni A. Rabino
Associate Professor of city and territorial systems, DIAP – Dipartimento di Architettura e Pianificazione, Polytechnic of Milan, Piazza Leonardo da Vinci 32, 20133, Milan, Italy.
Email: rabino@polimi.it

Asaf Ros
Graduate student, Department of Geography and the Human Environment. Researcher at ESLab – Environmental Simulation Laboratory, Porter School of Environmental Studies, Tel Aviv University. Email: asaf@eslab.tau.ac.il

Lena Sanders
Senior researcher, director of UMR Géographie-cités, 13 rue du Four, 75006 Paris, France. Email: lena.sanders@parisgeo.cnrs.fr

Ferdinando Semboloni
Researcher; currently Professor of Analysis of urban and regional systems. Department of Town and Regional Planning, and Center for the Study of Complex Dynamics, University of Florence, Via Micheli, 2 50121, Firenze, Italy.
Email: semboloni@urba.arch.unifi.it

Philip J. Steadman
Professor of Urban and Built Forms Studies, Bartlett Graduate School, University College London, 1-19 Torrington Place, London WC1E 6BT, UK. Email: j.p.steadman@ucl.ac.uk

Karin Talmor
Graduate student, Faculty of Architecture and Town Planning, Technion, Haifa. Researcher at ESLab – Environmental Simulation Laboratory, Porter School of Environmental Studies, Tel Aviv University. Email: karint@eslab.tau.ac.il

Harry Timmermans
Professor of Urban Planning, Eindhoven University of Technology, Director of DDSS – Design & Decision Support Systems Research Programme. PO Box 513, 5600 MB Eindhoven, The Netherlands. Email: h.j.p.timmermans@bwk.tue.nl

Paul. M. Torrens
Assistant Professor, Department of Geography, Arizona State University, PO Box 870104, Tempe, AZ 85287, USA. Email: torrens@geosimulation.com

Roger White
Professor, Department of Geography, Memorial University of Newfoundland, St. John's,
Newfoundland A1B 3X9, Canada. Email: roger@mun.ca

Yichun Xie
Professor of GIS, Planning and Geography, Director of the Institute for Geospatial Research and Education, Eastern Michigan University, 125 King Hall, Ypsilanti, Michigan 48197, USA. Email: yxie@emich.edu

Acknowledgments

This book is the outcome of a three-day international workshop on the study of complex artificial environments that took place on the island of San Servolo, Venice, during April 1-3, 2004. Following the workshop, the participants elaborated on their contributions and extended them into full-scale papers.

The publication of this book owes a great deal to many people: to the contributors, for their stimulating presentations and discussions during the international workshop and for transforming their presentations into the illuminating papers that make the content of this book; to the staff at Venice International University at San Servolo, specifically Mrs. Francesca P. Nisii, for their help during the workshop; to Ran Goldblatt and Michael Winograd from ESLab, for spending long hours on the editorial work of this book; to Karin Talmor and Rivka Fabrikant, who worked hard on the organization of the workshop; to my colleagues Itzhak Benenson and Itzhak Omer and the people of the ESLab, for contributing to the success of the workshop. Finally, I would like to express special thanks to Professor Hermann Haken – his initiative and support were central to the book's publication.

This book and the Venice workshop, upon which it is based, mark the climax of the first four years of ESLab's operation – the Environmental Simulation laboratory that was founded by the Charles. H. Revson Foundation in 2001 at Tel Aviv University, in association with the Porter School of Environmental Studies. I'm grateful to the Charles. H. Revson Foundation, in particular to Mr. Eli Evans, its former president, for his enthusiasm and vision, and to the current president, Mrs. Lisa Goldberg, for her support.

Introduction

Juval Portugali

The notion of *complex artificial environments* (CAE) refers to theories of complexity and self-organization, as well as to artifacts in general, and to artificial environments, such as cities, in particular. The link between the two, however, is not trivial. For one thing, the theories of complexity and self-organization originated in the "hard" science and by reference to natural phenomena in physics and biology. The study of artifacts, per contra, has traditionally been the business of the "soft" disciplines in the humanities and social sciences. The notion of "complex artificial environments" thus implies the supposition that the theories of complexity and self-organization, together with the mathematical formalisms and methodologies developed for their study, apply beyond the domain of nature. Such a supposition raises a whole set of questions relating to the nature of 21st century cities and urbanism, to philosophical issues regarding the natural versus the artificial, to the methodological legitimacy of interdisciplinary transfer of theories and methodologies and to the implications that entail the use of sophisticated, state-of-the-art artifacts such as virtual reality (VR) cities and environments.

The three-day workshop on the study of complex artificial environments that took place on the island of San Servolo, Venice, during April 1-3, 2004, was a gathering of scholars engaged in the study of the various aspects of CAE. The aim was to share experiences and to discuss both the issues noted above, as well as the more specific questions concerning the detailed structure of the models used in the study of cities, their association with the knowledge about human behavior in natural, artificial and virtual environments, as gained in cognitive sciences, and finally their use as planning tools. Following the workshop in Venice, the participants elaborated on their contributions and extended them into full-scale papers. This book is the outcome.

The book is divided into five parts. The first deals with general aspects of CAE; the second, with specific experiences of laboratories that in the last decade or so have specialized in CAE; the third part focuses on cellular automata and agent base models, which are currently the main approaches to urban simulation models; the fourth deals with cognition related to real, electronic and virtual environments, that is, cognitive aspects related to the various urban simulation models, while the fifth part concentrates on planning.

Part I: General aspects of complex artificial environments

The discussion in this section commences with an overview by the editor of this book of the scope of complex artificial environments (CAE). The paper identifies three major issues that signify CAE of the 21st century: a social change, the essence of which is that cities and urbanism are growing in significance, to the extent that one can speak of a *second urban revolution*; a theoretical change, viewing cities as complex, self-organizing, artificial environments and examining the behavior of urban agents from the first principles of humans' cognitive capabilities; and finally, there is a methodological change that is revealed in the use of a three-part general purpose support system – GPSS (Portugali, this volume) as the main tool for studying cities as self-organizing complex environments. A typical GPSS is composed of agent base (AB) and cellular automata (CA) simulation models, standard GIS and advanced virtual reality simulators of the kind described below in the papers of Bodum, Portugali and Kwartler.

The notion of CAE entails a question of definition: What are "artificial environments" and how can they be distinguished from their antonym notion "natural environments"? The second paper by Hermann Haken treats this issue by looking at it from the perspective of pattern recognition by humans and computers. From the point of view of pattern recognition, suggests Haken, artificial environments appear to be far more regular compared to natural environments. This difference shows up very clearly when processes of pattern recognition are implemented by the so-called synergetic computer. The paper goes on to discuss the significance of semantics in pattern recognition tasks associated with artifacts. It examines languages as artificial semantic networks and invokes them as a paradigm for a study on the semantics of cities. This is done by reference to the basic concepts of synergetics, such as "order parameter" and the "slaving principle."

The most prominent manifestation of humans' pattern recognition capabilities concerns *visualization*, and each of the three-part GPSS noted above is, to a large extent, a visualization device: AB/CA simulation models allow one to define the rules of the urban game and then to observe the resultant city as it evolves; GIS gives the user the option to map and observe different data configurations, while VR models create virtual environments, such as cities, within which one can virtually walk, drive, fly and see in real time. In the third paper, Michael Batty, Philip Steadman and Yichun Xie deal with the emerging role of visualization in the study of complex artificial environments. Visualization, they suggest, revolutionizes our notion of doing science: The twin goals of parsimony and verifiability that have dominated scientific theory since the Enlightenment give way to theories and models that "look right," often irrespective of what the statistics and causal logics tell. They then define three forms and purposes of visualization, and illustrate them by reference to specific models: *pedagogic visualization,* which makes things explicable, is illustrated by reference to the classical von Thünen model of land rent and density; *exploratory visualization,* which enables one to explore unanticipated outcomes is illustrated by the DUEM urban development model and *predictive visualization, which* enables end users such as planners to engage in using the models for prediction, prescription and control is illustrated by reference to a pedestrian model of human movement.

Complexity theories originated mainly in the natural sciences. Employing and applying them to the study and modeling of cities thus implies creating a bridge between the natural and social sciences. In her paper, Denise Pumain discusses this issue and suggests that this bridging entails an opportunity and a challenge for urban modelers. Borrowing concepts and tools from formalized disciplines may help to build more satisfying expressions of urban theories that will lead to a better understanding of the abstract processes behind urban dynamics. However, the specific features of social systems, as well as the urban knowledge accumulated in past studies, should not be neglected nor underestimated.

Part II: Specific experiences

Part II of the book deals with specific experiences of the laboratories that have been specializing in CAE in recent years. Two specific experiences are discussed. The first, by Lars Bodum, tells the story of the Centre for 3D GeoInformation established at Aalborg University in 2001. The paper focuses on the GRIFINOR, which is a general object-oriented system for real-time 3D visualization of geographically based virtual environments, developed by the Aalborg's center. The author describes this system in its present state, which allows for the generation of static physical elements, such as buildings, and also depicts the planned phase that will allow visualizing traditional geoinformation, such as socio-economic attribute values, on "top" of the virtual environment.

The second paper of Part II tells the story of ESLab – the environmental simulation laboratory established at Tel Aviv University in January 2001. The paper shows how the reality of 21st century urbanism, together with the somewhat abstract notions associated with complexity theory and its methodological tools and technologies, take a concrete form in the structure and activities of ESLab. The paper illustrates this by describing in some detail the ESLab's GPSS and its use as a planning support system (PSS), community support system (CSS) and research support system (RSS).

Part III: Urban simulation models

The discussion on urban simulation models starts with two interrelated papers. The first, by Paul Torrens, attempts to clarify the confusion associated with the current use of cellular automata (CA) and multi-agent systems (MAS). The author clarifies the similarities, and specifically the differences, between CA and MAS models and shows how confusing between the two may have important implications on the use of the models as applied tools. The paper then introduces a geographic automata system (GAS), which provides a spatial framework for urban simulation with automata tools.

The next paper, by Itzhak Benenson, Slava Birfur and Vlad Kharbash, further elaborates the GAS paradigm in relation to the latest version of object-base environment for urban simulation (OBEUS). The essence of this elaboration concerns mainly the treatment of time in models describing collectives of multiple interacting autonomous urban objects. The authors demonstrate how GAS and OBEUS can serve as a universal, transferable framework for object-based urban simulation.

While the focus of the previous AB models was on the movement of urban agents into, inside and out of the city, Ferdinando Semboloni's multi-agent urban simulation model, CityDev, looks at the way the internal economic interrelations between the agents, goods and markets affects the development of a city. Each agent (family, industrial firm, developer, etc.) produces goods by using other goods and trades the goods on the city's markets. When agents produce goods and interact in the markets, the urban fabric is built and transformed. Two innovative features of the model are the fact that it runs on a 3D spatial pattern organized in cubic cells and that it is implemented interactively on the web.

Roger White's paper turns the attention from the urban scale that has typified the AB/CA models discussed so far to the modeling of land use changes in large regions or whole countries. Modeling on such a scale, White argues, can be approached in two ways: When dealing with urban-centered regions with numerous and functionally coherent statistical areas, it is useful to build a combined model in which spatial allocation of population and economic activity is handled by a spatial interaction model that then drives a CA land use model. But when the areas are few and polycentric, an alternative approach is preferable, namely, to treat the dynamics at both scales using a single CA. This approach eliminates "boundary effects" and several other problems inherent in the conventional approach.

Part IV: Cognition and VR

The issue of cognition has already been featured in most of the topics described so far: The cognitive process of pattern recognition by associative memory provided the point of departure to Haken's discussion on natural versus artificial; Batty et al's contribution deals with visualization, which is one of the central themes of cognitive science; the SIRN approach to cognitive mapping and urban dynamics formed the core of my discussion on the scope of CAE, and was the basis of large parts of my description of ESLab's specific experience: InfoCity and CogCity are explicit cognitive models. The same holds true to Section III on urban simulation models: All such models make implicit and explicit assumptions regarding agents' cognition and behavior. Part IV further elaborates on the issue of cognition of real and virtual environments.

Part IV is composed of six papers dealing with three issues. The first three papers consider the various cognitive dimensions of AB and CA models. The following two papers deal with the electronic environment of the Internet and with cities as VR environments. The last paper examines geographical environments – real and virtual – as small world networks.

The first paper by Theo Arentze and Harry Timmermans makes explicit the link between the discussion in Part III and the issue of cognition. Based on research projects conducted in the DDSS research programme at Eindhoven University of Technology, the paper examines five multi-agent models referring to activity-travel behavior, pedestrian movement, land development, learning and adaptive behavior under conditions of uncertainty and information search, and finally, computational models linking cognition, choice set formation, activity travel behavior and land use dynamics. All these models deal with spatial cognition and put special emphasis on learning in relation to complex choice behavior in urban environments.

Modeling of the dynamics of settlements and settlement systems is commonly developed at different geographical scales: at the micro-level of single cities, at the meso-level of cities and regions and at the macro-level of hierarchical and spatial structures. In her paper, Lena Sanders shows that using a multi-agent system (MAS) approach to simulate the various scales raises an ontological question that concerns the cognitive status of "agent." At the micro level of cities, the notion agent refers to households or entrepreneurs, and thus makes intuitive sense. But what happens when the modeler moves to the meso and macro levels? Can a whole city be treated as an agent that takes decisions and executes them? Referring to results from SimPop – a MAS model that simulates the emergence and evolution of a settlement system in a period of 2000 years – Sanders' answer is yes!

Sylvie Occelli and Giovanni Rabino examine urban modeling from a new perspective: They treat each of the various urban simulation models as an action, learning and communication (ALC) *agent*. Each such model/agent is capable of performing a certain course of action (permitting a certain learning ability), which, because of its cognitive mediating role, communicates with other kinds of agents (other models). Each model must, therefore, contain three main components: a syntactic component, accounting for the mechanisms underlying the functioning of the modeled system; a representational-semantic component, specifying the kind of urban descriptions conveyed by the model, and a purposive investigation project component, referring to the activity simulated by the model.

The next paper by Stephen .C. Hirtle shifts the discussion to the issue of navigation – a central theme in the domain of spatial cognition. Humans come to the world equipped with spatial information processing capabilities that allow them to navigate and perform wayfinding tasks. Such capabilities, suggests Hirtle, might become efficient tools to navigate in the electronic environment of the Internet. He then presents a tripartite theory of navigation based on cognitive studies of navigation in physical spaces, and shows its applicability to navigation in VR environments and in the abstract information spaces of the Internet.

Software that enables the construction of virtual cities can be regarded as the most advanced technological tool currently available for the study of complex artificial environments. No wonder, therefore, that most studies in this domain tend to concentrate on the technological aspects of such models. Itzhak Omer, Ran Goldblatt, Karin Talmor and Asaf Roz's paper discusses spatial orientation and wayfinding difficulties that users experience when navigating in VR cities. Based on an empirical study of Tel Aviv residents' urban image, they introduce a wayfinding support system (WSS) that enhances legibility in the virtual city of Tel Aviv.

As we have just seen, complex artificial environments such as real and virtual cities or the Internet, are commonly described as networks of interacting objects – physical objects such as buildings or roads, and human agents meaning individuals, firms etc. Small world phenomena refer to the specific nature of networks, namely, to networks in which the distance between any two randomly chosen objects (e.g. yourself and the president of your country), is about six persons away, so called "six degrees of separation." This, in contrast to regular or random networks. In the final paper of this section, Bin Jiang demonstrates how geographic environments might be represented as small worlds, and discusses the implications thereof, among other things, on various cognitive tasks, such as search and navigation in real and virtual environments and in the Internet.

Part V: Planning

Planning is the domain where complexity and self-organization theories, AB/CA simulation models and technologically sophisticated VR environments meet the challenges of the real world. Two papers discuss this theory-practice interface. The first, by Michael Kwartler, tells the story of the Environmental Simulation Center, Ltd. (ESC) in New York, one of the few planning bodies that actually use the tools described so far, and one of the even fewer to use these tools for citizens' participation purposes. In his paper, Kwartler examines three projects undertaken in the last three years. The case studies demonstrate how to obtain citizen input as to their values and group identity through their participation in designing the place in which they would like to live. Fully integrating 3D/Geographic Information System-based simulations and visualizations into the visioning process makes it possible for citizens to better understand their choices at both a policy and experiential level and arrive at a consensus for the future of their communities.

One of the core assumptions of PSS is that these computer-based systems can be applied and found useful in actual planning situations. In the closing paper Richard Brail examines the performance of various PSS that have already been applied to the practice of planning. His main conclusion is that planning support systems have not yet reached their full potential in assisting public sector decision-making. The reasons vary but the consistent ones seems to be that some of the components of PSS, such as the simulation models that require mathematical sophistication, are still too abstract and as such offer less concrete application. On the other hand, data issues in supporting PSS appear to be manageable. However, visualization and VR tools are attracting attention and thus provide a way into PSS. As in many other areas, here too, education seems to be the key.

Part one:
General aspects
of Complex Artificial Environments

The Scope of Complex Artificial Environments

Juval Portugali

Abstract. This paper discusses the social significance, theoretical rationale and methodological and technological characteristics of complex artificial environments of the 21st century. It starts by identifying a social change, the essence of which is that cities and urbanism are growing in significance, to the extent that one can speak of a *second urban revolution*. The paper goes on to explore urban theory and the view that cities are complex, self-organizing, artificial environments, and that the theorization of their dynamics should start from the first principles of humans' cognitive capabilities. Finally, the paper considers the methodologies of agent base, cellular automata and advanced virtual reality simulators as tools for studying cities as self-organizing complex environments.

1. Introduction

The last decade has witnessed the emergence of laboratories, research centers, planning and design organizations, characterized by a configuration that includes the following features: a focus on artificial environments, in particular on cities and their dynamics; perception and study of cities in terms of theories of complexity and self-organization; use of cellular automata (CA) and agent base (AB) models to simulate the dynamics of cities; reliance on cognitive models as a source for agents' behavior and decision-making; intensive use of geographical and spatial information systems; use of virtual reality (VR) to visualize three dimensional (3D) urban environments and to simulate movement in them.

This paper aims to examine this new configuration – its social significance, theoretical rationale and methodological and technological implications. The thesis suggested in the discussion below is that this new configuration reflects several major changes that characterize 21st century society: First, a social change, the essence of which is that cities and urbanism are becoming especially significant, to the extent that one can speak of a second urban revolution. Next, a theoretical change emerges in the view that cities are complex, self-organizing, artificial environments, and that theorization of their dynamics should commence from the first principles of humans' cognitive capabilities. Finally, there is a methodological change that shows up in the prevalence of AB and CA models, as well as advanced VR simulators, as tools to simulate and study cities as self-organizing complex environments.

2. The second urban revolution

Cities have existed for more than 5,000 years, so what is suddenly so special about them? Let me answer by reference to two studies: One is a seminal paper published in 1950 by Gordon V. Chile entitled "The urban revolution," and the second is a monograph published 20 years later, in 1970, by Henri Lefebvre entitled *La Révolution Urbaine*, (The Urban Revolution). In his paper, Childe suggests regarding the first appearance of cities some 5,500 years ago in Mesopotamia as one of the most important revolutions in the history of humankind. This view was widely accepted, among other things, because cities emerged hand in hand with the invention of writing, which, in its turn, marks the transition from prehistory to history, and thus what is often called "the rise of civilization."

Unlike Childe's thesis, however, Lefebvre's was rather surprising, and as a consequence, controversial. He suggested that, while cities have existed for more than 5,000 years, a genuine urban society is only now emerging; in other words, we are on the verge of an urban revolution that is taking place in front of our very eyes. Lefebvre published his work in 1970, and at that time his ideas and predictions seemed prophetic and not very realistic. It is of no surprise that they have attracted criticism. The criticism of Lefebvre's student Castells (1977), in his *The Urban Question,* is probably the most famous. But the interesting part of the story is that today, more than 30 years after *La Révolution* was first published, his predictions seem to be being realized one by one (Smith, 2003): For the first time in human history, the number of people living in cities is reaching 50% of the world's population, cities such as Mexico City, Bombay (Mumbai) and Sao Paolo grew from 8.8 million, 6.2 million and 8.3 million respectively in 1970 to over 18 million, over 16 million and again over 18 million today. We speak today about *world cities*, or *global cities*, that form the centers for the globalization process; but Lefebvre was already writing about world cities in 1970 (according to him the term was coined by Mao Tse Tung).

All the above figures and signs indicate the more fundamental change of the urban revolution: According to Lefebvre, its essence is that urbanism is replacing industrialization as the dominant force in society. My view is that the essence of this change is that urbanism is replacing nationalism as the generative order of modern society.

The notion of *generative order* is taken from Bohm as an aspect of his theory of *implicate order* (Bohm 1980, Bohm and Hiley 1993), while the view of nationalism as a generative order, comes from Portugali's (1993) *Implicate Relations.* Intuitively, and in the context of society, a generative order can be described as an ordering principle, according to which individuals and communities conduct their affairs. The notion is close to what in social theory, especially Marxist theory, is often called *mode of production.* It also bears similarities to what in Haken's (1983) synergetic approach to complex systems is termed *order parameter* (see below).

The notion of generative order is related also to Wirth's (1938) classic "Urbanism as a way of life." In that paper, Wirth elaborates a theory of urbanism that is based on *population size*, *density* and *heterogeneity* as the key determinants of social and individual life in big cities. But the significance of Wirth's thesis lies, to my mind, in what it implied: that the specific spatial organization of society is not only a product of social, economic or political processes, but a force that can shape people's perception, behavior and action, and by implication their value judgment, ideologies and action. In other words, it can become a generative order. It is in this respect that one can say that urbanism as a way of life is replacing nationalism as a way of life.

The theory put forward in *Implicate Relations* is that nationalism has become the generative order of modernism and modern society in two respects. First, in the sense that its *information content* – the ideology of nationalism – has become the only ideology accepted by all otherwise rival positions: socialists, Marxists, liberals, capitalists, democrats all conform to the basic principles of nationalism. Second, in the sense that throughout most of the 20th century, its *material content* – the nation state – emerged as the most dominant and legitimate political structure.

The second urban revolution doesn't mean the disappearance of nationalism, however; it means its urbanization. There are several dimensions to this process: First, the decline of the welfare nation-state and the process of privatization that accompanies this decline. Second, the concurrent process of the emergence of a civil society that takes over many of the past duties and functions of the nationalist welfare state. Third, the quantitative growth of cities and urbanism noted above entails a new reality in which crucial problems of many (post)modern nations are no longer classical national problems (e.g. national self-determination, national boundaries and so on), but rather the problems of cities. This is so in a country such as France, in which a major threat and challenge to the national identity comes not from the outside, from an external nationalist "enemy," but rather from the inside – from the poverty-ridden Islamic quarters of its cities. This also applies in the Third World, where the urban questions of poverty, crime, sanitation and urban guerrilla warfare in the big cities override the "classical" national questions. Fourth is the process of globalization that represses the nation states, on the one hand, while making some world cities more dominant than the states within which they exist, on the other. Fifth, the events of September 11 and the ensuing wars in Afghanistan and the Middle East are tragic indications that the *urbanization of nationalism* is associated with a parallel process of the *urbanization of war*. As the current events in Iraq show, the "ordinary nationalist war," by which the American army invaded Iraq and conquered it, was only an introduction to the real urban-global war that followed.

3. What is a city?

Everybody knows what a city is. The workshop on which this book is based took place in the city of Venice; each of the participants came to Venice from his or her home city: Paris, Stuttgart, New York, London, Tel Aviv, Shanghai, etc. But what is a city? The history of the many attempts to answer this question is rather confusing: Whenever a definition was proposed, it was always possible to falsify it by putting forward cities that do not comply with the definition. In other words, the various definitions could never pass Popper's test of falsification.

The main reason for the failure to define cities is that the various attempts to do so were always made with reference to what in cognitive science is called *classical categories*. That is, groups composed of entities sharing some necessary and sufficient conditions that define them as a category and distinguish them from other categories. Students of urbanism have implicitly treated cities as classical categories, and yet, cities are not classical categories.

The study of concepts and categories as developed by cognitive scientists in the last three decades (Rosch, 1999) shows that one has to distinguish between "classical categories," which are sets of instances sharing some necessary and sufficient conditions, and sets of instances that form a category due to what Wittgenstein (1953) has termed *family resemblance*. Cities belong to the second group, and as a consequence, attempts to define them in terms of a classical category were not successful (Portugali, 1999).

A family resemblance category becomes a category not when its elements share some common denominators, but when they form a *network* of partial links and similarities. Further research and experiments into the cognitive dimension of categories and categorization have found that many family resemblance categories have a *core-periphery* structure, in the sense that some instances of the category are more *prototypical* of the category than others and they thus form its center while the rest of the instances form the category's periphery (Rosch, ibid; Johnson, 1987; Lakoff, 1987).

The city is a good example of a family resemblance category with a core periphery structure. On the one hand, there are no common elements between the "first" and "last" city but the name. On the other hand, the first city of some 5,500 years ago had space-time links and similarities with subsequent cities, which in turn had common elements with subsequent cities, and so on until the global city of today. The result of this process is that cities form a huge space-time family resemblance network extending in time and space from the ancient cities of 5,500 years ago to the cities of today. In this network, one can identify space-time moments during which certain cities became more characteristic or prototypical of the category than others. Such cities have temporarily captured the center of the category city, pushing to the periphery the rest of the instances, only to be replaced in subsequent space-time moments by other prototypical cities, other centers and other peripheries.

Cities are very large *artifacts*. As a consequence, in addition to being a member in the category "City," each single city has a category-like structure of itself: Each can be described as a network of entities (buildings, parks, roads, humans, cars, etc.) loosely connected by partial links and similarities. In each such city network, one can further identify elements and entities that are more (proto)typical of the entire city than other entities. One thus tends to speak of a typical Parisian café or a typical New Yorker person or street, and so on. As in the case of regular categories, here too, the prototypical elements form the center of the image of that city, while other elements its periphery. One implication from the above is that many cities have fuzzy boundaries, in the sense that their spatial extent is not clearly defined – a property that has immediate implications on the various attempts to quantify and model them.

4. Self-organization and the city

How do each single city and the category "city" evolve? The answer suggested in *Self-Organization and the City* (Portugali, 1999) is "by means of self-organization." *Self-organization* is a central property of open and complex systems. While the concept had already appeared in the 1940s, its modern use was pioneered in the 1960s by people such as Haken, (1983) with his theory of *synergetics*, Prigogine with his notion of *dissipative structures* (Nicolis and Prigogine 1977) and others (see review in Portugali, 1999, chap. 3.1). Such systems are typically in "a far from equilibrium condition" and exhibit phenomena of chaos, fractal structure and the like. For a long time, the term "self-organization" was used also as an umbrella name for these theories; nowadays, however, the umbrella name that is commonly in use is *complexity theories*.

The notions of self-organization and complexity originated in the sciences, specifically in physics, as a property of natural systems. But from the start they were associated with the artifact city: "An appropriate illustration would be a town that can only survive as long as it is a center for inflow of food, fuel … and sends out products and wastes." – So wrote Nicolis and Prigogine (1977, 4) in their introduction to *Self-Organization in Nonequilibrium Systems* in order to convey to their fellow physicists what they mean by "self-organization."

Physicists such as Peter Allen (1981) and Weidlich (1999) took this urban example seriously and showed that towns and cities are not just metaphors, but genuine self-organizing systems. Allen did so from the perspective of Prigogine's dissipative structures, while Weidlich from the perspective of Haken's synergetics. Their projects were important in that they demonstrated that the phenomena of self-organization and complexity extend beyond the domain of nature into the world of the artificial, and thus typify natural and artificial systems alike. They also provided a much more general and sound theoretical basis for the study of society, cities, urbanism and their dynamics.

Three and a half decades of intensive research with complex systems have shown that these theories indeed apply to natural and artificial complex systems in a variety of domains and phenomena, ranging from physics and life sciences to human behavior, cognition and society.

Self-Organization and the City (Portugali, 1999) explores the city as a complex self-organized and self-organizing system. It does so from the perspective of Haken's (1983) synergetic theory of complex systems and by means of FACS (Free Agents on a Cellular Space) – a family of agent base and cellular automata (AB/CA) simulation models specifically designed for this purpose. A discussion of AB/AC and FACS simulation models is given in the next section; here the focus is on synergetics.

At the core of synergetics is the view that the synergy between the many parts of a complex system, driven as it is by an internal or external *control parameter*, gives rise to several configurations of movement that enter into a competition. This competition is solved by means of the so-called *slaving principle* – when one (or a few) of the competing configurations "wins," in the sense that it "enslaves" the other parts of the system into its specific movement. This winning state, termed *order parameter,* can be likened to an attractor that governs the dynamics of the system.

In the case of cities, the "parts" of the system are the many urban agents – individuals, households, firms, public and private planning agencies and the like. Triggered by a certain control parameter – a demographic process, for instance – these urban agents enter into intensive synergy and interaction, which give rise to an order parameter that enslaves the behavior of the agents. Governed by this order parameter, the city then evolves at a steady state until a new control parameter enter the scene, triggers a new urban dynamics, phase transition, a new urban steady state, and so on.

In *Self-Organization and the City,* it is shown that this process characterizes the dynamics of cities and urbanism at a variety of scales: the slow, five-millennia, global process of urbanism and its outcome – the family resemblance nature of the category city; the mezzo scale of the faster urban processes that involve changes in land-use and socio-economic spatial configurations, as well as the very fast micro level that refers to the dynamics of the daily routines of cities. In all these scales we see the above noted self-organized and organizing evolutionary process at work.

5. The science of cities[1]

It is common practice to study complex systems, including cities, by means of simulation models. In the last decade or so, agent base (AB) and cellular automata (CA) models have become very popular devises for this purpose. A typical CA model commences with a cellular space in which, from iteration to iteration, the properties of each cell are redetermined by reference to the properties of the cell's neighboring cells. AB models add to the picture agents that are assumed to partly mimic living entities. For example, our FACS urban simulation models are built as a superposition of a CA submodel representing the dynamics of the urban infrastructure, and a superposition of an AB submodel representing the behavior, movement and action of the urban agents (Portugali, 1999).

There are several reasons why AB and CA models are very attractive tools for urban simulation: The fact that, in reality, the properties of many urban objects are determined by that object's relation to its neighbors, as in the original CA models, is one reason; simplicity is another.

Simplicity allows a scientific approach to the study of cities. It shows up in that most, if not all, CA/AB urban simulation models are bottom-up in their structure – they start with simply behaving agents, in simple local circumstances, that, by means of their simple-local interactions, give rise to the global complexity of the city. By so doing they follow, albeit implicitly, Simon's (1969/1999) thesis re-

[1] This section is based on Portugali, forthcoming.

garding *The Sciences of the Artificial*. The title of this section rephrases, as one can see, the title of Simon's book. I refer below to the 1999 addition, which takes into consideration the notion of complexity.

In his book, Simon develops an approach that, according to him, can transform the study of artifacts from a "soft" hermeneutic study into a "hard" science. His starting point is the study of nature as developed in the sciences. What made the study of nature scientific, he suggests, is the discovery that very few and simple natural laws govern an interaction between simple elementary parts, which then give rise to the enormous complexity that we observe in nature.

Unlike the few and simple natural laws that govern the observed complexity of nature, notes Simon, the causes governing the artificial world are, on the face of it, complex: humans' aims, plans, intentions, needs, policies and so on. But this observed complexity, he claims, is only an external appearance of innately simple behaving entities: Similarly to simple animals, we humans as "behaving systems, are quite simple. The apparent complexity of our behavior over time is largely a reflection of the complexity of the environment in which we find ourselves" (p. 53).

AB/CA urban simulation models are built in line with Simon's logic. They typically start with agents having a few simple aim(s) "in mind." These agents come to the city and enter into a local interaction with a cell, its neighboring cells and neighboring agents. This interaction gradually gives rise to an urban system, which from iteration to iteration becomes increasingly complex. As the urban environment becomes more complex, so does the observed behavior of the urban agents; but essentially it is not. Complexity is thus a property of the global system as a whole, but not of its individual parts.

The above explanatory model of "simple cause, complex effect" has a long history, of course. AB/CA urban simulation models have inherited it from location theory and quantitative human geographies of the 1950s and 1960s, which inherited it from neo-classical economic theory, which inherited it from the "sciences" as the genuine scientific method. This historical chain gives AB/AC urban simulation models extra theoretical and methodological strength; but there is a catch here: Several empirical and theoretical recent studies falsify the above model, and by implication Simon's view (Portugali, 2002, 2004). First, empirical studies show that rats' exploratory behavior, for example, (Golani et al., 1999) is innately complex – a finding that is in line with pragmatist and ecological approaches that currently dominate the study of brain, cognition and behavior (Gibson, 1979; Freeman, 1999; Varela et al., 1994 and further bibliography there).

Second, from *Self-Organization and the City* it follows that, similarly to languages (see below) the city is a *dual* self-organizing system: the city as a whole is a complex self-organizing system and each of its parts – the urban agents – is a complex self-organizing system too. This property reflects a fundamental difference between complex and simple systems in the domains of life and society: The initial conditions of simple systems are relatively few independent parts within a system that is itself isolated from its environment, while those of complex systems are relatively large number of complex interacting parts, linked by a complex network of feedback and feedforward loops, within a system that is open to, and thus part of, its environment.

Third, the conventional "simple cause, complex effect" scientific method collapses in the case of cities for yet another reason. A central property of complex systems is the process of *circular causality* that typifies also the dynamics of cities: The interaction between the local/micro urban agents gives rise to the global structure of the city, which in turn feeds back and prescribes the behavior, interaction and action of the agents, and so on. AB/CA urban simulation models are excellent tools to simulate the first part of this loop – the way local interactions give rise to a global structure – but they fail to describe the second, feedback part of the loop.

Fourth, AB and AC models were originally designed for systems whose parts are *simple* and *local*. The human agents that form the parts of a city are complex and global. Complex, in the sense noted above that every individual agent is itself a self-organizing complex system, and global, in that human agents have the cognitive capability to construct what might be called *c*- and *s-cognitive maps*, that is, conceptual and specific cognitive maps that refer to the global structure of the city. The outcome of this is that agents never act in the city *tabula rasa* (Portugali, 2004). The notion of SIRN introduced next is an attempt to take these properties into consideration.

6. SIRN cities

SIRN (*Synergetic Inter-Representation Networks*) is a theory that attempts to capture the process of the production of artifacts, and for two complementary purposes (Haken, 1996; Portugali, 1996; Haken and Portugali, 1996; Portugali, 2002): One, in order to understand cognitive processes in general and their role in the dynamics of cities and similar artifacts, in particular. Two, to understand the production and evolution of artifacts – small ones, such as laptops, and big, such as cities, metropolitan complexes and other artificial environments.

The notion SIRN is a composition of two terms: synergetics, which, as introduced above is the name assigned by Haken (1983) to his theory of complex, self-organizing systems, and *IRN (*inter-representation networks*),* which is an approach to cognitive geography (Portugali, 1996) suggesting that the production of artifacts plays a central role in the very process of cognition. The commonly held view in cognitive science is that artifacts are products of specific cognitive processes. To this view, IRN adds that in many cases the cognitive process doesn't end at this stage; rather it continues and has feedback effects, the essence of which is that artifacts function as an extension of the mind, and thus can be regarded as an integrative element in the very cognitive system.

Three decades of research on the synergetics of brain functioning and cognition (Haken, 1991/2004, 1996 and further bibliography there) have established the view that processes of cognition evolve as complex self-organized systems. According to SIRN, which combines synergetics with IRN, in many cognitive processes, some of a system's parts emerge as internally represented brain constructions, while others as externally represented artifacts. The interaction between the internal and external parts of such systems gives rise to order parameters that enslave the other parts and so on.

The cognitive validity of the SIRN model was discussed and exemplified in some detail in previous studies (Haken and Portugali, 1996; Portugali, 1996, 2002, 2004; see also Kitchin and Blades, 2002). Here it is of interest to note that the urban simulation model *CogCity* (Portugali, this volume) is built in line with the SIRN logic. It starts with agents that come to a city with an image of a city in mind – this image is termed *conceptual* or *c-cognitive map*. With this image in mind they observe the externally represented information afforded by the city. The interaction between these internal and external forms of information gives rise to the *specific*, that is, *s-cognitive map* according to which the agent then takes location decisions and action in the city. The latter action entails some changes in the city and in the information it affords to the next agent and so on in a process of circular causality.

The SIRN view comes close to the notion of *situated cognition*, which is currently dominant in the domains of cognition, *AI and A-life* (Clancey, 1997). Situated cognition emphasizes and elaborates on the active role played by context and environment in the overall process of cognition. SIRN accepts the situated cognition view but focuses its attention on the dual nature of artifacts, that is, on the interplay and circular causality between artifacts as the products of cognitive processes and artifacts as elements of cognitive systems.

According to SIRN, artifacts as elements of the cognitive system can take two basic forms: bodily artifacts and stand-alone artifacts. An utterance or a dance performed by a person is an example of bodily artifacts, while a sculpture, a painting, a map, a plan or text are stand-alone artifacts. All such artifacts come into being by means of the SIRN play between internal and external representations. SIRN further makes a distinction between intra-personal cognitive processes, for which the above artifacts are typical examples, and collective inter-personal SIRN processes. Every human language is an example of a bodily collective, inter-personal artifact, while a city is a typical example of a stand-alone, collective, interpersonal artifact.

7. On cities and languages

In *Knowledge of Language: Its nature, origin and use*, Chomsky (1983) makes a distinction between external and internal languages (E- vs. I-languages respectively). E-languages are the spoken languages (Hebrew, English, French, Chinese, etc.), while I-language is the innate universal language with which, according to Chomsky, every human being comes to the world, and by means of which he or she acquires specific E-languages. Evaluating these two concepts of language from his cognitive scientific approach to the study of language he writes the following:

The notion of E-language has no place in this picture. There is no issue of correctness with regard to E-languages, however characterized, because E-languages are mere artifacts. . . the concept appears to play no role in the theory of language. . . The technical concept of E-language is a dubious one in at least two respects. In the first place ... languages in this sense are not real-world objects but are artificial, somewhat arbitrary, and perhaps not very interesting constructs. In contrast ... statements about I-language ... are true or false statements about something real and definite, about actual states of the mind/brain and their components ... (p. 1986, 26-7, italics added).

Chomsky here suggests that E-languages are artifacts and therefore "somewhat arbitrary, and perhaps not very interesting constructs." I agree with the first part of his suggestion, namely, that languages are artifacts; but I disagree with the second – that they are "not very interesting constructs" – not only because artifacts, in general, and the artifact "city" in particular, form our topic of interest in this book, but because, the association between Chomsky's theory of language and the study of cities has already produced some interesting results. Another reason is that very much like cities, the artifacts E-languages are *complex systems*. Let us start with the first reason.

Possible links between Chomsky's theory of language and urban theory have already attracted students of cities. Two prominent examples are Hillier's *space syntax* (Hillier and Hanson, 1984; Hillier, 1996) and Alexander's *pattern language* (Alexander et al., 1977). Both employ the association to language to shed light on the nature of artificial environments: "In natural languages," write Hillier and Hanson (1984, p.50) "… a syntactically well-formed sentence permits meaning to exist, but neither specifies it nor guarantees it. In a morphic [architectural] language … a syntactically well-formed sentence … guarantees and … specifies … the meaning of a pattern." Hillier and Hanson further suggests three sets of laws, according to which artificial environments such as cities evolve: Laws regarding the relations between spatial objects, "laws from society to space" and "laws from space to society" (Hillier and Hanson, ibid).

Alexander (1979, 49-50) makes the link to Chomsky's theory of language even more explicit. The patterns of doors, buildings, neighborhoods and whole cities, he writes, are natural entities that are "actually there in peoples' heads and are responsible for the way the environment gets its structure." And in an interview with Grabow (1983, 184-5), Alexander emphasizes the difference between "his" pattern language and Chomsky's language: A natural spoken language … has a set of elements (words), a set of rules defining the possible relations between words and "the complex network of semantic connections, which defines each word in terms of other words." The pattern language of artificial environments is still more complex in the sense that "each pattern is also a rule which describes the possible arrangements of the elements – themselves again other patterns."

The second reason, as noted, is that cities and languages are complex systems. Chomsky's conceptualization of I- and E-languages is founded on the distinction between the "natural" and the "artificial" discussed above – in his words: between "real-world objects" and "artifacts." I-Language, claims Chomsky, is a natural, real-world object and thus lent itself to scientific inquiry, while E-Languages are artifacts, and as a consequence their study must take place outside the realm of the sciences (probably in the Humanities and other hermeneutic, non-scientific disciplines). This view of Chomsky echoes, of course, Snow's (1964) thesis that the "sciences" and the "humanities" form two unbridgeable scientific cultures and is related to our discussion above regarding "the science of cities."

The interesting discovery stemming from the various theories of complex systems is that, as we've seen above, they apply to natural and artificial systems alike. Not because natural and artificial systems are identical, but because, despite the genuine differences that exist between them (Haken, this volume), they share the property of complexity. E-languages and cities are thus similar to each other in that both are complex self-organizing systems; both are open to their environment, both have emerged as highly ordered systems out of a spontaneous complex interaction between a huge number of human agents and as complex systems both lent themselves to scientific inquiry – just like I-language.

Cities and languages as complex systems resemble each other in yet another way: both are *dual complex systems*. This is because the elementary parts of both are human agents and because each agent is itself a complex system. Historically, complexity theory was developed in light of phenomena in physics, such as the Bénard experiment, which acted as paradigmatic case studies. In the Bénard experiment, for example, the local parts are simple entities (atoms, molecules etc.) and complexity is the property of the emerging global system. Furthermore, complexity theory's main focus of interest has always been in the processes of emergence that take place in the global system. As a consequence, only little attention was paid to the intricacies of the circular causality phenomenon that typifies complex systems. Namely, to the way emerging properties in the global system react and trigger changes in each of the local parts. The main exception here is Haken's (1983) notion of the slaving principle, which is central to his theory of synergetics.

Haken and co-workers have studied processes of slaving that typify individual agents – with respect to pattern recognition and planning, for example (Haken, 1996, 1998), and also processes of slaving that are typical of whole cities and societies (Haken and Portugali, 1995, Weidlich, 1999). The challenge is to combine the two processes and theorize about them as two facets of a single system. The notion of SIRN introduced above is an attempt to meet this challenge.

But cities are not languages. For one thing, their products are stand-alone objects such as buildings, roads, bridges, etc. that can exist and survive independently of their producers. The products of languages are humans' voices and gestures that have no existence independent of their producers. Cities, in this respect, are akin to writing and texts – the external, stand-alone, representations of languages. The appearance of cities, some 5,500 years ago, hand in hand with writing, is, to my mind, not accidental.

A second difference concerns planning. To the best of my knowledge, there are no language planners (the attempt to "plan" the international language of Esperanto ended in failure), but there are many city planners; some of them contributed to this book. Moreover, unlike language, the city is full of planners and planning: professional planners and official/formal plans as well as non-professional and non-official planners and plans. Every agent operating in the city (person, family, company) is a planner on a certain level. But not one of the many planners can fully determine the final form and structure of the city. They are all *participants* in a big city-planning game (Portugali, 1999, Part III).

8. Forms of planning

One outcome from the above is a distinction between top-down, *global planning* versus bottom-up, *local planning*. The first refers to a planning process implemented by professionals – city planners, architects, engineers, etc. – while the second to planning as a basic human capability (Das et al., 1996).

There are two aspects to the above distinction. One that concerns the similarities, differences and links between local and global planning, while a second concerns the way local planning participates in the overall process of urban dynamics. The first aspect relates to the general process of specialization and division of labor by which general human capabilities become professions. The second aspect relates to the nature of the city as an open, complex and self-organizing system. Due to the non-linearity inherent in the behavior of such systems, there is always a possibility that local plans will be as effective as, or even more effective than, global plans in determining the city. A case in point is the cityscape of Tel Aviv.

From its early days the city of Tel Aviv has been a city of many balconies. People used to spend long hours sitting on their balconies, especially on summer evenings and nights. One day, probably at the end 1950s, a resident of Tel Aviv decided to enlarge his/her apartment by closing the balcony and making it a "half-room." He/she made a small plan, hired a builder and implemented the plan. One of the neighbors saw this was a good idea and did the same. A process of innovation diffusion (very much in line with Hägerstand's theory) started and before long the vast majority of balconies in the country as a whole was closed. At this stage, the municipalities decided to intervene and started to tax all balconies as if they are a regular room. In response, developers started to build buildings with closed balconies. For several years no balconies were built in Tel Aviv and other Israeli cities. But then, with the arrival of postmodern architecture, balconies became fashionable and architects started to apply for permits to build balconies – not to sit on them, as in the past, but as a decorative element. Equipped with their past planning experience and the wish not to lag behind the advancing (post)modern style, the city planners gave architects and developers permits to build open balconies but in a way that would not allow them to be closed as in the past. The result is the "jumping balconies" so typical nowadays in Israel's urban landscape.

A comparative empirical study on "urban pattern recognition," which took place in the early 1990s at Tel Aviv University and involved cities from Europe, America and East Asia, found that the most prototypical architectural patterns in the cityscape of Israel are one: the closed balcony, and two: the jumping balcony (Reuven-Zafrir, not published).

The co-existence of global and local forms of planning sheds new light on the notion of public participation in planning. The latter is based on an implicit assumption that there exists only one form of planning – global planning, and, as a consequence, on a sharp dichotomy between the planners and the planned. Public participation is the outcome of a common view among planners that in order for planning to be more democratic and just, planners have to give more say to the public, above and beyond the say given them via the standard political process.

The fact that global and local planning co-exist and interact in the dynamics of cities, and that in many cases local planning can be more dominant and effective in the overall urban process than global planning, implies that it must be perceived not as a reactive force, but an important source for planning ideas and initiatives. The role of public participation and planning democracy are thus not just to be more generous to the people affected by the planning, but also to allow the huge amount of planning energy to go bottom-up.

The second outcome of the above is a distinction between two forms of global planning: *mechanistic* or *engineered* or *entropic planning* versus *self-organized planning*. The first refers to a relatively simple "closed system" planning process, closed in the sense that it is, or rather should be, fully controlled. The second refers to a relatively complex "open system" planning process, which like other open and complex systems exhibits phenomena of non-linearity, chaos, bifurcation and self-organization. The planning of a bridge or a building is an example of the first form of planning, while city planning is an example of the second.

Once the planning of a building or bridge or any other engineered object is completed and implemented, their fate is predictable: Due to natural and artificial processes of decay and deterioration, they will eventually disintegrate, that is to say, they will reach a state of maximum entropy. However, once a city plan is completed and implemented, the story just starts – it triggers a complex and unpredictable dynamics that no one fully controls. This is true with respect to master plans, development plans and other forms of large-scale city planning, but it is also true for the global effect and role of small-scale plans implemented in the city: the effect of a new building or a bridge on the urban system as a whole is neither predictable nor controllable. Similarly to large-scale plans, they become participants in the urban self-organized planning game.

9. PSS – Planning Support Systems

AB/CA urban simulation models, which, as noted above, have become standard tools for simulating cities as complex self-organizing systems, are currently being proposed as the engines of Planning Support Systems (PSS) – the state-of-the-art planning tools that make use of the most advanced technologies currently available (Brail and Klosterman, 2001; Brail, this volume). A standard PSS is a three-part system composed of a set of simulation models (usually AB/CA), a Geographical Information System (GIS) and a set of 2D, 3D and VR visualization devices. The AB/CA simulation models are assumed to enable the planners to simulate future scenarios representing current trends, and also to envision the impact of various plans and policies; the GIS provides the data base for such scenarios, while the visualization systems provide the means to see the results at a high level of realism. As elaborated by Batty (this volume), visualization is gradually becoming a major analytical tool and evaluation device.

The enthusiasm currently surrounding PSS is reminiscent of the excitement that followed the appearance in the 1950s and 1960s of the *rational comprehensive planning* and its arsenal of quantitative planning tools. "This is an exciting time for simulation modeling and visualization tools in planning and public policy," writes Brail (this volume) and continues: "Planning support systems (PSS) have moved from concept to application. Is this future so bright ... ?"

10. Planning and the Prediction Paradox

AB and CA simulation models are becoming common tools in the context of planning as a means to predict future scenarios. This is their main role in the context of PSS, as we have just seen. But there is a problem here that can be described as the *prediction paradox* of self-organizing systems, in general, and of cities in particular. There are three interrelated facets for this paradox. First, the nonlinearities that typify cities imply that one cannot establish predictive cause-effect relationships between some of the variables. Second, many of the triggers for change in complex systems have the nature of mutations (Allen and Strathern, 2004). As such, they are unpredictable, not because of lack of data, but because of their very nature. Third, unlike closed systems, in complex systems, the observer, with his actions and predictions, is part of the system – a point made by Junsch (1975, 1981) more than two decades ago and largely ignored since then. In such a situation, predictions are essentially feed-forward loops in the system, important factors that affect the system and its future evolution with some interesting implications that include *self-fulfilling* and *self-falsifying* or *self-defeating* predictions.

The notion of self-fulfilling prediction is well recorded. A familiar example may start with a prediction that a given bank is about to face financial difficulties. Following this prediction, customers withdraw their money from the bank, which then indeed does face financial difficulties, as predicted. Self-falsifying or self-defeating predictions are less familiar. They refer to a situation by which a good prediction leads to its negation. A hypothetical case of self-falsifying prediction might arise if at 8 am all drivers hear in the radio that at 9 there is going to be a major traffic jam at a certain junction. If they trust the prediction, they avoid the junction and as a result there is no traffic jam.

A real case of self-falsifying prediction followed the immigration wave from the former Soviet Union to Israel in the early 1990s. The prediction of professional planners was that Israel was approaching a housing shortage and that the government should therefore purchase a large number of mobile houses and locate them on the outskirts of towns and cities. The implementation of this policy ended in failure due to spontaneous initiatives by a large number of individuals (latent planners), who, as a consequence of the predicted shortage and the prospect of making money, transformed existing non-residential buildings into residential ones.

This they did mainly in city centers, which, from the point of view of newcomers, are the most attractive places. The outcome was that the vast majority of apartments prepared by the latent planners were rented, while many of the mobile houses prepared by the government's planning bodied were left unwanted and unoccupied (Alfasi and Portugali, 2004).

On the face of it, a possible solution to the prediction paradox would be to make a second prediction that takes the first one into consideration. But this will entail the same result and so on in succession until infinity.

The way to untangle this paradox is to be aware of the fact that "the map is not the territory," that is, that our models are not one-to-one representations of reality but tools that allow us to study some aspects of it. Such tools cannot predict mutations or other "unpredictable events." What they can do is give us some indications as to the probabilities of the city evolving along certain courses if its current structure remains the same. Given the fact that once self-organized, cities tend to be rather stable systems, such information is significant to the various urban agents acting in the city as top-down professional or bottom-up latent planners.

11. Three-dimensional urban simulation models

The city is a 3-dimensional (3D) spatial structure (to which one might add a multi-dimensional socio-spatial structure). And yet, similarly to urban studies in general, the vast majority of urban simulation models are 2-dimensional (2D) in their structure and logic. Certainly, there are several good reasons for this tendency: First, the third dimension of cities is negligible compared to their first two dimensions. Second and related to the above, perceptually speaking, humans have a strong tendency to perceive the city 2-dimensionally. Third, and also related to the above, similarly to perception, theorization is essentially a process of interpretation and, as such, involves some form of generalization and information compression.

However, the two-dimensionality of urban theory and modeling has its drawbacks. First, cities are 3D structures, and by ignoring this in our modeling, we ignore an important aspect of the urban dynamics. Second, via the process of planning, this tendency feeds back on the very structure of cities: plans that are 2-dimensional in their structure, like many if not most urban plans, tend to entail 2D solutions (e.g. zoning) and as a consequence to produce 2-dimenasional real cities, that is, *flatcities* – a term that echoes *flatland* – Abbot's beautiful little book from the 1880s. This outcome has been specifically damaging in US cities – as was shown with great talent already in 1961 by Jean Jacobs in her classical criticism on urban theory and planning – *The Death and Life of Great American Cities*.

Third, the recent wave of urbanism that is strongly related to globalization and to the very rapid processes of urban growth and spread that typify cities around the world, is associated also with a tendency to go up: the number of high-rises in cities is increasing, their height is increasing and with them there emerges a whole complexity of urban processes that takes place in the third dimension of buildings and cities. In this new urban reality a 50-story building with some 250 units (apartments, offices, etc.) is already a kind of 3D neighborhood. An agent's location decision, in such circumstances, might be influenced more by who lives in the building as a neighborhood, on top, below and around, than by the properties of its neighboring buildings (as is common in 2D urban simulation models). Furthermore, the variability within such buildings in terms of property value, rent and land-use, might be of great significance, to the extent that describing them as "mixed land-use" (as is common in 2D land-use plans) is no longer sufficient. The three-dimensional urban simulation models – Semboloni's (this volume) CityDev and Portugali's (this volume) 3Dcity – are attempts to respond to this need.

12. VR Cities

The basic proposition of SIRN is that the cognitive system as a complex system must be distinguished from the brain as a complex system in that, unlike the brain, the cognitive system can include artifacts in the world. Such artifacts therefore become extensions of the mind. A city map, for instance, can be regarded as an external representation of a city, as perceived by a person or a group of persons, and as such participate in a SIRN process of city building. One can say that modern VR technologies are of this nature too: they are artifacts, and as such, are an external representations and extensions of the mind, like maps.

But VR cities are external representations in a very special way: Unlike maps that are essentially interpretive tools as they conceptualize the city and thus compress information, VR cities do not necessarily conceptualize the city nor compress information, and thus do not necessarily help interpret the city. Their quest for high realism in many cases entails a VR city that is as complex and unclear as the real city from which it was derived and in several ways even more complex, problematic and unclear than the original. This shows up very clearly in the experience of real-time navigation in a VR city: since there is no gravitation here, one very often literally "loses ground." One solution for this is to add to the system "*VR gravitation*" that will limit the movement to the required mode (pedestrian, driver, flight and so on) and in this way will partly overcome the situation of zero gravitation. However, this implies limiting the potential of the VR city. One can thus think of the V-city as a world in itself, or as an extension of the real city; that the more it will be in use, the more it will require exploration of its specific properties and experiences.

The above somewhat philosophical considerations have immediate implications on the next generation of urban simulation models and city planning tools. Already, VR technologies form the major visualization component in standard PSS (see the articles of Batty, Bodum, Portugali, Kwartler and Brail in this volume) and as indicated above and to be illustrated in more detail below (Portugali, this volume), they allow the development of genuine 3D urban simulation models.

13. Concluding notes

According to Lefebvre, we are witnessing the first genuine urban revolution in human history. The essence of this revolution, according to him, is that urbanism is replacing industrialization as the dominant force in society. The view in this paper is that we are witnessing the second urban revolution and that the essence of this change is that urbanism is replacing nationalism as the generative order of modern society. Nationalism does not disappear, but becomes urbanized and thus loses its primacy and dominancy as a generative order.

The increasing significance of cities and urbanism in modern life raises the old question of "What is a city?" What has been suggested above is that the appropriate place to search for an answer is the discourse on concepts and categories as developed in the last two decades within cognitive science. Examined from this perspective, the city appears as a non-classical, space-time family resemblance category and concept, a collective artifact and a self-organizing system.

The specific point of view of cognitive science, to my mind, is also the key to understanding the very dynamics of cities. We have tried to capture this by the notion of SIRN (Synergetic Inter-Representation Networks). According to SIRN, the dynamics of cities can be described as a synergetic self-organization process that involves interplay between internal representations created in the minds of urban agents and the city as an externally represented product of their behavior and action in the city.

On the face of it, perceiving cities as complex self-organizing systems entails doubts regarding the very rationale for planning: If cities organize themselves, planning becomes superfluous. The discussion above shows the exact opposite, however: Approaching the city as a complex system enriches our view of planning in that it gives rise to several new planning forms. In the discussion above, we examined global vs. local planning, entropic vs. self-organized global planning, linear vs. non-linear local planning and authoritative vs. democratic planning.

AB and CA simulation models have become common tools in the study of self-organizing systems, in general, and cities, in particular. In the above discussion, it was further noted that while AB and CA models are good urban simulators, they need further development in order to adapt to the specific peculiarities of cities. Namely, to the fact that cities are dual self-organizing systems, in that the agents that form their elementary parts are themselves complex systems. As suggested by SIRN, urban agents are complex, among other things, in the sense that the cognitive system is a complex system that in many tasks extends beyond the brain to include stand-alone objects in the environment. The construction of cognitive maps by urban agents is one such task. As shown elsewhere (Portugali, 2004), and below (Portugali, this volume) the cognitive mapping capability of urban agents implies that AB and CA urban simulation models should be adapted to allow top-down decision and action processes on the part of urban agents.

One aspect of the second urban revolution noted above concerns the third dimension of cities. Cities are not only becoming larger than ever before, but also higher. So far, most urban theory and urban simulation models have been 2-dimensional in their treatment of cities. The result has been "flatcities." In the past, this 2D approach was fully justified by the need for abstraction and information reduction; today, in light of the new urban wave, it becomes problematic and questionable. What we see today, therefore, in urban theory and urban simulation modeling, is a tension between flatcities and *3Dcities*: between the need for abstraction and information reduction and the growing importance of the third dimension of cities.

Another tension concerns the new VR technologies that enable us to produce virtual urban objects, small objects such as buildings, roads and bridges, and large ones such as neighborhoods and whole cities. But what exactly are these new VR cities? Are they ordinary artifacts? Are they representations of real cities that, like maps, enable us to better understand cities? Are they a new kind of artifact – independent virtual worlds that are as complex as ordinary artificial cities? Can they be integrated with ordinary artifacts and function as extensions and components of real cities?

These questions, in particular the ones concerning the new tools that enable high realism, bring to mind two pieces of work that, while originally produced independent of each other, together nicely conclude what has been said above. The first is Jorge Luis Borges' beautiful story about "precision in science" in his *The Garden of Forking Paths* and the second, McLuhan and Fiore's (1967) little book *The Medium is the Massage*. Here is Borges' short story *Precision in Science* in full.[2]

> *.... In that empire, the art of mapmaking has reached such perfection that the map of one region was as big as a whole city while that of the empire big as the whole region. But these huge maps still did not satisfy people. Mapmakers then started to prepare a map of the empire on a 1:1 scale with the empire itself and identical to every element in it. Later generations thought that this huge map was superfluous and abandoned it to the cruelty of the hot sun and cold winters. In the western desert, one can still find a few fragments of that map, together with wild animals and beggars. In the country as a whole, however, no remains were left from the art of mapmaking.* (From Suares Miranda's Travels of the Heroes, book four, Chap. 45, Larida 1658).

The question this little story poses is this: To what extent does the new media introduced in this book – simulation models, GIS, 2D and 3D visualizations and the PSS as a whole – enable a deeper insight into, and a better interpretation of,

[2] This is my personal translation from the Hebrew version. The discussion from this point onwards is partly based on Portugali, 2002a.

the world in which we live and intend to plan, and to what extent does their ultra-realism and sophistication obscure reality by leaving no room for imagination and as a consequence interpretation? The answer to the question can be found in McLuhan and Fiore's book:

> *Societies have always been shaped more by the nature of the media by which men communicate than by the content of the communication.*

The new media we are discussing and developing do not necessarily provide society with a clearer and more accessible description of the world. Rather, the new media *create* a new virtual world, several parts of which are complicated or even obscured for the non-expert, while others are yet unknown and unfamiliar.

The new media we are developing and discussing are essentially models. Every model is, implicitly or explicitly, a statement or a theory about the dynamics of cities, regions or environments. And the question of whether the theory, and by implication the model, is true or false is not automatically related to the sophistication, beauty or realism of the model. A model might be a beautiful mathematical construction and still false. It might be visually realistic and even correspond nicely to quantitative data and still be a false representation of the forces that shape reality. One of the principles of model building is that the more parameters you add to your model, the less you understand what your model is actually doing and why. The catch is, however, that since a good (controllable and thus understandable) simulation model should be based on as few parameters as possible, it often happens that different sets of parameters (social, economic, cultural, cognitive etc.) give rise to the very same urban or environmental scenario. Hence, the answer to the question "Which set of parameters is appropriate?" stands outside the model itself – exactly as the theory behind the model. Given this nature of urban simulation models, the ambition to make them a standard tool at the hands of researchers, planners and users who are not experts in the inner structure of the simulation models is problematic, to say the least.

As noted at the opening of this paper, today we are witnessing the emergence of a fascinating new domain of research that attempts to integrate several high-tech media hitherto developed in relative autonomy. On the one hand, the attraction and fascination of this new medium results from its foreseeable promise: the integration of the various media (GIS, simulation, visualization, etc.) has the potential to create a new medium stronger than the sum of its individual components. On the other hand, it is attractive and fascinating precisely because of its unforeseeable properties – the fact that we are entering a new domain with new properties and rules is associated with curiosity and motivation to enter and study this new domain. As testified by this workshop and book, scientists and planners are already studying this new domain. At the moment, however, the number of people engaged in its study is relatively small and thereby so is the direct impact of this study on planning and society. What will be its fate in the future? Will it continue to be closed in experts' ivory towers or will it overflow to society at large to play a role in the larger scale processes of globalization and urbanization? It remains to be seen.

References

Alexander, C., Ishikawa, S. and Silvestein, M. (1977). *A Pattern Language*. New York: Oxford University Press.

Alexander, C. (1979). *The Timeless Way of Building*. New York, Oxford University Press.

Alfasi, N. and Portugali, J. (2004). Planning just in time versus planning just in case. *Cities* 21, 1, 29-39.

Allen, P. A. (1981). The evolutionary paradigm of dissipative structures. In *The Evolutionary Vision* (E. Jantsch, ed.), pp. 25--71. Boulder: Westview Press.

Allen, P.M. and Strathern, M. (2004). Complexity: The integrated framework for integrated models of urban and regional systems. A talk delivered at a conference on The Dynamics of Complex Urban Systems, November 4-6, 2004, Monte Verita.

Bohm, D. (1980). *Wholeness and the Implicate Order. London*. London: Routledge & Kegan Paul.

Bohm, D., and B. Hiley. (1993). *The Undivided Universe*. London: Routledge.

Borges J.L. "precision in science" in his *The Garden of Forking Paths*.

Brail, R.K. and Klosterman, R.E. (Eds.) (2001). *Planning Support Systems*, ESRI Press,

Castells, E. (1977). *The Urban Question,* London, Edward Arnold; trans. Of *La Question Urbaine* (1972) Paris, Maspero.

Clancey, W. J. (1997). *Conceptual Coordination: How the mind orders experience in time*. Lawrence Erlbaum Associates, New Jersey, London.

Childe G.V. (1970). The Urban Revolution. *The Town Planning Review* 21, 3-17.

Chomsky, N. (1983). Knowlegde of Language: Its nature, origin and use. New York: Praeger.

Das, J.P., Kar, B.C. and Parrila R.K. (1996). Cognitive Planning: the psychological basis of intelligent behavior. Sage, New Delhi.

Freeman, W.J. (1999). *How Brains Make Up Their Minds*. London: Weidenfeld & Nicolson.

Gibson, J.J. (1979). *The Ecological Approach to Visual Perception*. Boston: houghton-mifflin.

Golani, I., N. Kafkafi, and D. Drai. (1999). Phenotyping stereotypic behaviour: collective variables,range of variation and predictability. *Applied Animal Behaviour Science* 65:191-220.

Grabow, S. (1983). Christopher Alexander: The search for a New Paradigm in Architecture. London: Oriel Press Stockfield, Boston Henley.

Haken, H. (1983). *Synergetics, An Introduction*. Heidelberg: Springer.

Haken, H. (1991/2004). *Synergetic Computers and Cognition*. Heidelberg: Springer.

Haken, H. (1996). *Principles of Brain Functioning*. Heidelberg: Springer.

Haken, H. (1998). Decision making and optimization in regional planning. In Beckmann M.J., Johanson B., Snikars F. and Thord, R. (Eds.) *Knowledge and Networks in a Dynamic Economy*, Springer, Berlin.

Haken H. and Portugali, J. (1995). A synergetc approach to the self-organization of cities. *Environment and Planning B: Planning and Design* 22, 35-46.

Haken, H., and Portugali, J. (1996). Synergetics, Inter-representation networks and cognitive maps. Pp. 45-67 in *The construction of cognitive maps*, edited by J Portugali. Dordrecht: Kluwer academic publishers.

Hillier, B. (1996). *Space in the Machine*. Cambridge: Cambridge University Press.

Hillier, B. and Hanson, J. (1984). *The Social Logic of Space*. Cambridge: Cambridge University Press.
Jacobs, J. (1961). The Death and Life of Great American Cities: the failure of town planning. Harmondsworth, Penguin.
Jantsch, E. (1975). *Design for Evolution*. New York: George Braziller.
Jantsch, E., ed. (1981). *The Evolutionary Vision*. Boulder, CO: Westview Press.
Johnson, M. (1987). The Body in the Mind: the bodily basis of meaning, imagination, and reason. Chicago: The university of Chicago press.
Kitchin, R. and Blades, M. (2002). *The Cognition of Geographic Space*. (I.B.Tauris, London)
Lakoff, G. (1987). Women, Fire and Dangerous Things: what can categories reveal about the mind. Chicago: The University of Chicago press.
Lefebvre, H. (1970). *La Révolution Urbaine*. Paris: Gallimard. Translated into English in 2003 as *The Urban Revolution*.
McLuhan, M. and Fiore, Q. (1967). *The Medium is the Massage*, Penguin books.
Nicolis, G. and Prigogine, I. (1977). Self-Organization in Nonequilibrium Systems: From Dissipative Structures to Order Through Fluctuations. New York: Wiley-Interscience.
Portugali, J. (1993). *Implicate Relations: society and space in the Israeli-Palestinian conflict*. Dordrecht: Kluwer academic publishers.
Portugali, J. (1996). Inter-representation networks and cognitive maps. Pp. 11-43 in *The construction of cognitive maps*, edited by J. Portugali. Dordrecht: Kluwer academic publishers.
Portugali, J. (1999). *Self-Organization and the City*. Heidelberg: Springer.
Portugali, J. (2002). The seven basic propositions of SIRN (Synergetic Inter-Representation Networks). *Nonlinear Phenomena in Complex Systems* 5(4) 428-444.
Portugali, J. (2002a). A book review on *Planning Support Systems* by Brail, R.K. and Klosterman, R.E. (Eds.) 2001, *Planning Theory* 1(3) 271-279.
Portugali, J. (2004). Toward a cognitive approach to urban dynamics, *Environment and Planning B, Planning and Design* 31, 589-613.
Portugali, J. (Forthcoming). Complexity theory as a link between Space and Place. *Environment and Planning A*.
Rosch, E. (1999). Reclaiming concepts. *Journal of Consciousness Studies* 6, No. 11-12, 61-77.
Simon H. A. (1969/1999). *The Sciences of the Artificial*. MIT Press, Cambridge.
Smith, (2003). Introduction to *The Urban Revolution* by Henri Lefebvre.
Snow, C.P. (1964). *The Two Cultures and a Second Look*. Cambridge: Cambridge University Press
Turing (1936). "On computable numbers, with an application to the Entscheidungsproblem". *Proceedings of the London Mathematical Society*, Series 2, 42, 230-265.
Varela, F.J., E. Thompson, and E. Rosch. (1994). *The Embodied Mind*. Cambridge Mass.: The MIT press.
Weidlich W. (1999). "From fast to slow processes in the evolution of urban and regional settlement structures: the role of population pressure". In *Population, Environment and Society on the Verge of the 21st Century* (J. Portugali, ed.). A special theme issue, *Discrete Dynamics in Nature and Society* **3** 2-3, 137-147
Wirth L. (1938). Urbanism as a Way of Life. *American Journal of Sociology* 44,1-24.
Wittgenstein, L. (1953). *Philosophical Investigations*. Oxford: Blackwell.

Recognition of Natural and Artificial Environments by Computers: Commonalities and Differences

Hermann Haken

Abstract. In this paper I point at specific differences between natural and artificial environments with respect to recognition procedures by computers. In general the artificial environment appears far more regular than the natural environment. Basic computer procedures for pattern recognition are outlined. Special emphasis is laid on the semantics of a city where the study of languages as semantic networks is invoked as a paradigm where basic concepts of synergetics such as order parameters and the slaving principle are used.

1. Use of Computers

As it is witnessed by the various papers in this book, computers are more and more used in modeling and recognition in all sorts of problems of geography. Quite often modeling is based on cellular automata or multi-agent systems. In my paper I will be primarily concerned with recognition, though, of course, feedback effects between modeling and recognition must be taken into account also. More precisely, I will deal with pattern recognition.

> *One of the most interesting aspects of the world is that it can be considered to be made up of patterns. A pattern is essentially an arrangement. It is characterized by the order of the elements of which it is made rather than by the intrinsic nature of these elements.*
> *Norbert Wiener*

In general, each individual element of a pattern is itself a pattern, that is, an arrangement of still smaller elements and so on. Thus, we will have to consider also hierarchies of patterns. A standard approach to pattern recognition consists in using the concept of an associative memory. A simple example is provided by a telephone book. When we look up the name of a person it tells us his or her telephone number. Another example is provided by meeting a person where we want to associate a name with his or her face. Thus an associative memory is a completion of an incomplete set of data. From this follow, at least in principle, two basic tasks: The task of *learning*, the aim of which is to collect sets of complete data and the task of *recognition* which is a process of data completion. The recognition may be done by looking up tables (e.g. by humans or search machines) or by dynamical (synergetic) systems (e.g. human brains, neural networks, synergetic computers). Actually, and ultimately, all mentioned processes must be realized by a dynamical system (including the brain!), provided we are looking at the fundamental level.

When applying pattern recognition we may have different objectives in mind as is listed in the following Table 1. Please note that "pattern" can refer to quite abstract relations.

handwritten characters • printed letters
phonemes •letters
objects • designation, name
incomplete, deformed patterns • prototypes
movements, gestures • designation, meaning
complex scenes • "understanding" spatial relations
time-dependent patterns • recognition of temporal evolution, causal relationships
written or spoken texts • meaning, orders for specific action

Table 1 some objectives of pattern recognition

In order to get some insight into what computers can do, we need to discuss, at least superficially, how computers for pattern recognition work.

2. Typical computer procedures

Let me remind the reader of typical computer procedures to deal with associative memory. Usually a pattern, for instance in the form of a picture, is decomposed into its pixels. This "test pattern" may be incomplete or deformed, etc. It is then to be compared with stored "prototype patterns". An example is provided by face recognition (Fig.1). The data set of prototypes comprises faces with their names (Haken, 2004a).

Fig. 1. Prototype patterns (faces with letters encoding for family names)

The prototype to be stored depends on the recognition purpose. In the case of Figure 1, for example, it is sufficient to store only the front view; for other tasks, other views must be stored also. The prototype patterns serve so-to-speak as templates and one speaks of template matching of a test pattern. This matching is done by means of similarity measures where also deformations can be taken into account as well as specific transformations such as shifts in space, rotations and scaling. An important task consists in the learning of prototypes by means of the computer. In the learning process, a series of faces is randomly and repeatedly shown to the computer. The learning process works even when the examples shown are partly hidden, provided the square is positioned at random location (Fig. 2).

Fig. 2. learning of partially hidden faces

Another approach rests on feature extraction e.g. where lines or other significant parts are extracted, i.e. parts that contain high information. Such parts are usually those in which strong local changes of color or shades occur and they indicate edges, corners, etc. of objects. A major task for the computer (or better for its design) is to apply or develop methods to correlate the individual parts in such a way that a meaning can be attributed to this arrangement. (E.g.: eyes, nose, mouth → face).

3. Natural vs. Artificial

Here I come to the main objective of my contribution, namely a comparison between the natural and artificial. First of all we must observe that all objects are undergoing temporal changes, with the implication that our comparison will strongly depend on the time scales of our observations. In nature, consider for instance a forest. When we walk through it, it appears as a static object with its arrangements of trees (except for a stormy day). Over a period of half a year, some forests change their appearance (loosing leaves, for instance). The period of several years may show other effects: A newly planted forest grows. Another forest is macroscopically in a steady state: While its individual trees come, grow and vanish, it retains its macroscopic shape. The corresponding remarks hold, of course, for cities, as an example for an artificial environment.

The main issue I am proposing is this: when we consider static objects or environments there is a pronounced difference between natural and artificial in that the natural is, at least in general, far more structured and irregular. Figs. 3-6 may illustrate my point.

Fig. 3. oak tree

after: Kiedrowski R. Bäume dieser Welt. Augsburg: Naturbuch-Verlag, 1997

Fig. 4. prawn

after: Werner B. Farbatlas Meeresfauna. Bd.1. Stuttgart: Ulmer, 1993

Fig. 5. Hongkong Bank

after: Thiel-Siling. S (ed.) Architektur! München, London, New York: Prestel, 1998

Fig. 6. houses in Florida

after: Thiel-Siling. S (ed.) Architektur! München, London, New York: Prestel, 1998

Table 2 compares natural versus artificial entities. As Table 2 reveals, there are pronounced differences between the natural and the artificial, but there are also a number of counter examples. In fact, my observations must be taken with a grain of salt. Here the spatial scale plays a role and I treat objects seen by a human observer or navigator at his or her "natural" scale. The question is, of course, how important the differences between natural and artificial as well as the counter examples will be in special cases of practical interest. In my article I will take the typical differences as granted, however.

a) "static"

Natural	**Artificial**
irregular	regular
trees, plants	buildings, streets
forests, mountains	furniture
boundaries of objects	straight lines, rectangular
fractal nature	vertical, horizontal, but: perspective plain surfaces VR: limited computer capacity
structures at many scales	structures at few scales

Counter examples

desert	paintings
sea? but waves at many scales!	megacities

b) "dynamic"

depends on time and space scales of observation, e.g. traffic

ants	pedestrians
birds	cars
fish	lorries
	telephone calls, …

Table 2: Comparison between the natural and the artificial

4. Pattern recognition

Let me discuss some consequences for pattern recognition starting with the natural (Table 3). Because of the highly structured patterns there is much information per pixel conveyed. This makes it extremely difficult to apply the method of template matching because e.g. each tree looks quite different. So basically the number of prototypes is tending to infinity so that this approach becomes more or less unfeasible. Of course, one may try to "smear out" specific patterns, e.g. to return to types of trees but again this is a very difficult job. More details will be discussed in section 5.

In the artificial we expect little structured objects, i.e. they carry little information per pixel. We have so-to-speak a degeneracy of pixels. This means here one may apply the method of feature extraction, e.g. to focus on edges, corners or at larger scales on landmarks. Here I may refer to the work by Lynch (Lynch, 1960) and by Haken and Portugali (Haken & Portugali, 2003). But what will be further needed is more semantics. Details will be provided in sections 6, 7 and 9.

Natural	Artificial
highly structured	little structured
much information per pixel	little information per pixel "degeneracy" template matching (possibly waste ful)
template matching?	feature extraction:
number of prototypes →∞	edges, corners landmarks
"smearing out" types of trees	Lynch (Lynch, 1960) H. & Portugali (Haken & Portugali, 2003)
more details in sect. 5	↓ further needed semantics more details in sects 6, 7, 8

Table 3: Pattern recognition of natural versus artificial objects

5. Pattern recognition vs. pattern formation

5.1. Synergetics

Let me remind the reader of the definition of synergetics (Haken, 2004b). Synergetics is an interdisciplinary field of research that deals with the self-organization of structures and functions in systems that are composed of many individual parts. These systems may belong to both the animate and inanimate world. Synergetics aims at a unifying point of view, by focussing its attention on those situations, where qualitatively new features of the complex system appear at its macroscopic level. As is shown, close to these "critical points", the dynamics of a complex system is governed by, in general, few variables, the so-called *order parameters* that determine the behaviour of the individual parts of the complex system via the "slaving principle". In short, macroscopic structures, their growth and final establishment are determined by a competition between order parameters. In this way also pattern completion connected with an associative memory can be understood. Pattern recognition is nothing but pattern (or structure) formation. This is the basis of the synergetic computer. Figure 7 illustrates a process of face recognition implemented on the synergetic computer (For details see Haken, 2004a).

Fig. 7. face recognition by synergetic computer

5.2. Fractal Theory

There is, however, also a phenomenological approach to cope with complicated structures we are observing in nature, namely fractal structures (Mandelbrot, 1982). The principle is easily explained by means of Koch's snow flake (Fig. 8), where the same procedure is again and again applied at various scales of space.

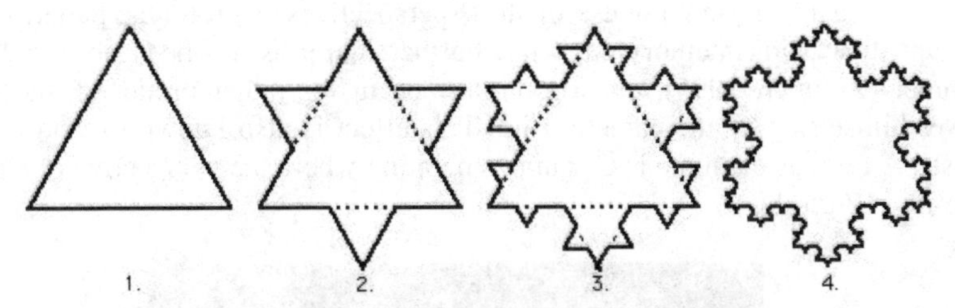

Fig. 8. Construction principle of Koch's snow flake

Mandelbrot's work (Mandelbrot, 1982) is well known here, e.g. the application of the same rule leads to the structure of a fern (Fig. 9), to tree-like structures (Fig. 10) or bizarre mountains. Such simple rules can lead to the formation of plants according to an algorithm originally introduced by Lindenmayer (1968) in a different context. Perhaps what is still lacking in such an approach is the deduction or explanation of these rules by means of a microscopic dynamics. It is an open question whether these phenomenological rules can be exploited for pattern recognition. Probably, at least in general, they are not exploited by humans because their analysis would require too much time but one may ask whether high-speed computers would be able to do so.

Fig. 9. fern

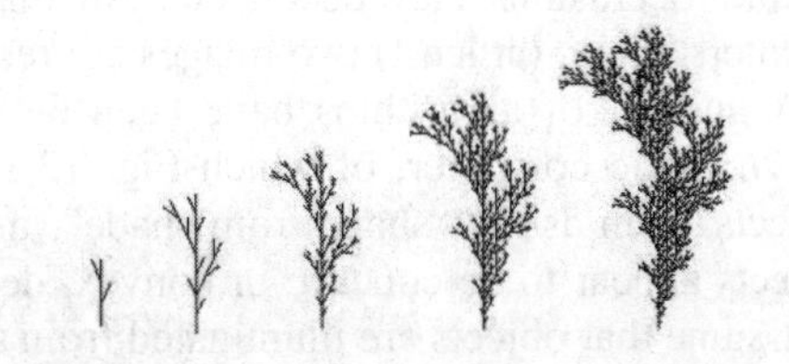

Fig. 10. „Construction" of a plant by use of the Lindenmayer algorithm

6. VR and its inverse

In order to mimic reality, virtual reality (VR) must take care of perspectives. For example, when we are driving through a city, the perspective changes all the time. In pattern recognition we must solve the inverse problem, namely, to restore, for instance, the front view of a house that is shown to us only in perspective. In fact, quite obviously, storing a house in all its perspectives as prototype patterns would amount to wasting memory capacity. Further more, as has been observed by A. Zimmer (Zimmer, 1995), distortions may occur, or rather produced, by the observer himself or by the camera used. This effect is also known to, and used by, artists. A famous example is Cezanne's painting where the perspective of a plate is distorted (Fig. 11).

Fig. 11. Section of a painting by Cézanne
after*:* Düchting H. Paul Cézanne. Köln: Taschen 1990

Inverse problems are difficult in mathematics: they may be not uniquely solvable, occlusions may occur, etc. An important task is also 3-D vision by computers. Here, (at least) two images are required to make use of the disparity effect. A number of algorithms have been developed including our own based on the synergetic computer, of which Fig. 12 is an example (Haken, 2004a). Other effects, such as the "shape from shade", may also be taken into account. Here, objects appear to be concave or convex, depending on shades, where we intuitively assume that objects are illuminated from above.

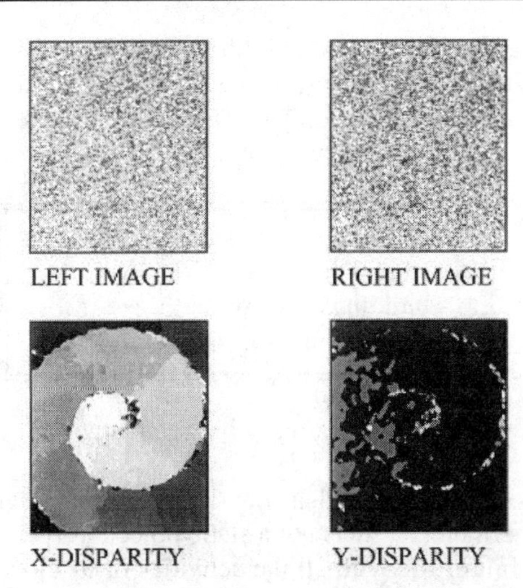

Fig. 12. 3D-recognition of a spiral
upper part: Julesz random dot patterns for the left and right eye, respectively
lower part: disparity maps (Haken, 2004a)

7. Semantics of the City

A good starting point for a discussion of cities as semantic networks is the fundamental work by Kevin Lynch (1960) "The image of the city". Lynch's work has recently been carried on in several ways. I will emphasis mainly two (Figure 13): One approach, by Haken and Portugali, (2003) in their paper "The face of the city is its information", deals with the interplay between Shannonian and semantic information. The second approach concerns the interpretation of the city as a language. Examples here are Alexander's "Pattern language" (Alexander et al, 1977) and Hillier's "space syntax" (Hanson and Hillier, 1984). A further, important line of thought is that of self-organization of a city (Portugali, 1999).

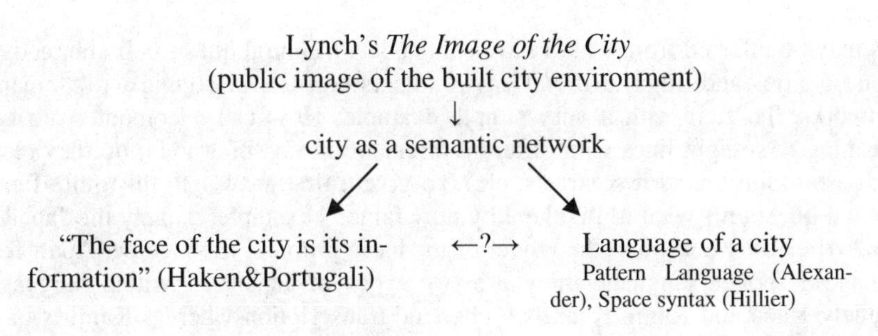

Figure 13 Recent elaborations of Lynch's *The Image of the City*

Because "the face of the city is its information" was published elsewhere, and is described in some details below (Portugali, Part II of this volume) here I just want to point at it and to focus mainly on the analogy between language and city. Here the analogies are rather obvious (Table 4):

city • text, article, book
street • sentence
buildings • words
parts of building •letters, syllables

Table 4. Some analogies between language and city

One should note, however, that the language of a city must be multi-dimensional. Furthermore, a city is not a static object, as (perhaps) some architects would see it. It is full of life with all the activities of its citizens including traffic, trade, sports, etc. Before turning to my main task which is to explore the relation between synergetics and understanding a language, I want to mention some other artificial environments. Here a few catchwords may suffice:

Interior of buildings (navigation & action)
Factory (production process)
Hospital (care of patients)
Apartment house (living together)
Flat (household)

These questions are not only of concern for architects, but also for design computers (CAD) and for all kinds of robots used in these environments. For instance household robots are becoming more and more fashionable though their "intellectual" capabilities are still extremely limited.

8. Semantic networks

As may be inferred from what I have said above, a central question in the recognition of cities and other artificial environments consists in coping with semantic networks. To start with a very simple example: How can a computer attribute meaning to straight lines with different orientations, e.g. in how far do they represent a building, windows, streets, etc? To give a first answer to this quite fundamental question I want to invoke a by now famous example, namely the Japanese typewriter. Just to remind the reader of the basic problem in this case. Apart from Japanese spoken language there are two types of Japanese written languages, namely Kana and Kanji. Kana is a phonetic transcription whereas Kanji is based on Chinese characters. An example is given in Table 5.

The basic difficulty of transforming Kana (or Romanji) into Kanji consists in the fact that there are numerous homonyms in Japanese. Or in other words, a phoneme may refer to many quite different meanings. (An example of a homonym in English is given in Figure 14).

English transcription	Roman transcription	Kanji
“kee”	“ki”	木
“neehon”	“nihon”	日本

Table 4: The Japanese written languages, namely Kana and Kanji.

In order to make this mapping from Kana to Kanji unique one has to take care of the context, i.e. one has to relate words that occur in the same sentence. Thus one has to apply a rather complicated semantic network, which can, however, be established by means of the new efficient computers. The learning phase consists in the establishment of tables. This can be done by programmers or by machines that operate on the frequency of pairwise occurrence of words in sentences, where, however, ultimately meaning is provided by people. The Japanese computer has been developed under the guidance of Ken Ichi Mori and is considered a major achievement nowadays. An example is given in Figure 15). Clearly what I have in mind here is rather obvious, namely can we apply a similar procedure to the city as a semantic network? In other words, we must relate different objects to each other and build networks where, of course, both human experience as well as computer power come into an interplay.

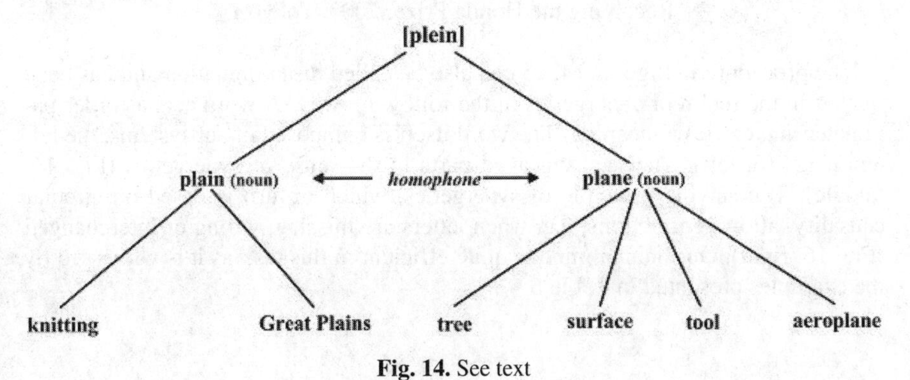

Fig. 14. See text

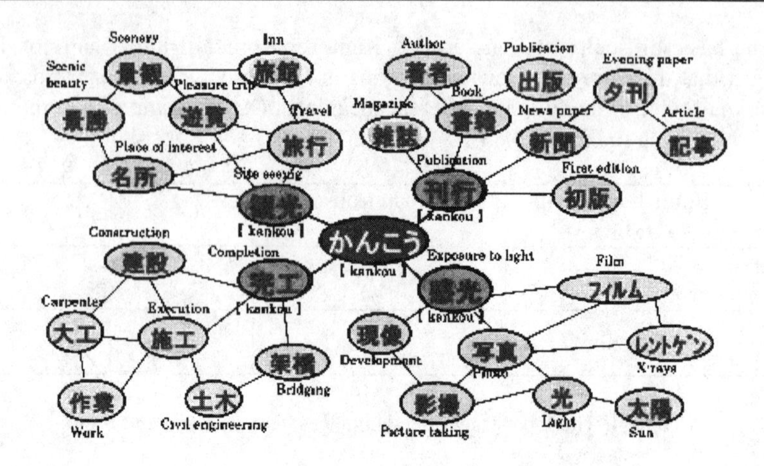

For example, homonyms corresponding to a *kana* expression "kankou「かんこう」" include 敢行(resolute action), 慣行(habitual or customary practice), 観光(sight-seeing), 刊行 (publication), 完工(completion), 感光(exposure to light) 官公(government and municipal offices), 勘考(taking in consideration) etc. However, if there are words such as travel, place of interest and excursion in the same sentence, the probability of hitting a right homonym will be higher when the word 観光(sight-seeing) is chosen from the list of homonyms corresponding to "kankou「かんこう」". This relationship is called "a relationship of co-occurrence between words.

Fig. 15. (After Ken Ichi Mori in his Commemorative Lecture on the Occasion of Receiving the Honda Prize, 2003. Tokyo.)

The procedure in Figures 14, 15 can also be called disambiguation and has been treated in the realm of synergetics in the following way: A word acts as order parameter that carries a meaning. The word itself is composed of subsystems, the letters (Fig. 16, left). They are enslaved parts in the sense of synergetics (Fig. 16, middle). The slaving principle of synergetics, which in turn is based on circular causality, allows corrections, e.g. when letters are missing, wrong or interchanged (Fig. 16, right). Our human mind is quite efficient in this task as it is witnessed by the examples presented in Table 6.

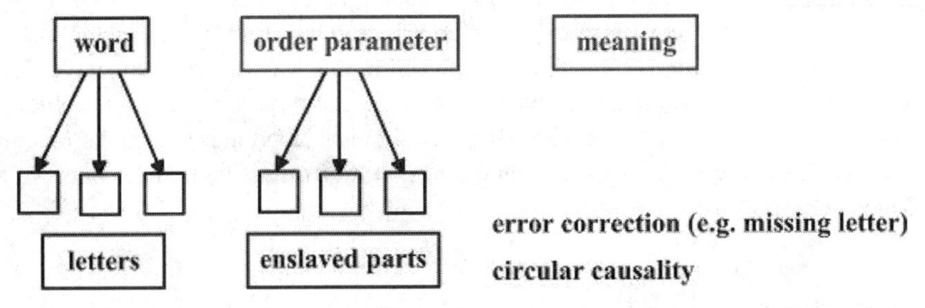

Fig. 16. A word conceived as order parameter conveys a meaning.

Spelling Errors

Aoccdrnig to a rseearch at an Elingsh uinervtisy, it deosn't mttaer in waht oredr the ltteers in a wrod are, the olny iprmoatnt tihng is that frist and lsat ltteer is at the rghit pclace. The rset can be a ttoal mses and you can sitll raed it wouthit porbelm. This is bcuseae we do not raed ervey lteter by it slef but the word as a wlohe.

Preosllnay I tinhk its cmolpete Bkolcols

Gmäeß eneir Sutide eneir elgnihcesn Uvinisterät, ist es nchit witihcg in wlecehr Rneflogheie die Bstachuebn in eneim Wrot snid, das ezniige was wcthigg ist, ist daß der ester und der leztte Bstabchue an der ritihcgegn Pstoiion snid. Der Rset knan ttoaelr Bsinöldn sien, tedztorm knan man ihn onhe Pemoblre lseen. Das ist so, wiel wir nciht jeedn Bstachuebn enzelin leesn, snderon das Wrot als gseatems.

Ehct ksras! Das ghet wicklirh!

Und jtzet prboiert es ssbelt!

Table 5: Our human mind is quite efficient in this task as it is witnessed by these spelling errors in English and German (see Fig. 15)

Words as order parameters carry some symmetry or, in other words, they have multiple meanings, e.g. the English word "bank" is ambiguous and may either be connected with "river" or with "money" (Fig.17). This internal symmetry must be broken by means of additional words such as "riverbank" or words occurring in the same sentence (Fig. 17). The whole text can thus be understood as sequence of symmetry-breaking events through which a hierarchy of order parameter runs so that eventually a complex meaning is built up. In this way one may conceive a city as a hierarchy of order parameters that undergo symmetry-breakings and, of course, a future task will be to realize this kind of approach by means of computers. Here, of course, not only static configurations must be treated, but also all kinds of activities.

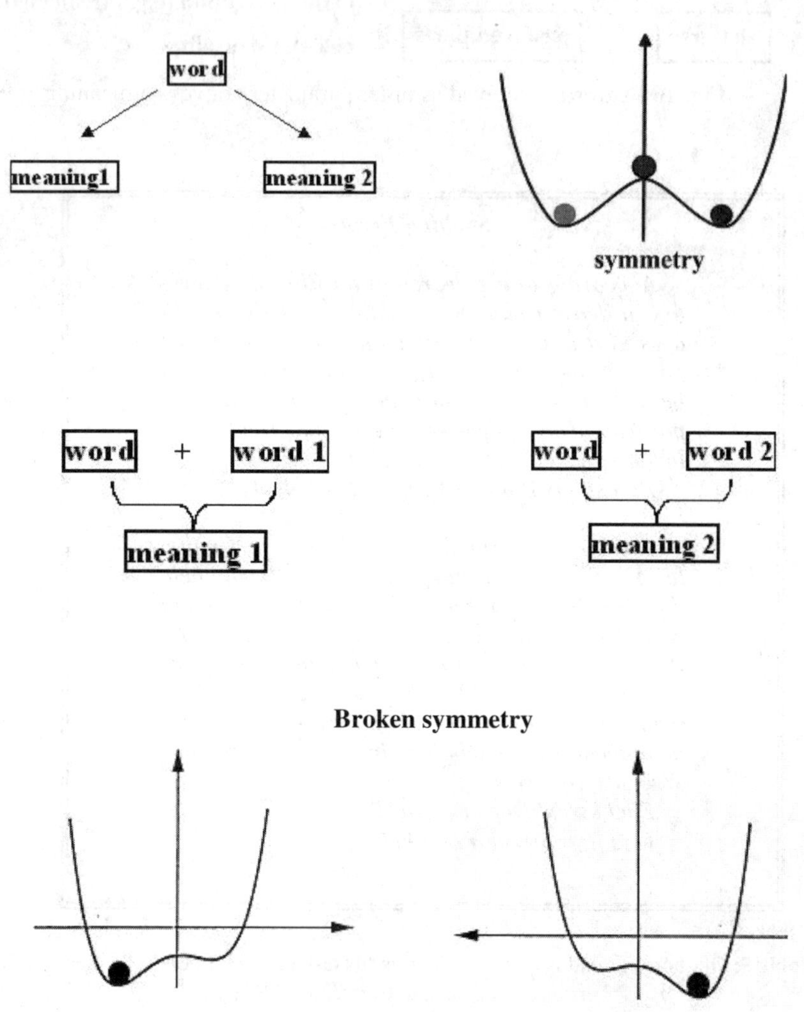

Fig. 17. *Upper part*: a word carrying two meanings can be interpreted as an order parameter that can, in principle, occupy two stable positions (valleys) in a "potential" landscape with symmetry. *Middle part*: by adding a second word, this symmetry can be broken, and a unique meaning results. *Lower part:* Visualization of broken symmetry by deformed potential landscapes

9. By means of associations and analogies to a new insight

In my contribution I have tried to stress the important role of *associations* in recognition processes. It may become clear from what I have said above that also *analogies* play an important role. So let me briefly discuss some further developments that one may expect. A comparatively simple network is that of a tree, which we may read in two directions, namely in the direction of more and more branches or in the opposite direction of a convergence of the individual branches (Fig 18).

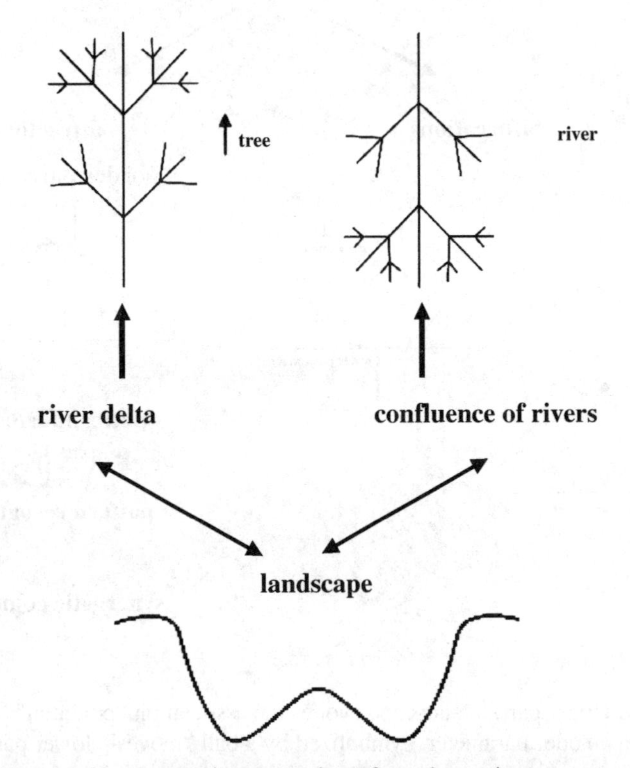

Fig. 18. Upper part: interpretation of graphs as rivers or trees

This is still a rather static picture. Now these pictures are not only reminding us of trees and their growth but also of rivers. In the one case of a river-delta, in the other of rivers formed from smaller rivers. But what causes these forms of rivers (in a self-organized fashion!)? They are caused by the landscape that itself is formed by the flowing water. With my bias coined by synergetics, the landscape evokes the association of a "potential landscape". In it a ball symbolizing an order parameter may move (Fig. 19 upper part).

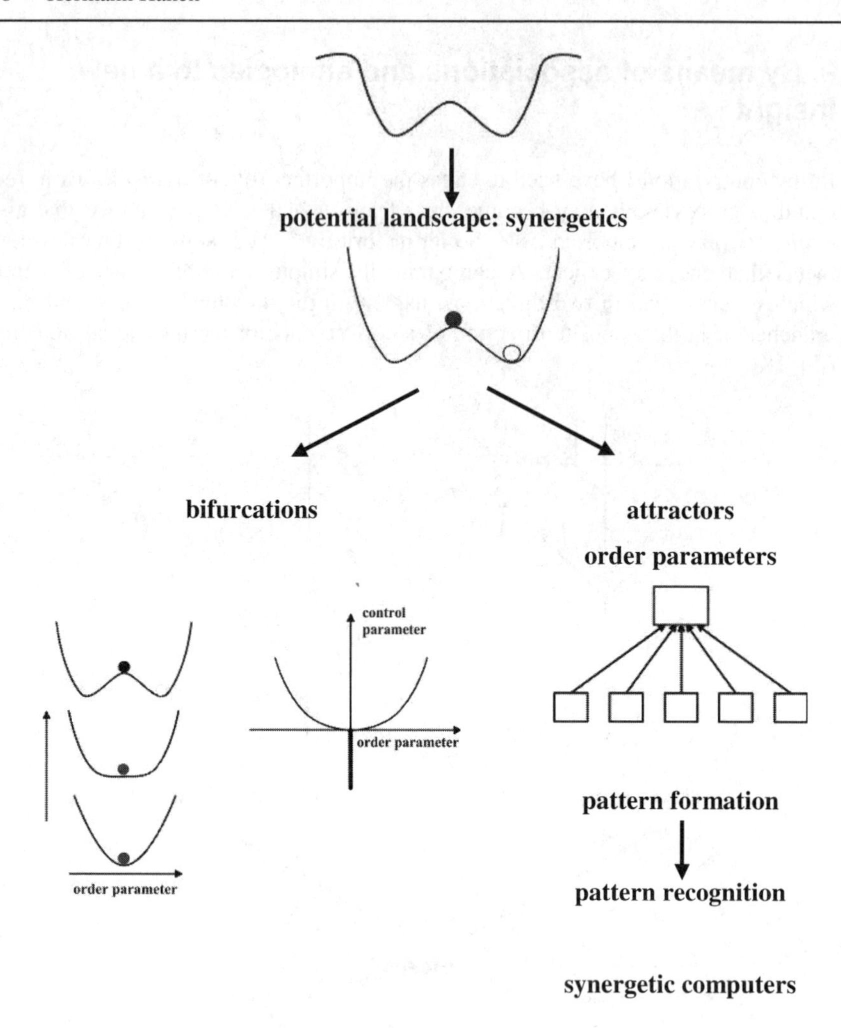

Fig. 19. Upper part: a landscape evokes the association "potential" landscape, in which an oder parameter, symbolized by a ball "moves". lower part: two different interpretations, cf. text

This figure allows us to understand the meaning of "bifurcation". Consider the left hand side of the lower part of Fig. 19. Here, three different potential landscapes are shown that are deformed (in the sequence from bottom to top) when an external control parameter (an externally given condition) is changed. Clearly, in the potential landscape at the bottom there is only one stable state of the order parameter ball. But in the double valley landscape at the top, there are two possible values of the order parameter. Thus, when we plot the displacement versus the size of the control parameter, a branching, i.e. a bifurcation occurs. This interpretation is, of course, well known, but it lies at the root of the tree-like structures I will discuss in the context of Fig. 20. Finally, going on the right hand side, lower part, of Fig. 19 upwards, we see a system of individual parts running into a collective state described by one or several order parameters. Thus, in the lower part of Fig. 19, left hand side, going the tree up we are reminded of bifurcations, going the opposite way we feel that the system runs into specific attractors. In the latter case I am led to the analogy between pattern formation and pattern recognition, an analogy which forms the basis of the synergetic computer on the one hand but also leads to the establishment of meaning on the other hand. Clearly, the steps I just described can be iterated which leads to Figs. 20 and 21. These schemes allow for many applications and interpretations. In the present context, we apply them to language (as a metaphor for cities). In Fig. 20, the upward axis indicates the increase of a control parameter, which may be time on the scale of millennia. The branching indicates the increasing complexity of languages. Clearly, much more should and could be said here, but because of lack of space these remarks must suffice here. In Fig. 21, by climbing up the order parameter hierarchy, we find the emergence of meaning of words, sentences, texts, but also, more generally, the meanings of patterns. But also from this point of view a central problem of a universal grammar might become visible: What are the building blocks of the structure of Fig. 21? Chomsky's work as well as the X-bar theory of linguistics can be mentioned here (Fig. 22).

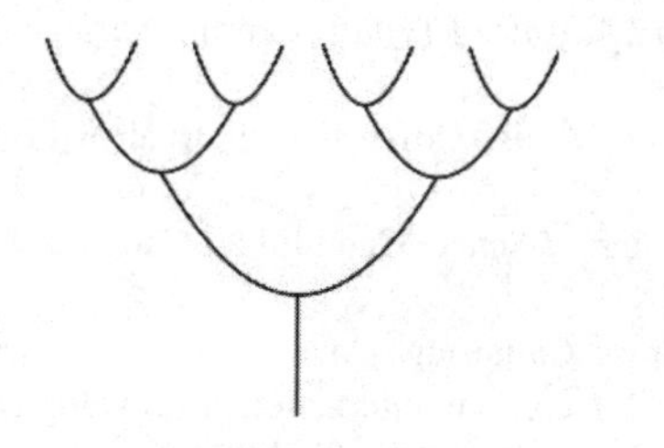

Fig. 20. Bifurcation tree

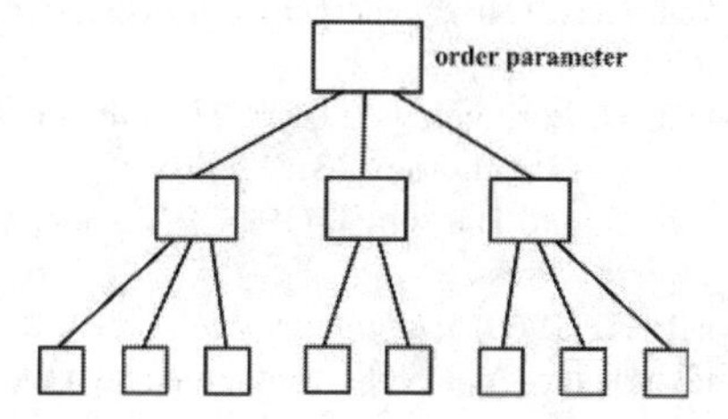

Fig. 21. Order parameter hierarchy

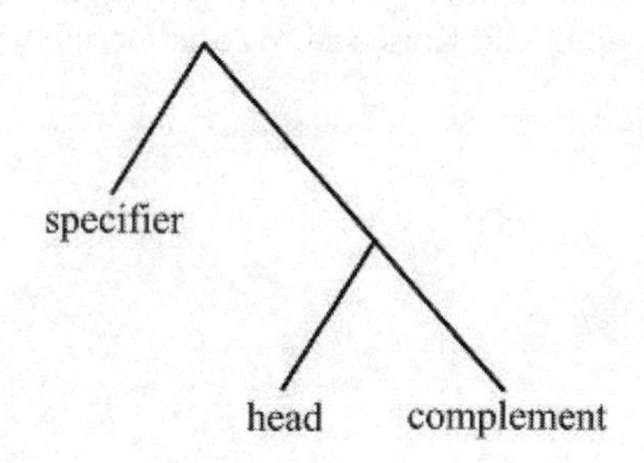

Fig. 22. Basic building block of a universal grammar

The application of these schemes to the study of cities, especially in the context of computer applications as generalizations of the synergetic computer must be left to further research. In my opinion recognition, cognition or the search for meaning by humans and computers, will be an open ended story. Perhaps we should adopt the perspective that the *journey is the reward.*

In other words, depending on our specific aims, we (or the computer) will have to climb up and down in the schemes of Fig. 20 and 21, thereby also switching between theses schemes. Eventually we are led to the following program: Can we devise a dynamical system so that from specific inputs in each case a specific meaning results. This concept was the foundation of the synergetic computer but it can be carried much further as we see by now, namely by following the individual steps.

Acknowledgement

I wish to thank Prof. Juval Portugali for valuable discussions.

References

Alexander, C. (1977). *A Pattern Language*. Oxford University Press.

Haken, H. (2004a). *Synergetic Computers and Cognition*, 2nd ed., Springer, Berlin.

Haken, H. (2004b). Synergetics. *Introduction and Advanced Topics*, Reprint, Springer, Berlin.

Haken, H, Portugali J. (2003). *The face of the city is its information*, Journal of Environmental Psychology **23**, 385-408.

Hillier, B. and Hanson, J. (1984). *The Social Logic of Space*. Cambridge: Cambridge University Press.

Lynch, K. (1960). *The image of the City*. MIT Press, Cambridge, MA.

Lindenmayer, A. (1968) Mathematical models for cellular interaction in development I. Filaments with one-sided inputs. Journal of Theor. Biol. **18**, 280-289.

Mandelbrot, B. (1982). *The Fractal Geometry of Nature*, W.H. Freeman & Company.

Portugali, J. (1999). *Self-organization and the city*. Springer, Heidelberg.

Zimmer, A.C. (1995). Multistability - More than just a Freak Phenomenon. In: *Ambiguity in Mind and Nature* (P. Kruse and M. Stadler, eds), pp.99-138. Springer, Berlin.

Visualization in Spatial Modeling

Michael Batty, Philip Steadman, and Yichun Xie

Abstract. This chapter deals with issues arising from a central theme in contemporary computer modeling – visualization. We first tie visualization to varieties of modeling along the continuum from iconic to symbolic and then focus on the notion that our models are so intrinsically complex that there are many different types of visualization that might be developed in their understanding and implementation. This focuses the debate on the very way of 'doing science' in that patterns and processes of any complexity can be better understood through visualizing the data, the simulations, and the outcomes that such models generate. As we have grown more sensitive to the problem of complexity in all systems, we are more aware that the twin goals of parsimony and verifiability which have dominated scientific theory since the 'Enlightenment' are up for grabs: good theories and models must 'look right' despite what our statistics and causal logics tell us. Visualization is the cutting edge of this new way of thinking about science but its styles vary enormously with context. Here we define three varieties: visualization of complicated systems to make things simple or at least explicable, which is the role of pedagogy; visualization to explore unanticipated outcomes and to refine processes that interact in unanticipated ways; and visualization to enable end users with no prior understanding of the science but a deep understanding of the problem to engage in using models for prediction, prescription, and control. We illustrate these themes with a model of an agricultural market which is the basis of modern urban economics – the von Thünen model of land rent and density; a model of urban development based on interacting spatial and temporal processes of land development – the *DUEM* model; and a pedestrian model of human movement at the fine scale where control of such movements to meet standards of public safety is intrinsically part of the model about which the controllers know intimately.

1. Defining visualization

Visualization is a term that gained widespread currency in the mid-1980s when for the first time, computer graphics were linked to supercomputer processing, particularly in scientific contexts such as astrophysics where it was essential for the results of such processes to be absorbed and understood visually (Kaufmann and Smarr, 1993). There was explicit recognition that many large data sets, whether produced for input to computer models or as outputs or simply as raw data captured by digital instruments, needed to be understood holistically using the synthetic properties of the mind and eye in unaided form. In parallel, our interaction with computers was becoming more visual and although this began prior to the micro-revolution, it was spurred on by the immediacy of interaction which the PC enabled. The ultimate outcome is that we now interact with computers almost exclusively using graphical user interfaces (GUIs) and this, in itself, has broadened the concept of visualization to most aspects of human-computer interaction.

Visualization is so broad a term that to define its role in spatial modeling, we first need to stand back a little and examine the kinds of models that the visualizations that we present here pertain to. Classifications of models go back to the 1950s and 1960s when the term became popular and it is instructive to note, for example in papers such as Ira Lowry's (1965) "A Short Course in Model Design", that the starting point was defined as continuum of models from iconic to symbolic embracing analog along the way. Iconic models are physical versions of the real thing, usually scaled down to toy-like proportions such as architects' block models, while analog models represent the system of interest using another but different system which may be either physical or digital. Good examples represent the movement of pedestrians in streets using analogies with hydrodynamic flow theory. Symbolic models are those which replace the physical or material system by some logical-mathematical structure, usually algebraic, with computer, hence digital, representation and manipulation a central feature of such simulation. Models in all these senses are of course simplifications where key features of their system relevant to their users are emphasized, often to the exclusion of many other features.

These issues are being rapidly influenced by a sea change which is occurring in how we view models in science and social science. Fifty years ago, modeling was parallel to classical science in that it was based on implementing good theory in models in the most parsimonious way possible. Good theories and models were those that could explain the data in the simplest, most efficient way, notwithstanding the fact that there were often critical issues which most would agree form part of the system being modeled, left out. This is now changing in every discipline and domain. The cutting edge of theory and modeling in the spatial-geographical domain, particularly where this involves human systems, is embracing ever more diverse and richer model structures. These structures are never likely to be validated in their entirety against data, they are too rich, and the data required for their testing too poor for any complete assessment. Many of the models that we introduce here follow this tradition in that there may only be a few points at which their data and processes touch the real world in terms of the data available. Science is now much more comfortable with this notion of a theory or model which is only partly testable, than it has been hitherto. There is increasing recognition that our systems of interest are intrinsically complex and must be handled rather differently from those on which classical science has been founded.

A further twist to the visualization paradigm involves the way the model is represented. As digital computation has become all pervasive, symbolic models no longer represent the sole focus. Iconic and analog models are also increasingly digital with the key issue being the mix of symbolism, analogy and iconic representation that can be employed for a single system where the simulation involves passing between any of these styles of modeling. The best examples involve relatively real renditions of spatial systems based on digital modeling of the physical appearance of the objects of interest. In this case, the appearance may only be used as the visual container in which analysis takes place. Such is the case in 3-d GIS (geographic information systems) where the 3-d container is a digital version of the physical fabric. Increasingly 3-d is being used is display patterns which can then be projected back onto the digital iconic model, or even onto a physical model of the system itself as in the *Tangible Media* projects at MIT (http://tangible.media.mit.edu/). Moreover once a digital iconic model is developed, it can be aggregated into various forms, put into other digital contexts in semi-recursive fashion, and even used to manufacture actual physical models as hard copy print versions in traditional wood or plastic. An early version of such a mix of media is embodied in the hypothesis that Vermeer used a simple camera to generate some of his paintings (see Steadman's (2001) book *Vermeer's Camera* at http://www.vermeerscamera.co.uk/). Animating such a mix and using this digitally as part of the story line is what Tufte (1997) calls a 'visual confection'.

In this paper, we will concentrate on what has come to be called 'scientific visualization' whose main focus is on the inputs, processes, and outputs associated with symbolic or mathematical models, in this case urban and spatial systems focused on the human-built environment. In the next section, we will introduce a generic modeling process and show how this can be linked to planning, management and action. It is this nexus of explanation, simulation, forecasting, design and control which provides the wider context for visualization, and we will thus identify the key types that map onto this spectrum of possibility. We will then develop a generic form for visualization in the spatial modeling field, which tie these various possibilities together in what we call the 'space of visualization'. This sets the scene for three distinct demonstrations: the first is pedagogic and focuses on an explanation of a well-known theory of land use due to von Thünen, the second enables a model of dynamic (temporal) urban development to be explored, and the third shows how important it is to develop models with strong visual content which enable designers and decision-makers to use models to generate effective designs and policies.

2. Defining spatial modeling

Computer models are structured in many different ways but the standard sequence follows digital processing which involves manipulating a series of inputs which drive the model to a series of outputs, thus reflecting the various functions that tie the elements of the model together. This sequence reflects the logic chain that any model is built around, with inputs defining the exogenous or independent variables that dictate how endogenous or dependent variables are conditioned. Many models involve elaborate causal chains which may be activated many times, recursively through different kinds of time which in turn enable the model's outputs to become stable. This *model processing* is usually nested within a wider process of *model fitting*, estimation, or calibration which enables parameters of the model – macro variables that usually have global significance – to be tuned to values that connect the inputs to the outputs in the most satisfactory way. This kind of prediction is enabled so that confidence in the way the model reproduces a known situation as reflected in the input data and the independently corroborated data associated with the model's outputs or predictions, is assured. The third process is *model use*, in conditional prediction which often mirrors calibration, but is part of a wider phase in which the model can be used in design, management, and control.

Each of these processes can be represented using different forms of visualization. Indeed every aspect of model operation and use can be visualized as the ultimate structure of the model is digital whose location in computer memory can be mapped in some way to the 2-dimensional screen. As all our models are spatial, hence some of their inputs and outputs are mappable, then associations between inputs and outputs with respect to map pattern provide an obvious form. There are many ways in which such inputs and outputs can be linked - offline or online in terms of showing how inputs are converted into outputs and a classic strategy of visual comparison is to array these maps as separate and comparable layers, as 'small multiples' in the manner suggested by Tufte (1990). 3-dimensional forms can be widely exploited too as in the standard manner where such representations are portrayed using the three Euclidean dimensions. But at the level of abstraction used in this style of modeling, the third dimension is more likely to be employed for scientific visualization of the phase space of model solution rather than for more literal, or iconic visualizations associated with built form or rural landscapes.

These visualizations almost assume that what is being modeled is static in structure where outputs occur at a single point in time but many spatial models are dynamic and thus sequences of inputs and outputs need to be visualized. This is the space-time process and although small multiples are useful, animation in 2-d or 3-d is often employed. We will return to this when we deal with calibration Low; but animation also constitutes a way of linking inputs to outputs, thus revealing model functions or processes. There are also different dynamics, from the routine where the focus is on showing how objects move in space, to longer-term migrations where the focus is in comparing map patterns at different points in time. Movement is most simply conceived of in terms of animation but there are a variety of ways in which such animations can be linked to more abstract properties of the map patterns and the space-time movement of related objects.

A time-honored strategy for spatial visualization is to use multiple windows with different phenomena in each, some spatial in 2-d or 3-d but some aspatial or non-spatial; and to hot-link these windows so that change in each can be related. The process of linking inputs to outputs involves the functioning of the causal chains which form the core of the model. If the model is dynamic, visualization may be built around space-time in the literal sense but if the process is recursive for the model to converge on stable outputs, this too might be visualized in much the same way. In cases where the model is both temporally dynamic and recursive in terms of its path to solution, then a combination of both is possible.

Spatial models are often built around aspatial or non-spatial processes which although touching the spatial system at some point, can be represented using visual traditions very different from the 2-d map or 3-d surface. For example, the spatial economic model that we first introduce below is conceived in terms of demand and supply curves and only then mapped onto a simplified spatial landscape. In fact, one of the great powers of scientific visualization is to make such links between non-spatial, aspatial and spatial representations, as much for pedagogic purposes as for use in more practical contexts. However such mapping from one visual medium to another in terms of representation always needs to be determined before visualization takes place. All this means that visualization is a creative process. It is only as good as our imagining of how different elements of a model relate to one another and to the wider context in which they sit.

The second environment in which spatial models are formed involves the process of *model fitting*. This connects up directly to searching for pattern in data but in particular for pattern in the input data which is exploited by tuning the model's processes and functions to explain as much of this pattern as possible. This is usually accomplished using a process of trial and error fitting with successive improvement to the best fit. To state the long standing analogy, the process might be visualized as climbing a hill where the surface terrain represents the different performance of the model with respect to different input parameters. The process of model fitting is then the process of climbing this hill and reaching some global optima, ensuring that the process does not get stuck in some local optima, some hillock in the landscape of hills. In many traditional models which are parsimonious in the extreme such as those embodying spatial interaction (Batty, 1976), the structure is such that a unimodal performance surface can be ensured if the model is formulated with mathematical correctness. Standard procedures can then be used to reach the global optima. In fact, visualizations of this process showing the climb across the terrain have been used quite widely since the 1960s.

Most of our models are much more complex in that the phase space which embodies this terrain can no longer be mapped due to the very large number of parameters that contemporary spatial models contain. This is especially true of simulation models incorporating new notions of cells and agents. There the process is often much more partial, in that the calibration might be visualized using exploratory procedures which do not aim to find any global optima. In such cases, there still needs to be structure to the process. In fact, the calibration phase of spatial modeling is pushed one stage back in these more exploratory models due to the fact that the very formulation of the model itself comes under scrutiny as soon as spatial data begins to be explored. Exploratory spatial data analysis (ESDA) which became popular after the first wave of scientific visualization had been established in the early 1990s, threw up the notion that the model should emerge naturally from an exploration of pattern in data. Although there have been many demonstrations, the focus has been more on inductive generalization than on the development of explicitly deductive models which arise from such analysis. Pattern and error in the data is a key consequence of such ESDA but there have been very few examples where such analysis has then proceeded to build models round these patterns, other than those which are based on inductive statistics. Although visualization may help generate new models, this is likely to be over a much longer time span than the modeling process itself and although the notion that the model emerges naturally from such exploration is an attractive one, almost Eureka-like in its impact, this is unlikely to be the case in most modeling efforts. Usually the model is proposed in advance and all the focus is on tuning it to a real situation, understanding that situation, and using the model predictively and prescriptively to solve some problem or implement some new design.

The third environment in which spatial models sit involves their practical use in prediction and policy. In the simplest sense, similar visualizations might be built around prediction and prescription as around data exploration and model calibration. Early software embodying GIS functionality in standard cross-sectional urban models illustrates this principle where standard sets of visualization functions apply to any of the four model stages – data assembly and checking, analysis, prediction, prescription (Batty, 1992). However this process raises the question of purpose and engagement in terms of what the model for, who is it for, and how is it to be used? In short, models are rarely for the indulgence of the model-builder or scientist, more likely for the persons who commissioned it in the first place for practical use. In our examples below, we present three models; the first is for educating ourselves, while although the second is for exploration of urban development processes, it is conceived as being applicable to practical problems and policies involving urban sprawl. The third is quite definitely for the stakeholders involved in solving problems in the local environment.

In these applied and policy contexts, visualization is likely to be a little different from the kinds of scientific visualization we have been implying so far. In fact it is likely to reflect a much looser interpretation of model inputs, processes and outputs and may be linked to media that do not form part of the model in the first place. This is no less rigorous but it does change the kind of engagement that modelers and stakeholders have in the process. Visualization thus becomes an essential part of communicating complex ideas to a non-expert clientele, and in this sense, it probably involves developing procedures for involving this clientele at different stages of the modeling process. The notion that the model is delivered, scientists explore and tune it using visualization, and its outputs are then pictured in conventional scientific form, is not necessarily the most appropriate procedure in situations where stakeholders are involved in using models directly. We will present such applications as part of our third example below.

3. The space of visualization

Visualization as a style and set of activities is almost impossible to classify for every aspect of computer modeling and its application can be subject to representation using digital pictures. Nevertheless it is useful to begin to organize the field with respect to the models we will illustrate here, if only to show how the kinds of visualization employed depend intimately on purpose, the system of interest, and the environment in which the application exists. We will define a generic space which is organized on two levels: first with respect to the purposes for which the model, thence its visualization is designed, and second, in terms of the key techniques used to implement its visualization. We could have developed a third level based on different media but in all the cases we conceive of here, the media are conventional pencil and paper and their digital equivalents. The panoply of VR and tangible media have not yet been invoked in any of the models that we present here although this is an important direction in which these more abstract models should be developed.

We define four distinct purposes: education, exploration, explanation, and engagement. These purposes are not mutually exclusive, nor are they arrayed orthogonally; more likely a model and its visualization tend to stress these four purposes in different ways, often with one purpose dominating. For example, visualization for education can be both narrow and wide although in the sense used here, we will be taking a narrower view. Of course all model building and applications involve education of ourselves and of our clients but in this context, we are specifically thinking of visualization for the prime purpose of getting the message over of how a model actually works. In this sense then, we see visualization as enabling an understanding which would not be possible without pictorial help. This kind of visualization makes the operation and meaning of the model much clearer than any other form of communication.

Exploration is more geared to investigating how model structures translate inputs into outputs. This is an essential quest in learning about how the model works. The more complicated the simulation, the more likely that exploration is required to test the limits of the model, and to enable researchers to be sensitized to the impacts of their scientific decisions. All modeling involves some exploration but in the development of models separate from immediate practice, then exploration is of the essence, especially where causal structures cannot be analyzed using mathematical formalisms. Exploration might simply be trial and error based on trying to find out how the model behaves, or it may be more systematic as, for example, in the process of calibration.

In contrast, explanation involves using visualization to confirm or falsify some theory which is embodied within the model, and the usual processes of comparing pattern in the input and output data is central to this. A tricky issue however is to visualize how the various processes linking inputs to outputs match what we know about the operation of such processes in real life. Often visualization as explanation is rather partial, being based solely on a comparison of outputs from the model with those that observed in the real world. Every model which is built afresh, requires some sense of how well it explains the reality to which it is being applied, although this is more likely to be to the fore in applications which are removed from practice.

The last purposive activity we define is engagement. Rather than define purposes which involve forecasting for policy-making, forecasting to test design impacts, management or control, we prefer to simply note that models which are developed for purposes other than science *per se*, involve the engagement of non-modeling experts. In fact, as models are often built by large teams whose expertise differs markedly between team members, then it might be supposed that visualization might be used to engage the team in assembling the best model. This is indeed the case but here the contrast is greater between scientists and non-scientist stakeholders who need to be involved in the process in rather different ways. Essentially engaging stakeholders and non-scientific experts involves different kinds of visualization and dissemination which probably requires more non-scientific information to be assembled and related to the model and its application. Furthermore, the process of using visualization tools becomes significant when diverse groups are involved in this kind of communication.

The second level of visualization we define is based on a limited number of techniques. Much ingenious visualization is one-off and cannot be classified generically. This is because visualization tends to involve some insight which is produced idiosyncratically and then pictured in some meaningful way. There are no formulas for creating such graphics although there are some simple and obvious methods for taking spatial media and representing them visually. We define the 2-d map and the 3-d icon – surface, iconic physical shape, and so on – as being key elements of the way we visualize map pattern. Spatial process is harder to fashion but the notion of a process occurring in space and time can be illustrated using small multiples (of pictures) which provide a sense of change in space and time; animation is often simply arranging these multiples as frames in sequence. We also invoke the parameter space as being a vehicle which controls the operation of the model specifically through calibration. These five features can be arrayed against one another and combined in diverse ways to give real substance to the idea of modeling as visual confection. The use of hot-linking through multiple windows is simply one of the ways in which such visualizations can be animated and linked.

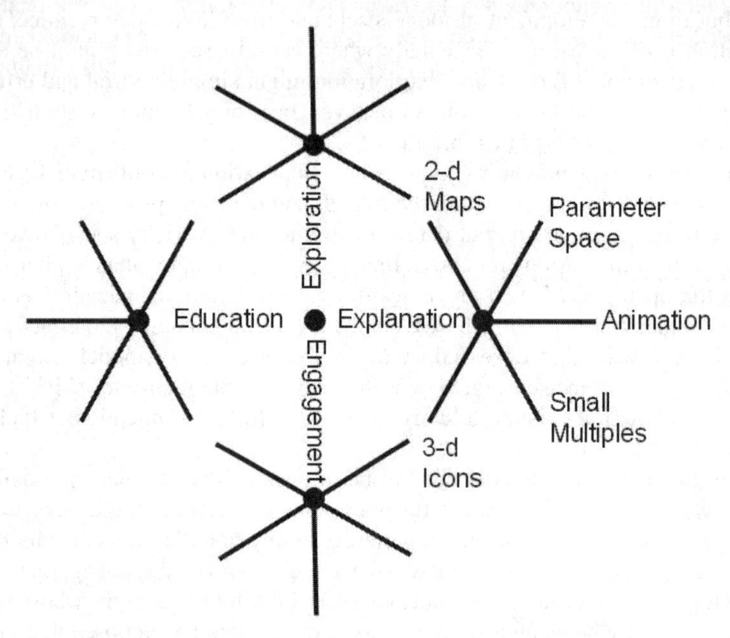

Fig. 1. Elements of Visualization: Purpose and Technique

These two levels of visualization by purpose and technique lead to a tree of possibilities that we have arrayed in Figure 1. This provides a simple means of classifying the types of visualization in any spatial model application and we will use it to illustrate the relative differences between the three applications which follow. In any full scale application, many branches of this tree will be invoked: models are defined to educate, explore and explain with these activities often taking place, even though the main purpose is to engage those stakeholders who control the policy and design process. Our first example below clearly illustrates this point although the three applications we have chosen are very different from one another.

4. Visualizing theoretical explanations: the von Thünen model of market land rent and transport

Count Johann Heinrich von Thünen is accredited with devising the first land use model in 1826 which explained the spatial distribution of crop cultivation (and/or animal husbandry), using evidence from his farm in East Prussia (Hall, 1966). His model essentially examines the trade-off between the productivity of agricultural land with respect to the yield and price associated with selling the crop at the market and the transportation cost necessary to move it there. In essence, this tradeoff fixes the level of rent that the farmer is able to pay and the use of the land is thus determined by the crop that generates the highest rent. The model involves comparing several crops and determining, with respect to one market center, the actual crop cultivation on each land parcel at different distances from the market. If there is more than one market center, the allocation depends on a comparison of possibilities but the solution is still stable for the selection of the land use simply depends on the maximum rent payable from whatever market.

For a single market with uniform transport costs which imply location on a homogeneous plain, land uses arrange themselves concentrically around the center. This can be easily extended to several centers. If physical distortions due to the transport network are introduced, then the pattern of location is affected, with fast transport routes having higher accessibility than the areas between them. This model is simple in form but the spatial outcomes from its generalization to many centers and to many different transport routes are often difficult to anticipate, hence the need for some simple demonstration. Moreover, when the price of the good or its composite yield, and the relative transport costs are changed, the pattern of land uses shifts. This is the essence of the model for which we have produced a very simple but effective visualization. Of course, the importance of the model does not lie in its application to an agricultural or rural system but in its generalization to the urban realm where it forms the basic equilibrium model of the urban land market, underpinning contemporary urban economics (Alonso, 1964).

To illustrate how tricky the model is to non-mathematicians, we first define a location with respect to a single market center as j, the distance from the center to j as d_j, and the cost of transport for the commodity in question, k, as β^k. The quantity produced per unit of land or yield is assumed to be uniform – on a homogeneous plain, and is defined as Q^k with its price at the market as ρ^k. We will also assume a fixed cost of production for each commodity called c^k. The rent which a farmer cultivating a crop k at j is called R_j^k and is calculated as

$$R_j^k = Q^k[\rho^k - c^k] - \beta^k d_j \tag{1}$$

This is a linear equation, sometimes called a bid rent curve, in that it shows what a farmer cultivating crop k can bid for renting the land at different distances

j from the market center. In the model, we have built, we assume that one element of the yield, price and fixed cost in equation (1) can be varied and thus this equation can be written as

$$R_j^k = \alpha^k - \beta^k d_j \tag{2}$$

For a single market center, the use of each parcel of land j is computed as the maximum rent payable over all crops, that is $R_j^l = \max_k R_j^k$. For any pair of land uses/crops, it is also easy to compute points in the landscape where the rent payable is the same, that is the breakpoints between crops with respect to their distance from the market. This occurs where $R_j^k = R_j^l$ and this equation can be solved for any k and l to yield the break point $d_j(k:l)$ as

$$d_j(k:l) = \frac{\alpha^k - \alpha^l}{\beta^k - \beta^l} \tag{3}$$

Finally where there are two or more market centers called i, the land rents need to be computed using a modified form of the bid rent equations which are now indexed with respect to each center as

$$R_{ij}^k = Q^k[\rho^k - c^k] - \beta^k d_{ij} \tag{4}$$

We choose the land use for each location j which maximizes the rent as

$$R_j^l = \max_{ki} R_{ij}^k$$

The visualization necessary in this model is largely so that the user can understand how the land use equilibrium occurs. The software essentially enables the user to show how the bid rent curves for three land uses – milk, grain and livestock production – can vary in price and transport cost, thus changing the intercepts and slopes of the linear bid rent function, α^k and β^k. The guts of the visualization are a blank canvas – the map – onto which one can draw transport routes and locate market centers. It is also possible to define constraints such as unproductive land. The canvas is initially an homogeneous plain but background maps can be attached to it so that the user can draw on features that pertain to some real situation. The second type of canvas but within the same GUI, reflects the bid rent curves, one canvas for changing the slope of these, the other for the intercept. When these are changed, the distribution of land uses appears immediately and thus there is a direct association from the parameter space to the real space, with the parameter space in fact being constituted as a cross section through the hypothetical real space. The final feature of the visualization is the same canvas but with the distribution of land uses portrayed in 3-d as a wire frame or rendered image. There are two sliders to change the orientation of the *x-y* and *z* dimensions associated with this visualization.

This visualization is to educate the user through explanation. It was devised as part of the Open University Third Level Course on *Cities and Technology* and the software was distributed free to students registered on this course (Steadman, 1999). It also contains elements of exploration but it is strictly designed to explain and educate. In terms of our characterization in Figure 1, we show this form of visualization in Figure 2(a) (where we also compare this with the two other examples reproduced here) and this shows immediately that we are using minimal but effective visualization techniques. The essence of this is a small portable piece of software which embodies a kind of sketch explanation or sketch modeling. Only one window is used and there is no hot-linking but the control over the model is so quick and direct that this is an example of extremely parsimonious visualization which is pretty effective.

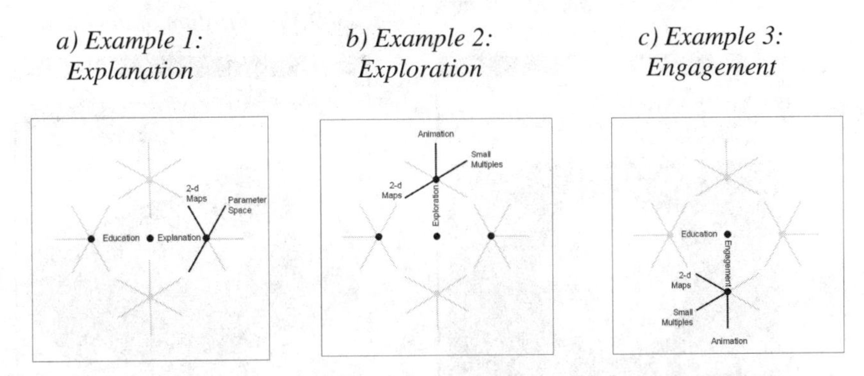

Fig. 2. Structure of Visualization in the Three Example Models

We show two examples of the use of this software in Figures 3 and 4. In Figure 3, we show the single market center with no distortions associated with transport routes. This is the homogenous plain example which appears everywhere in the location theory literature. The concentric symmetric ring structure around the market is clearly shown. In Figure 4, we have taken a map of Chicago and its railroads in 1861 from Cronon's (1991) magnificent book and use this to impress the fact that land use around Chicago is influenced by these routes. Note that we define Lake Michigan as 'unproductive land'. The resulting land use pattern shows the classic distortion posed by differential transport routes. The 3-d surface also shows the limits to the degree of distortion in that the picture is a little too confused. Nevertheless this does show how a model can be moved from theory to practice, from hypothesis to reality, albeit that the realities we choose are more caricatures of the real thing than the sorts of model reality we present below. There are other bells and whistles that are invoked in this software – fuzzy boundaries and precise distances, scaling of bid rent curves to reflect actual distances, and so on – but the essence of the visualization is shown in Figures 3 and 4 which provide a complete picture of this approach to explaining location theory and its relationship to the micro-economy in a spatial setting.

The Canvas:
The Homogeneous Plain

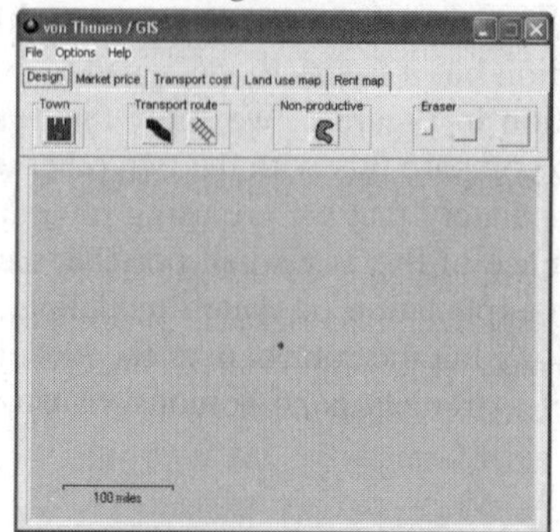

The Price-Transport Cost Tradeoff

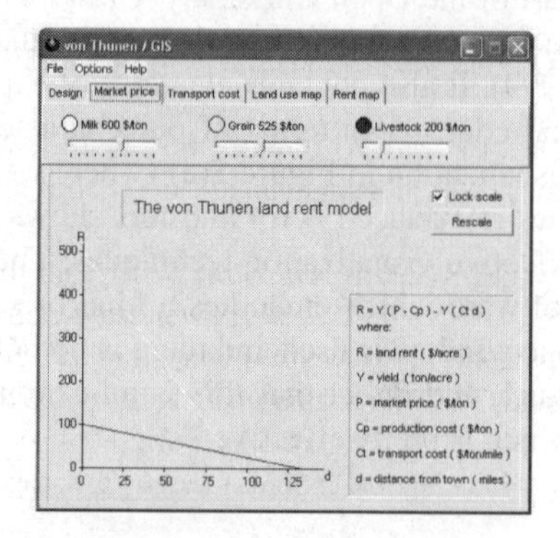

The Concentric Land Use Ring Solution

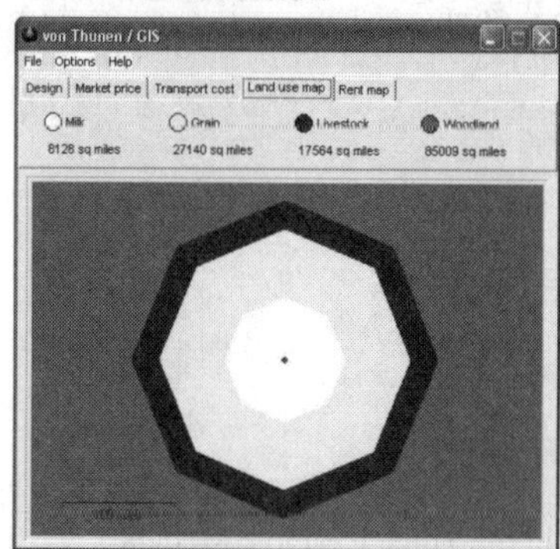

3-D Visualization of the Rings

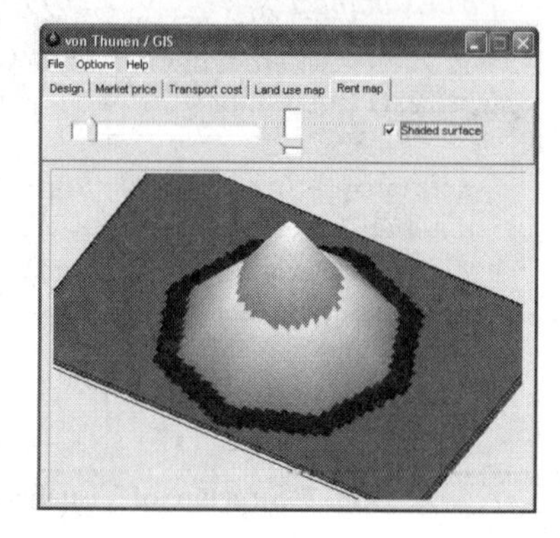

Fig. 3. von Thünen's Model

Simple software demonstrates how the tradeoff between product yield and transport cost gives rise to land use competition and stable spatial organization. You can download the software for this application from http://www.casa.ucl.ac.uk/vonthunen/

The Canvas: Chicago 1861 (from Cronon, 1991)

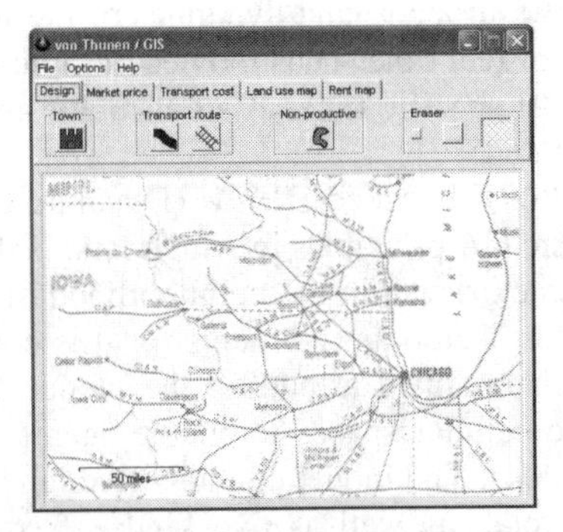

Railroads and Lake

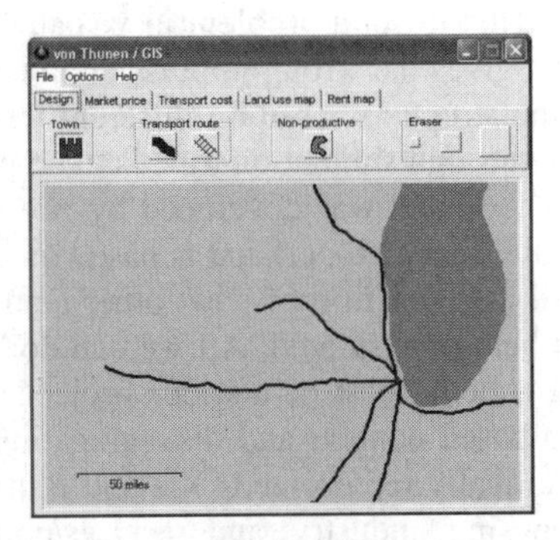

Land Use Solution: Distorted Rings

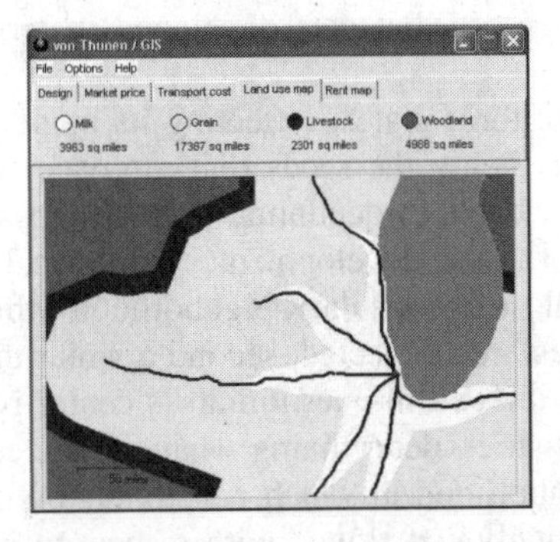

3-D Visualization

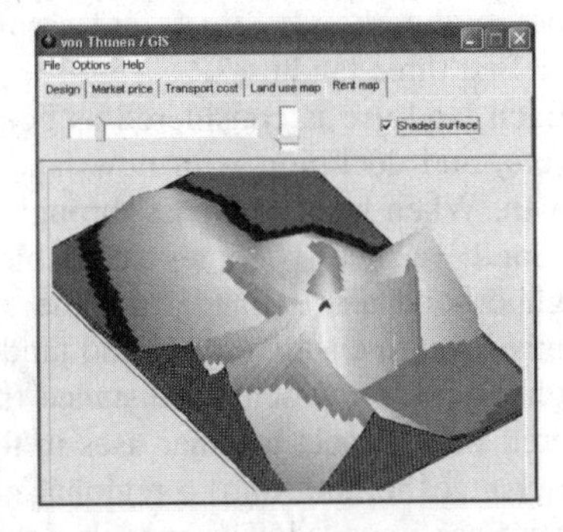

Fig. 4. Nature's Metropolis: How Railroad Structure Distorts the Concentric Pattern in 19thC Chicago and the middle west

5. Visualizing model exploration and calibration: urban development using cellular automata

Our second and third visualizations involve models whose outputs can be examined as they are executed. In a sense, this is true of the von Thünen model but as this is a comparative static structure, its operation is immediate. Models which can be examined as they run are usually temporally dynamic with the time simulation synchronized with simulation in computer time, notwithstanding any additional processing involving trial and error calibration during the simulation time itself. In the case of our first model of this genre – a dynamic model of urban land development using cellular automata principles – as the land area becomes more developed, more and more functions are invoked to examine distance relationships and comparative links relating different spatially specific land uses to one another. Thus the computer time taken slows in linear proportion to this additional processing. This is not a problem in visualizing the structure and dynamics of the model but it gives the wrong impression in terms of real time. Thus movies must be made of the structure based on different frames at specific times if an accurate impression through simulation time is to be presented.

The model was developed by Xie (1996). It is called *DUEM* (Dynamic Urban Evolutionary Model) and is based on classic CA principles in which land is developed as a function of what other land uses exist within the neighborhood of each site being considered. All we can do here is sketch its rudiments for it is complicated, being part of a wider project which is aimed at putting the model into a web-based context and disaggregating the cellular spaces to enable agents to be specifically represented (Xie and Batty, 2004). The model contains five land uses – housing, industry, and services/commercial, as well as two kinds of street – junctions and segments. Junctions are needed to connect segments, and housing, industry and commerce cannot develop without there being streets nearby, within some neighborhood. Streets are a function of what gets developed in terms of these first three land uses.

Each land use is considered as being in three states, reflecting its aging: new, mature, and declining with new land uses being the seeds that motivate further growth. When land uses pass through their cycle to declining, they disappear and the land vacated becomes available for new development. Three scales of neighborhood are defined: first the small strict cellular neighborhood which is mainly used to ensure streets and land uses are connected, second a wider district neighborhood in which the distance from the cell in question at its center is considered with respect to other uses in the district, determining what use the central cell changes to, and third a regional neighborhood in which constraints on development are imposed. The model is thus richly constituted with respect to its life cycling and the representation of spatial scale.

We have designed the interface in an entirely visual way based on two kinds of windows: the map canvas and four related windows which show the trajectories of growth and decline for housing, industry and commerce, and all three of these. The user can switch elements on or off in each of these windows. The key reason for arranging the visualization of urban dynamics in this way is to present two features: the capacitated growth of the spatial system which is reached when the map canvas fills up, thus illustrating logistic growth and oscillations around the equilibrium, with even the possibility, yet to be seen, of simulating chaotic growth; and the wave effects seen when land uses age through their cycle. Waves of change cycle across the map, eventually dissipating as the system gets older. Because these kinds of model are highly diverse and contain many parameters, visualizations must be highly tuned to particular purposes. In this case, we have not produced any visualization of the parameter space for our focus is more on showing how different morphologies of development can result from very different initial and boundary conditions. Again one of our quests is to show how scale makes a difference.

Behind these windows, access to the phase space within which the parameters are set is accessible using various dialogues, also called up as windows. Essentially by doubling clicking on each land use, we are able to bring up a series of dialogues with respect to how neighborhoods are configured, how distance influences the operation of state changes in each neighborhood, and how long different land uses remain in their different life cycle stages. We have not provided any means for visualizing how these changes in parameters might impact on the different morphologies produced. Offline, we can use any of the usual strategies to explore the parameter space using small multiples of maps and linking the parameter values in its phase space to these spatial differences. In fact, it is easy enough to make animations from the simulations which can be linked to different values in the parameter space but our focus here is more exploratory. The framework is so diverse in terms of scale that an obvious approach is to see how different kinds of initial conditions can be simulated. Comparisons of these thus become important, and again techniques such as small multiples and layering are clear ways to visualize these differences. Finally these kinds of models can be explored on-the-fly. As the model is running, we are able to explore changes to the morphology either directly or by stopping the model and changing parameters.

Initial Conditions: Random Seeds with Equal Proportions of All Land Uses

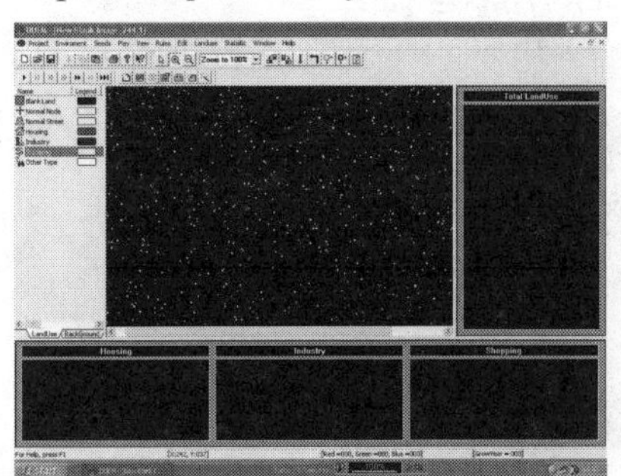

Growth to Time = 30: Early Exponential Growth

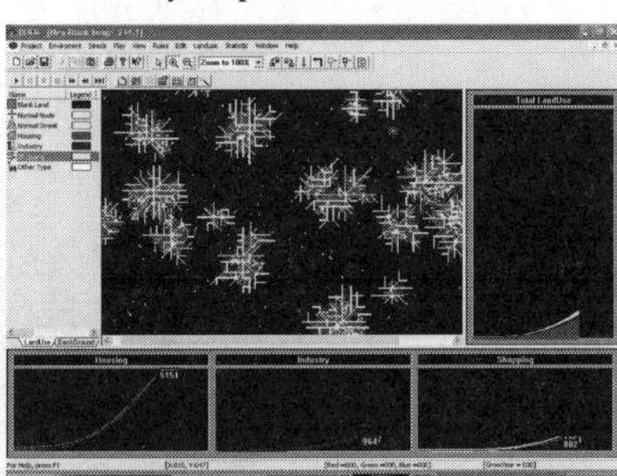

Growth the Time = 60: Exponential Growth

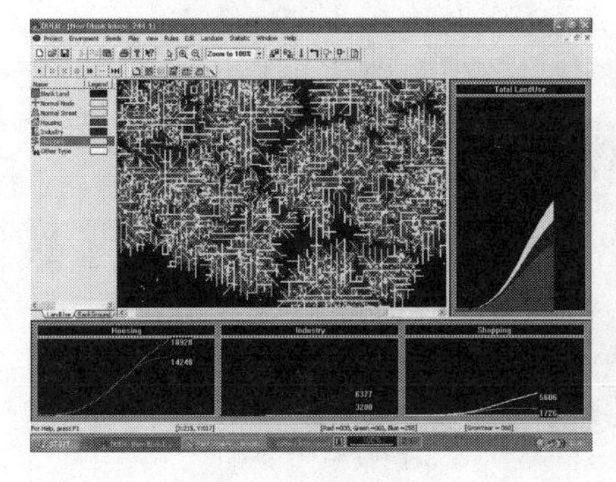

Growth to Time = 120: Capacitated Growth and Logistic Oscillations

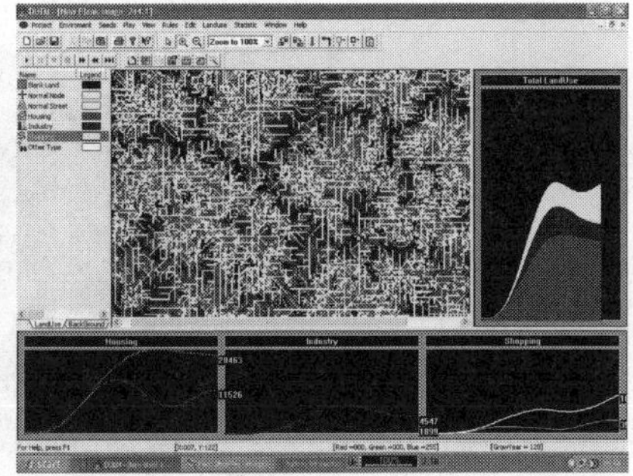

Fig. 5. Small Multiples of Urban Growth Processes

To show the power of this visualization which is charted in Figure 2(b), we present a typical run of the model in Figure 5. This shows the various windows and the development of a system of cities to the capacity of the map canvas as illustrated through the model's trajectories. The effect of different waves of growth in the residential sector can also be seen as distinct gaps in the development by the time the simulation reaches t = 120 but these traveling waves are much clearer in the animation. In Figure 6, we show what happens in the steady state, how the land use totals oscillate but also how industry gradually encroaches on the other land uses, thus indicating that the default rules, no matter however plausible, are badly specified. This is the classic finding of our exploration: that models although plausible can be quite unrealistic when pushed to their limit. This, of course, is an essential diagnostic in developing a more realistic model with visualization being essential to this process.

There are many elaborations on this system that we might make, especially as the visualization occurs in computer time and the location of activity can be changed on-the-fly. Within the package is a drawing capability, a little like that contained in the von Thünen model, and this enables the user to interact with the model in direct fashion. In fact, in all three packages we have used, there is drawing capability that lets the user interact with the model physically, notwithstanding that the data that drives these models is by and large numerical.

Oscillating Growth at Time = 240

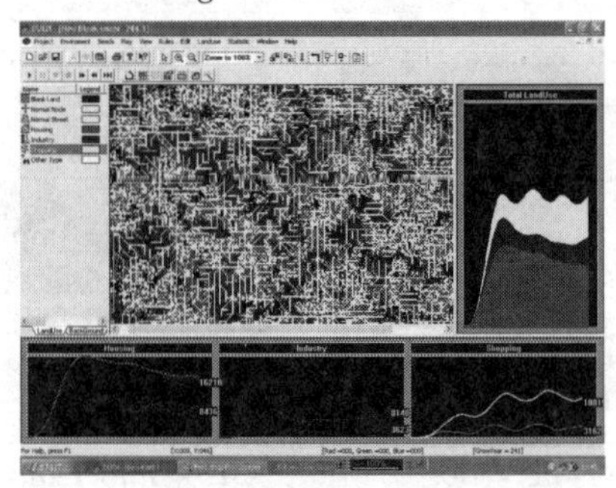

Exploring the Steady State to Time = 1500:
Industry Takes Over Indicating that the Default Rules are Badly Specified

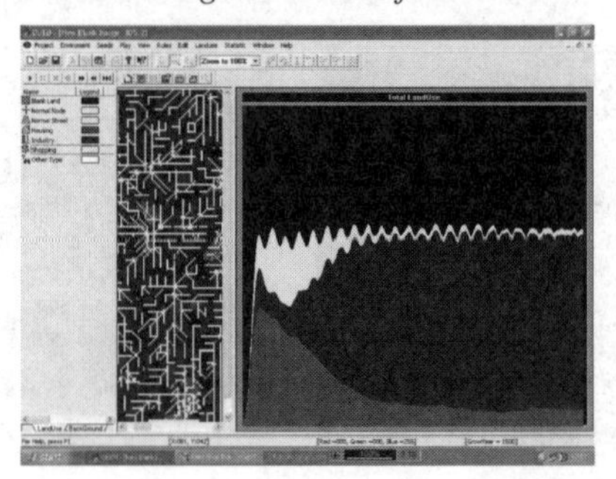

Fig. 6. Exploring Trajectories to the Steady State

6. Visualizing model operation with stakeholder involvement: pedestrian movement and public safety

Our last example builds on the cellular dynamics approach of the last section but develops a model at a much finer spatial scale with a real problem driving the application. We are working with several varieties of active walker model which combine agent-based with cellular modeling, using the cells to represent the landscape on which the agents – the pedestrians – move. The generic model essentially combines movement in the default direction of forward with some random perturbation and with obstacle avoidance, the direction being fixed according to a walker attraction surface which is formed from a synthesis of the multiple forces determining why people wish to walk and for what purpose (Batty, Desyllas, and Duxbury, 2003). In a shopping mall, for example, this surface would reflect the kinds of goods and their locations in shops that walkers wish to purchase.

We have applied this model to a situation of crowding which is associated with a major street festival, the Notting Hill Carnival, which is held once a year for 2 days in west central London. There are major problems of public safety associated with this event and the model has been built to show how people enter the area, flock to the attractions, namely the street bands and parade, visiting a series of events which are located in a small area of around 3 square kilometers. The model essentially walks visitors through the street system to the carnival attractions according to the existing controls on the event managed by the police and other services. The problem of crowding is severe in that there is a serious conflict between the parade and the street bands, the bands being inside the parade route which is circular and continuous. Visitors crossing into the area where the bands are located are in conflict with the parade and there are more general problems of crowding at different points within the area where streets are narrowest. Problems of crime have grown as the carnival has gained in popularity and reducing crowd densities is seen as a way of making the event safer.

The critical focus of this model is that it is designed to help in alleviating crowding by showing how streets might be closed, barriers erected, sound (band) systems moved and the route of the parade changed. All these elements can be controlled within the model and posed as 'what if' questions. Most models can be put into a 'what if' context but in the case of this particular model, the situation is so highly controlled already that it is impossible to think of the model as simulating some relatively uncontrolled situation and then adding controls to meet some objectives. This kind of problem is quite unlike the problem of optimal city design where it is assumed that most cities develop organically from the bottom up and that planning control is imposed to direct growth rather specifically in situations where such direction is lacking or ineffective. This is not the case in something like the Notting Hill Carnival as throughout the history of the event, there has been strong control and management.

This suggests that those involved in managing the event and who know it best be intimately involved not only in the use of the model but in its design. Moreover, in a situation where there is high control, it is useful to think of model calibration as reflecting various degrees of control, for example, by beginning with a relatively uncontrolled situation and then adding controls one by one. To do this effectively, stakeholders who know what controls are effective, should be involved so that the process of model calibration and use in problem-solving and plan-making is simply a natural extension of the model fitting process. In the case of the Notting Hill model, this process can be extended even further back with the data needed to operate the model being in itself simulated in cases where it is difficult to observe.

To illustrate these various stages, visual interfaces are necessary with the software being user friendly and interactive as in all the programs so far in this paper. However it is debatable as to whether stakeholders and non-experts should be involved in the software *per se* as the simulations can be captured as animations and pictured using small multiples. This may be enough to engage in debate although this is uncharted territory in so far as communicating the model to a wider set of non-scientific experts is concerned. The actual model developed begins with data relating to where people enter the carnival and are finally destined for. This is site specific and the first thing that is done is to model the tracks that pedestrians make from their entry points to the destinations at the carnival itself. These tracks can be found as shortest routes from entry points to attractions through the street system and a swarm algorithm is used to find these (Bonabeau, Dorigo, and Theraulaz, 1999). This is a rather technical stage of the model design but once completed, an attraction surface is formed from these shortest routes and the second stage invoked. This surface is used to direct walkers from entry points to the carnival attractions and it is at this point that crowding is assessed. We start with a situation of no control and then gradually introduce controls until safety levels are reached. This involves running the model through many stages. Ideally it is during this process that those who best know the carnival should be involved. In this stage, it is entirely possible that the current situation is replicated but in fact it is likely that the current situation will be found wanting in some way, as we know it is, hence the rationale for this style of modeling.

In short, this kind of modeling involves using a model in such a way that the expertise of those who know the problem best is gradually added into the simulation. This is why we do not define this as a planning model or even a forecasting model but a model which engages those who know the situation best. The model can thus be seen as the product of many decisions from those who know the event and this naturally leads to a rather different style of 'what if' analysis and a rather different kind of policy making process. We illustrate two of the stages of the model using small multiples – in Figures 7 and 8. In Figure 7, we show how the walkers swarm out of the carnival area in search of destination points which is part of the early stage of generating an appropriate and realistic data set for the model. In Figure 8, we show the second stage where the walkers climb the surface of attraction, generating crowds and leading to an analysis of key problems of safety which need to be resolved. Animations of these processes are essential in visualizing how crowds move and thus how they might be controlled and in this sense, the model has a usage in almost real time.

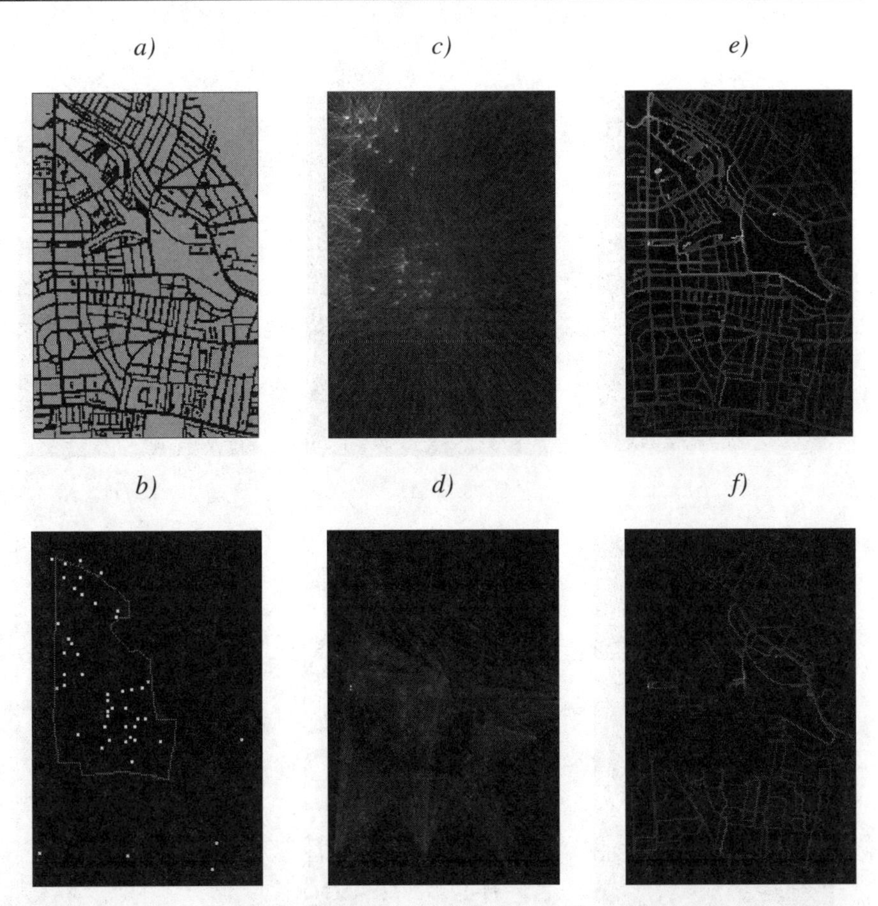

Fig. 7. Exploration of the Street System in Notting Hill. a, The street geometry b, The parade route (dark grey), the static sound systems (light grey) and the tube stations (med grey) c, Accessibility from parade and sound systems without streets d, Shortest routes to tubes without streets e, Accessibility in streets f, Shortest routes in streets.

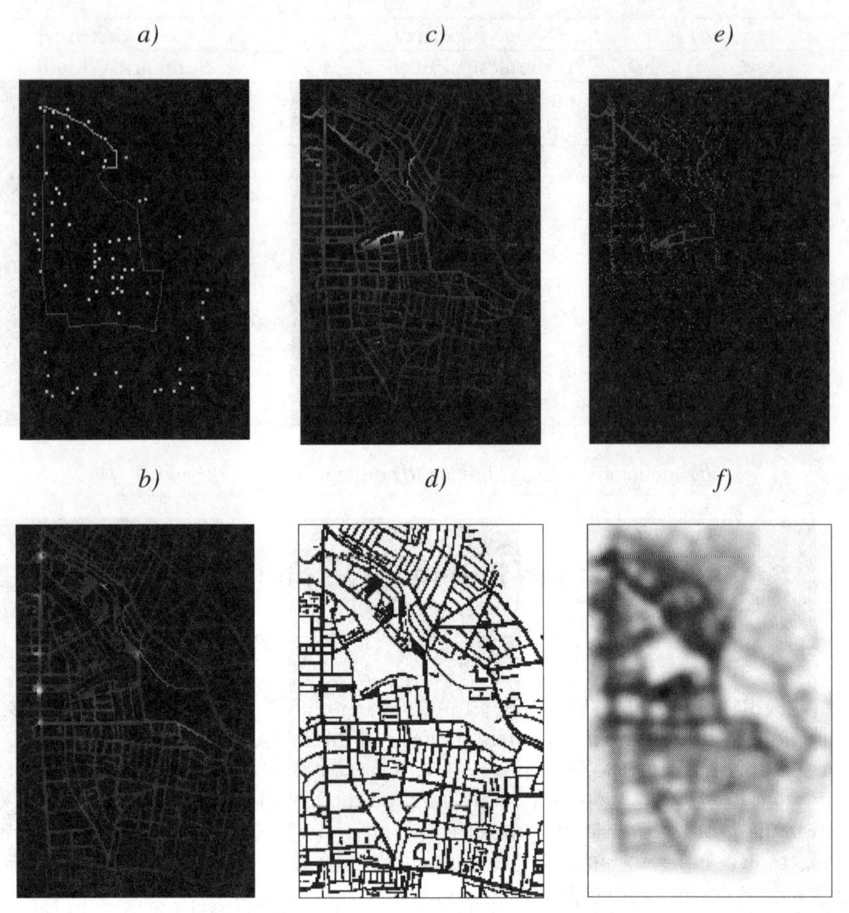

Fig. 8. The Full Modeling Sequence and Identification of Vulnerable Locations. a, The 2001 parade route (dark and med grey) with the proposed 2002 route in dark grey, the static sound systems (light grey) and the entry points (med grey) b, The composite accessibility surface from stage 1 c, Traffic density from stage 2 d, Areas closed by the police used in stage 3 e, Location of walkers in the stage 3 steady state f, Vulnerable locations predicted from stage 3.

7. Next steps: a paradigm for visual modeling

We are very conscious that we have not mapped out here a comprehensive framework for visualization in spatial modeling. This is largely because so much visualization is characterized by ingenious solutions which involve putting unlike pictures together, by large scale simulations that depend on very sophisticated software in the search for pattern, and by the very nature of the models themselves and how they are formulated. All this is influenced by the imagination we bring to bear on the pictorial world. Equally well there are many insights to be made using other intellectual media – verbal discourse and numerical reasoning. In getting a complete picture of how our models can be understood and best applied, all these strategies are required. In fact we have shown here that rather than visualization *per se*, what is evolving are visual models: models that cannot be designed in any other way than using the visual medium. In our last example, this was even more pointed in that to involve non-scientists in developing such models, the visual medium is essential.

What we have not addressed here but is something that needs to be pursued are the physical media for spatial modeling. As our models are digital, they can be manipulated in countless ways. For example, GIS is often used not for spatial analysis but for paper map-making with the physical product and its perfection being the rational for digital representation of the reality in the first place. In the same way, models of the built environment can be printed in 3-d using the new generation of hard copy printers, thus simply aiding the manufacture of iconic models in their traditional physical form. The notion of building 'models of models' – simulacra as Baudrillard (1994) refers to them – is also a useful way forward and in the examples illustrated here, some elements of this recursion do permeate the model building process. In short, scientific visualization is increasingly being informed by physical visualization but with the digital representation being central and stable to this entire medium. These are exciting developments and there is an urgent need to engage a debate about such possibilities in spatial modeling and analysis.

References

Alonso, W. (1964) *Location and Land Use*, Harvard University Press, Cambridge, MA.

Batty, M. (1976) *Urban Modeling: Algorithms, Calibration, Predictions*, Cambridge University Press, Cambridge, UK.

Batty, M. (1992) Urban Modeling in Computer-Graphic and Geographic Information System Environments, *Environment and Planning B*, *19*, 663-685.

Batty, M., Desyllas, J., and Duxbury, E. (2003) The Discrete Dynamics of Small-Scale Spatial Events: Agent-Based Models of Mobility in Carnivals and Street Parades, *International Journal of Geographic Information Science*, *17*, 673-687.

Baudrillard, J. (1994) *Simulacra and Simulation*, University of Michigan Press, Ann Arbor, MI.

Bonabeau, E., Dorigo, M., and Theraulaz, G. (1999) *Swarm Intelligence: From Natural to Artificial Systems*, Oxford University Press, New York.
Cronon, W. (1991) *Nature's Metropolis: Chicago and the Great West*, W. W. Norton and Company, New York.
Hall, P. (1966) (Editor) *Von Thunen's Isolated State*, Pergamon, Oxford, UK.
Kaufmann, W. J., and Smarr, L. L. (1993) *Supercomputing and the Transformation of Science*, Scientific American Library, No 43, W. H. Freeman & Co, San Francisco, CA.
Lowry, I. S. (1965) A Short Course in Model Design, *Journal of the American Institute of Planners*, *31*, 158-166.
Steadman, P. (1999) *Computing Booklet Part 1: Feeding the City: Von Thünen's Model of Agricultural Location*, AT308 Cities and Technology: From Babylon to Singapore, The Open University, Milton Keynes, UK.
Steadman, P. (2001) *Vermeer's Camera*, Oxford University Press. Oxford, UK
Tufte, E. R. (1990) *Envisioning Information*, Graphics Press, Cheshire, CN.
Tufte, E. R. (1997) *Visual Explanation*, Graphics Press, Cheshire, CN.
Xie, Y. (1996) A Generalized Model for Cellular Urban Dynamics, *Geographical Analysis*, *28*, 350-373.
Xie, Y. and Batty, M. (2004) Integrated Urban Evolutionary Modeling, in Atkinson, P. M., Foody, G. M., Darby, S. E., and Wu, F. (Editors) *GeoDynamics*, CRC Press, Boca Raton, FL. pp.:349-366.

Transferring Concepts for Urban Modeling: Capture or Exchange?

Denise Pumain

Abstract. The development of a theory of complex systems that establish bridges between disciplines of the natural and social sciences appears together as an opportunity and as a challenge for urban modeling. Borrowing concepts and tools from formalised disciplines may help to more satisfying expressions of urban theories and to a better understanding of abstract processes that are behind urban dynamics. However, the specific features of social systems should not be neglected or underestimated while operating these transfers. Proper adapted ontologies have to be preserved for urban entities and the new experiments should be able to become part of the previously existing urban knowledge, helping to revise it, not replacing it.

1. Introduction

The recent multiplication of initiatives for an academic institutionalisation of the sciences of complex systems (as for instance the creation in Europe of half a dozen of dedicated schools or networks that are more or less inspired from the Santa Fe Institute) is a good indicator of the large developments that have brought that research field on the front stage during the last fifty years. The announced emergence of a perhaps mythical "science of complexity" is now challenging all sciences. It is very often suggested that the dynamic processes that lead to the emergence of sometimes unexpected structures from individual interactions can be analysed by using a common variety of concepts and tools, as provided by self-organisation theories and non linear mathematics. According to the Springer Complexity publication program, "These deep structural similarities can be exploited to transfer analytical methods and understanding from one field to another". We would like to question that claimed cross-fertilisation between scientific disciplines by drawing attention towards a few difficulties that are in practice associated to such a trans-disciplinary approach, especially when trying to develop a theory of complex systems that would be transversal to sciences in natural and social world.

Our laboratory has experienced the use of different modeling frames aiming at formalising, simulating and predicting the development of urban and regional systems: models of non linear differential equations under the paradigm of self-organisation theories (dissipative structures and synergetics (Pumain, Sanders, Saint-Julien, 1989; Sanders, 1992), as well as multi-agents systems in the context of artificial intelligence and theories of complexity (Bura et al., 1996). Since more than thirty years, urban modeling has progressed mainly by conceptual or even paradigmatic borrows from physics, mathematics, information theory or computation theory (Allen, 1997; Batty and Longley, 1994; Portugali, 2000; Weidlich,

2000; Pumain, 1998). We want to explore to what extent such notional transfers have contributed to urban theory. After noticing a reversal in the scientific paradigm beneficial to the "historical sciences" within the contemporary building of a "theory of complex systems", we review a series of misunderstandings or hiatus in the experimentation of various concepts that were transferred and applied to urban systems dynamics. In some respect, the practice of transfer is very often perverted by the relative social and epistemological status of the disciplines. We emphasize the need for a broader circulation of concepts and a wider attention paid to meaning and signification in applications.

2. Complexity: a reversal in the dominant scientific paradigm?

Research on complex systems has lead to a paradoxical situation. For a long time, classical physics and mathematics have dominated the criteria of scientific work, as exemplified in the Popper model. Trying to be "scientific", social scientists had to resist to the simplifying hypothesis of this reductionism and could not meet the requirements of repeatability of experience and universality of results. Some sociologists as J.C. Passeron (1991) even imagine a specific way of reasoning for social sciences ("le raisonnement naturel"), because, even when falsifiable hypothesis can be formulated, their results never can be exactly repeated and abstracted from the context (historical or local) where they were established (or, the relevant contextual variables cannot be enumerated in an exhaustive way). Such a position is perhaps tactical and probably overestimates the effectiveness and rigour of the "hard sciences". Moreover, it seems to miss an essential turn in the recent evolution of a larger part of them, towards what is called "complexity".

New ideas that emerged in physical sciences with the development of dissipative structures (Prigogine, 1973) or synergetics (Haken, 1977) under the general label of complex systems dynamics have prepared a perhaps more general change in our way of thinking about systems. While these ideas were more and more applied to living species in the framework of an evolutionary thinking (Lewontin, 2003), or to adaptive cognitive systems in economy or social networks (Anderson, Arrow and Pine, 1988; Arthur, 1994; Arthur, Durlauf and Lane, 1997), the focus shifted from "self-organisation" of spatio-temporal patterns, from interactions between very large numbers of elementary particles in open systems submitted to external energy flows, towards the "emergence" of new structures and new properties stemming from the internal and/or external interactions between a limited number of heterogeneous elements or individuals, that may have reactive, adaptive and even cognitive behaviour, with sometimes a capacity for changing their interaction rules. The criteria in use for the definition of complex systems progressively evolved, including more and more aspects that were not often mentioned before. The systems "far from equilibrium" that are now under study, even in physical sciences, are explicitly considered in distinct contextual conditions of space and time. They come to meet the specific properties of the social systems, that were until now considered as diriment obstacles to any scientific formalisation: the irreversibility of temporal processes, the uniqueness of a system's trajectory in phase space, the non predictability of its future (Prigogine, 2001). All these features are now part of the theory of complex systems. New models have been developed for exploring fuzzy elements, strange attractors and uncertain events that were not considered before, and even human brains and thoughts have received a due attention in the new framework of cognitive sciences (Bourgine and Nadal, 2004).

It could be argued, as I did some time ago (Pumain, 1997), that social sciences should more than ever borrow their models from the sciences that are more advanced in formalisation, since these are now offering tools and concepts that no longer hurt the basic principles of research and knowledge about social systems. By applying ideas and models that have been developed within physics or mathematics of complex systems, we could learn more about the universe of possible evolutions framing the observed urban dynamics, and perhaps discover some abstract hidden processes (or formal theories) that could better explain the observed similarities appearing in urban structures and evolution, despite the overwhelming diversity of physical, economic, political and cultural forms that urban systems are exemplifying all over the world. I want to develop here the complementary idea that while the so-called (self-called?) "sciences of complexity" are evolving towards an attempt at unifying the analysis and, even if possible, "the" *theory*, of complex systems, social scientists should be keen on maintaining their previous knowledge as a most valuable input in the models that are now developed. Actually, that knowledge about social systems, although less formalised and incompletely integrated, already incorporates the principles that are today the distinguishing mark of complexity theory, and then should be recognized and integrated as such: heterogeneity of elements and their properties, diversity of interactions that are not only non-linear but often multi-scalar, dependence towards initial conditions and contextual variables, path dependency of the evolution, unpredictability of the future, irreducible role of intentional behaviour, intervention of the point of view of the observer in constructing the situation... Too often though, that specific knowledge was ignored, for instance by the enthusiastic promoters of "econophysics" (see for instance Durlauf, 2003) or "sociophysics" (fortunately there is nothing yet such as an "urbanophysics" ?).

Of course we are interested in the highly formalised concepts and powerful tools of complex systems sciences. But while experimenting new concepts and new modeling methods, we should keep in mind the objective of developing a relevant and sensible urban theory, that provides a really new contribution to knowledge in that field.

3. Quality of ontology and measurement for social systems

A first difficulty in applying the complexity paradigm to urban systems stems from the meaning of the word "complexity" itself. As the theory of complex systems aims at exploring how new entities, structures or properties emerge at a one observation level from the interactions between objects and behaviour or practices that are occurring at a lower level, a significant theory can be developed only if the objects, their attributes and their interactions that are under study are correctly identified. In the case of cities, this identification is not an easy task, for conceptual and for practical reasons.

On the theoretical side, there is a higher degree of complexity in social systems because of a higher difficulty in separating entities that would have clear limits and definite attributes, and because of the plurality of theories that frame the possibly relevant representations of these systems. One could for instance imagine to develop and refine many specific ontologies of the city as a complex system, that would define it either after its morphological properties as a progressive composition of buildings, or as a demographic aggregate of resident population constrained by the various and competitive needs of different age groups and professions, or as a portfolio of economic activities linked by agglomeration economies, or as the expression of the political and cultural organization of a society articulating co-operative and conflicting groups, or as a place where the accumulated knowledge gives rise to the emergence of innovations… Each discipline of the social sciences participating to urban research has developed its own dynamic models of the city as a complex system, including non linear feed- backs effects and using differential equations or agent based computational representations for their simulation. Though, when implemented in a model, any of these partial representations would require the consideration of some important features that were not included in the model, as soon as a confrontation with the real world is to be tempted: in most cases, the evolution of any observed city, even when restricted to a narrow disciplinary description, is controlled by more than one of the features mentioned above. For instance, the ecological resources are necessary to explain some industrial urban specialization, even if they are not part of a "pure" economic theory of the city; land values and urban densities can be related in a single model but applications will require precisions about collective values and land regulation policies… Even if these models could be conceived as "purely" theoretical, there is little doubt that they would miss most of the specificity of urban dynamics, that is precisely made of permanent adjustments between many of the possible determinants of urban evolution (as morphology, demography, economy, policy or culture).

According to J. Casti (1994), complexity is not an intrinsic property of systems but a subjective view of an observer confronted to the "surprise" of emerging properties (see also Batty and Torrens, 2001), and the degree of complexity of a system is directly proportional to the number of equivalent ways (models) in which the system can be described for explaining them. Following that view, we could then argue that the complexity of urban systems could be measured by the diversity of the several distinct representations of what a city is and how it functions. However, what are the "equivalent models" in the field of social sciences? Can they be assimilated to different explanations that are considered as satisfying by one particular discipline in a given theory within the field of urban research? Actually, these explanations are not "equivalent", since they do not give an alternative interpretation for the same reality, but they build a coherent view of a particular aspect of that reality. The various disciplinary approaches have to be articulated to reconstruct a comprehensive description of the city as a complex system. We can then suggest that, besides the definitions that have been proposed for complex systems, either mathematical or computational, social scientists could have their own complementary interpretation of complexity. The level of complexity of any situation (or dynamics) could be evaluated by enumerating the number of disciplinary concepts or points of view that would be necessary to provide an explanation of that situation, that can be considered both consistent and sufficient, according to an operational purpose or to a degree of precision of the description that is thought of as acceptable. That view is possible since social sciences built themselves and differentiated from each other by deepening the knowledge in one specific aspect of society, but while getting more and more insights in one direction they discover very often that they have to include within the description of the context of their study many other aspects that are developed by other disciplines. This is especially the case when cities are considered in the complexity of their evolution. A consequence is the need for periodically building new interfaces between disciplines of the social sciences (see section 4 below).

In this respect, the specific contribution of geography to the theory of cities as complex systems could be, not only in the traditional consideration of the phenomenological diversity of cities according to regions of earth space and historical times of societies, but also in the recognition of the multi-scalar character of urban systems. The seminal expression by B. Berry of "cities as systems within systems of cities" coins the ontological definition of urban systems by geographers. Of course that expression, that refers to the nested hierarchy as typical of the "architecture of complexity" as conceived by H. Simon at the beginning of the sixties, has to be questioned and updated, for instance with regard to modern communication systems, leading perhaps to a reformulation that would make a larger place to long distance and non hierarchical interactions in the former quasi-nested representation of two levels in urban systems. But in the building of a trans-disciplinary approach, we shall meet another difficulty. According to the principle of decomposability that, together with predictability and linearity, makes the difference between simple and complex systems, if there is some relevance in the concept of a "system of cities", and if that system is complex, then the consideration of a single city as disconnected from the whole system would change its dynamics. That connection between one city and the system of cities is missing in the economic theory of "the" city and even in the "new economic geography" (Fujita, Krugman and Venables, 1999). The lack of recognition of the constraint exerted by the existence of other cities on urban dynamics is probably a major weakness of the new urban economics, despite the important advances that were made in that field of research during the last decades.

So the acceptance of a point of view about cities as complex systems would mean that social sciences do co-operate in the elaboration of the theory. Even when it is well defined for one level of analysis, the theoretical conceptualisation of cities should include aspects of the context that are relevant for the dynamics under study. How many analyses of urban sprawl take for granted a description of properties of a "centre" and a "periphery" that are inspired by a cultural representation of American cities and derive corresponding attributes, without considering the values that are attached to the locations (as expressed for instance by the spatial distribution of land values, or urbanism regulations) within the country where the model is applied? The same carefulness should be required when defining the individual interactions that shape a system's structure: even if Shelling's model of social segregation provides a beautiful case of a non desirable unintentional collective result of intentional individual behaviour, it probably overestimates the intensity of segregationist practices, for instance by not allowing the residents that are satisfied to move, as the non satisfied do. Once again, the benchmark of their possible application to a diversity of observed situations seems to remain a necessary part of the construction of models and theories.

The same definitional accuracy should be applied to the apparently more trivial question of data collection for empirical analysis or model testing. We shall give only one example that could become a source of problem with the growing craze for the question of scaling (Pumain, 2004). Scaling processes are essential in complex systems dynamics, because they are probably rooted in very general constraints on the organisation, through the circulation of energy or information (West and Brown, 2004). Urban systems are very likely to arouse many attempts of conceptualisation through scaling laws, because many empirical regularities have already been observed and modelled by Pareto distributions or fractal geometry. The spatial distribution of residential population densities or land prices, as well as Zipf's rank size rule, or Christaller's central place theory, do suggest the relevance of scaling processes for explaining the urban density gradient or the persistent inequalities in city sizes and functions. However, discovering new expressions of scaling laws cannot merely result from adding a new experiment on any urban data. If scaling effects are suspected, the data in use must be relevant for the process under study and the quality of their measurement has to be very high. Of course, the question of the definition and delimitation of urban entities in space and during the course of time is very difficult and their comparison within and between countries remain a delicate exercise. But it is not sufficient to use existing definitions and data bases if they do not represent meaningful entities for the analysed process.

For instance, many authors use the urban data base that give the populations of the cities of United States as a benchmark for testing Zipf's of Gibrat's models. But the SMSAs that are included do not represent without bias an entire urban system: as an SMSA is defined around a centre (urbanised area or continuously built-up surface) that groups at least 50 000 jobs, a number of smaller urban centres, although still functioning as urban agglomerations, are neglected. This lack in information has been acknowledged recently (in 2000) by the Bureau of Census who decided to add "micropolitan" statistical areas (including centres with 10 000 jobs) to the set of SMSAs. Many tests of models that try to relate the distribution of city sizes and growth processes have nevertheless used that data base for model testing, as did for instance Gabaix and Ioannides (2004), and Spyros (2003). Although the models these authors develop are each very interesting, their conclusions cannot be totally reliable because of this bias in data. Moreover, it is well known that the United States are not a representative case for all systems of cities, because during the last two centuries of its development the US system mixes classical dynamic processes of distributed growth in a mature system of cities together with more specific processes of expansion through new frontier settlements. The results of such experiments cannot be generalised (for instance to Europe and Asia where urban systems have a pluri-secular and most of times more than millennial history) and cannot make definitive conclusions in terms of a model that would become a reference for every urban system.

Another example of the importance of a correct definition of geographical urban entities for measuring urban growth processes is given by comparing the studies of Batty (2003) on Britain and Bretagnolle et al. (2000 and 2002) for Europe and France. Both try to identify a trend in urban concentration or dispersion at the scale of a system of cities during a long time period, by adjusting a Pareto model to the distribution of city sizes at different dates. But while Bretagnolle et al. use the definition of urban agglomerations (that can expand in space over time), Batty refers to an exhaustive partition of Britain, measuring the evolution of population within the stable 459 municipalities of England, Scotland and Wales between 1901 and 1991. Of course there is in that last case a possible bias in measuring the variance in urban population size, since the largest urban agglomerations are not allowed to overcome the limits of their municipality, their growth may be underestimated, while at the other end of the distribution of town sizes, the urban agglomerations that became smaller than their municipality limits have their population overestimated. This difference in data is likely to explain, at least for a part, the differences in results showing, after the values of the slopes of the adjusted distribution, a trend towards deconcentration in Britain versus a reinforcement of the urban hierarchy in the European and French study. The problem of comparability of data in space and time has to be solved if one wants to rise correct conclusions about the observed evolutionary trend, in order to further elaborate about the theory of the dynamics of systems of cities (Pumain, 2000).

We should then make a plea for using and helping to develop more comparable urban data bases. We do regret that the most recent attempt by Eurostat (program named Urban Audit II, 2003) that includes a very carefully designed survey (more than 300 variables) will provide urban data that are meritoriously comparable in their statistical definition but absolutely not in the spatial framework of the urban entities under consideration: the delimitation of the urban entities according to the different European states varies in this document from political agglomeration (France), to NUTS3 (Spain) or NUTS4 (UK) regions... There were however previous successful attempts for providing comparative information at the scale of all European urban agglomerations, for historical periods (Bairoch et al. 1988; de Vries, 1984). At the world scale, the very exhaustive Geopolis data base prepared by F. Moriconi-Ebrard (1994) has been too rarely used and quoted as a powerful instrument for international comparisons using the best comparable definition and reliable delimitation of urban entities (Pumain and Moriconi, 1997). Another example of a cautious comparative attempt in urban comparison for scaling has been recently made by M. Guérois who used remote sensing data for comparing the shape of the urban field in different European countries (Guérois, 2003; Guérois and Pumain, 2004). Using the CORINE Land Cover data set, she was able to demonstrate that the urbanised areas are distributed around the main historical urban centres according to a dual density gradient, one rather steep corresponding to the urban agglomeration (with a radial fractal dimension between 1.7 and 1.9) and the other with much less contrasts representing the rural part of the functional urban area (automobile commuting zone, with a fractal dimension less than 1). More careful measurements like this are needed for a better understanding and significant modeling of the spatial expression of urban morphology and dynamics.

4. Cumulativity of knowledge

Another difficulty in the development of applications of new ideas and tools for complex systems to cities is in establishing connections between the ancient and that new knowledge. Knowledge accumulation, after remaining for a long time an academic and educational problem (UNESCO, 2003) and a preoccupation for archivists and museums, has become during the last decades a major political and economic issue. Being now considered as an important input in production, besides labour and capital, the scientific and technological achievements lead to the development of a new discipline, the "economy of knowledge". At the same time, the epistemological thinking tended to avoid the debate about what we call the "cumulativity of knowledge": it relates to the scientific and sociologic conditions that permit knowledge accumulation. The post structuralist deconstructivism as well as the postmodernist theories insisted on the plurality of "systems of knowledge" and the parallelism of theories that were alternative explanations and could not be cumulated.

To a large extent, many explorations that were conducted in the field of urban research for the sake of using concepts and tools of complexity theories did not try so much to contribute to knowledge accumulation in that field. Their main objective was not so much to connect their results to the existing state of knowledge in the domain than to underline the originality and novelty of their approach. We review here in detail one example, not as a criticism of that particular work that provides in other respects an excellent contribution, but as an illustration of the too limited use of what could be one of the most interesting and promising approach of the dynamics of urban networks. We refer to a paper by Anderson et al. (2003). They use an algorithm building « scale-free » networks for describing the distribution of land prices in an urban system. A "scale-free network" corresponds to a class of growing networks whose node degrees are power law distributed. In their model, the nodes of the network represent pieces of land which become over time more and more connected by edges representing exchanges of goods and services (actually the result of this trade is simulated by a trade benefit or financial investment directed from one node to another). The model proceeds by adding new links between already developed nodes, with a probability that is proportional to the relative size of the node in the total of nodes, and by selecting new nodes. The mean probability of developing existing nodes is significantly higher than the one attached to the development of new nodes. Spatial rules are added for specifying this selection process, according to hypothesis about a distance-decay interaction model. The model is calibrated in order to fit an impressive empirical data set about land values in Sweden (almost 3 millions observations). The paper demonstrates the ability of the model to reproduce the global statistical distribution (frequency of land squares according to land price) and its main parameter (Pareto exponent of 2.1). The authors assume a linear relationship between the value per unit of urban land and the size of urban population, so their model could be used as a starting point for fitting population data as well.

But the paper is not clear about the scale of application of the model: whereas referring at first to Zipf's law, which is a model of the interurban distribution of city sizes, it represents "systems of specialised trading activities" that "can be resolved to any resolution down to individual transactions", whereas the explanation of the model in "an urban economy context" seems to refer mainly to intra-urban land values formation (for instance, looking at different processes at the perimeter of urban areas and predicting the emergence of urban sub-centres). In any case, the model predicts a single and unified statistical distribution of land values at a country scale, making no distinction between the intra-urban gradient of land prices and the interurban distribution of land values. The model produces only a sharp break between rural and urban land values. To be coherent with the existing state of knowledge, the authors should have tested the variations of land prices inside the nodes (between centres and peripheries) as well as between the aggregated nodes. It could happen that the rather high level of inequalities they find between land prices is more linked with intra-urban inequalities than to interurban. Actually, when looking at he average housing, offices and land prices per urban area in Europe, one discovers that prices are surprisingly similar from one city to the next (low variance) and the correlation with city size (as measured by population figures) is rather low for the entire distribution (even if large cities as London or Paris have the highest prices). Meanwhile, the inequalities between the prices per hectare inside the same single city may reach a factor 10 and more, at the block level, and frequently 5 or 6 at the scale of neighbourhoods (Fen Chong and Pumain, forthcoming).

5. Conclusion: organizing a more symmetrical trans-disciplinary communication

Within the framework of the developing theory of complex systems, urban research is more and more open to the use of the large variety of concepts and tools that are imported from the more formalised sciences. While welcoming the appeal towards a general use of these references for urban modeling and theoretical elaboration, we have claimed for more caution and perhaps a better reciprocity in this process of transferring notions between disciplines. The modern paradigm of complexity being more and more inspired by conceptions that are emanating from social sciences, the methods that are now in use for research should not forget about the specific procedures for identifying and selecting relevant entities and processes that were specifically elaborated for the complex systems they are studying. As social scientists, we should not loss our specific expertise, the knowledge that was accumulated from past experiences using other methods but still valid and reusable, even if, as always, revisable.

We have underlined the originality of the representation of multi-scale urban systems that is built by urban geographers, and its high compatibility with the paradigm of complexity. However, in order to be recognised as more than a descriptive discipline among social sciences, there is a need for geography to build and communicate better formalised representations of its specific knowledge, by the means of basic theories and models. This theoretical approach did perhaps not progress enough since the seminal attempts by W. Bunge or W. Tobler. Why do we hesitate to formulate normalised geographical views of the city, or of systems of cities? Do we have to define an "homo geographicus"? Or can we borrow individual attributes and behavioural rules to other disciplines? Because economics was the first discipline in social sciences that undertook its formal and mathematical formalisation, a specific attention should be devoted to its approaches of urban systems. It is likely for instance that the Dixit-Stiglitz model of centre-periphery will become a building block for many urban models, as suggested by Fujita, Krugman and Venables (1999). However, the inter-disciplinary circulation of concepts and models should be two-ways, if one wants to avoid strange false innovations! For instance, Fujita et al. recommend to adopt the distinction made by Cronon of "first nature" and "second nature" advantages in location. The first correspond to advantages stemming from pure natural resources whereas the latter would be linked to man-made investments on the spot. Is that distinction really theoretically useful and necessary? Geographers have demonstrated for long that practically in all places the distinction is no longer possible to be made (as human intervention in modifying the quality of the site have been ancient and numerous), but that a very fruitful distinction in location advantages could arise when scale effects are recorded as site and situation. Improving the dialogue about such a question could be profitable to the two disciplines.

There is to avoid the periodical reinvention of the wheel, the misuse or neglect of former discoveries, the dilapidation of our intellectual heritage and to organise its preservation for the future ("to the generation before us", as said the dedication of the textbook by Abler, Adams and Gould). Something like sustainable development in science?

References

Allen, P. M. (1997). Cities and regions as self-organising systems: models of complexity. Reading, Gordon and Breach.

Anderson, P., Arrow, K. and Pines, D. (eds) (1988). *The Economy as an Evolving Complex system. Redwood City*, Addison Wesley.

Anderson, C., Hellervik, A., Hagson, A. and Tornberg, J. (2003). *The urban economy as a scale-free network*. ArXiv:cond-mat/0303535 v2

Arthur, B.W. (1994). Increasing returns and path dependency in the economy. Michigan University Press.

Arthur, W.B., Durlauf, S. and Lane, D. (eds) (1997). *The Economy as an Evolving Complex System II. Redwood City*, Addison Wesley.

Batty, M. and Longley, P. (1994). *Fractal cities: a geometry of form and function*. San Diego, California, Academic Press.

Batty, M. (2001). Polynucleated urban landscapes. Urban Studies, 38, 635-655.

Batty, M. and Torrens, P.,(2001). *Modeling complexity. The limits to prediction*. Cybergeo, 201, 24 p.

Batty, M. (2003). The emergence of cities : complexity and urban dynamics, CASA Paper 64, 16 p.

Bourgine, P. and Nadal, J.P. (2004). *Cognitive economics, an interdisciplinary approach*. Berlin, Springer.

Bretagnolle, A., Mathian, H., Pumain, D. and Rozenblat, C. (2000). *Long-term dynamics of European towns and cities: towards a spatial model of urban growth*. Cybergeo, 131, 17 p.

Bretagnolle, A., Paulus, F. and Pumain, D. (2002). *Time and space scales for measuring urban growth*. Cybergeo, 219, 12 p.(http://www.cybergeo.presse.fr)

Bura, S., Guérin-Pace, F., Mathian, H., Pumain, D. and Sanders, L. (1996). Multi-agent systems and the dynamics of a settlement system. *Geographical Analysis*, 2, 161-178.

Casti J. 1994, Complexification. Harper and Collins.

Durlauf S.N. (2003). *Complexity and empirical economics*. Santa Fe Institute, Working paper.

Fen Chong, J., Pumain, D. (Forthcoming). Real estate prices in the space of European cities.

Fujita, M., Krugman, P. and Venables, A. (1999). *The spatial economy: cities, regions and the international trade*. Cambridge, MIT Press.

Gabaix, X. and Ioannides, Y.M. (2004). The evolution of City size distributions. *Survey prepared for the Handbook of Urban and Regional Economics*, 4 (53), V. Henderson and J-F. Thisse eds, North-Holland, p.2341-2378.

Guérois, M. (2003). Les villes d'Europe vues du ciel, in (D. Pumain, M.F. Mattei M, dir.) Données urbaines 4, Paris, Anthropos, 411-425.

Guérois, M. Pumain, D. (2004). CORINE Land Cover and the urban field, forthcoming.

Lewontin, R.C. (2003). Four Complications in Understanding the Evolutionary Process. Santa Fe Institute Bulletin, 18, 1, 17-23.

Moriconi-Ebrard, F. (1994). Geopolis, pour comparer les villes du monde. Paris, Anthropos.

Passeron, J.C. (1991). Le raisonnement sociologique. L'espace non-poppérien du raisonnement naturel. Paris, Nathan.

Portugali, J. (2000). *Self-organisation and the city*. Berlin, Springer.

Prigogine, I. (2001). La fin des certitudes. Paris, Odile Jacob (1ère édition 1996).

Pumain, D. (1998). Urban Research and Complexity, in (C.S. Bertuglia C, G. Bianchi, Mela A., eds) *The City and its Sciences*, Heidelberg, Physica Verlag, 323-362.

Pumain, D. (2000). Settlement systems in the evolution. *Geografiska Annaler*, 82B, 2, 73-87.

Pumain, D. Sanders, L. and Saint-Julien, T. (1989). *Villes et auto-organisation.* Paris, Economica.

Pumain, D. and Moriconi-Ebrard, F. (1997). City Size distributions and metropolisation. *Geojournal*, 43 :4, 307-314.

Sanders, L. (1992). *Système de villes et synergétiques*. Paris, Anthropos.

Weidlich, W. (2000). Sociodynamics. Reading, Gordon and Breach.

West G. B., Brown J. H. (2004), Life's Universal Scaling Laws. *Physics Today* 57(9), 36.

Part two:
Specific experiences

Design Issues to be Considered for Development of an Object-Oriented System for 3D Geovisualization: The Aalborg Experience

Lars Bodum

Abstract. The Center for 3D GeoInformation at Aalborg University (DK) became a reality in 2001. Among the many activities in the center, there is onc that goes through all the others as an important red line. That is the development of a general object-oriented system for real-time 3D visualization of geographically based Virtual Environments, called GRIFINOR. This paper will reveal some of the considerations and aspects that have been discussed in the preliminary design of GRIFINOR. The system involves use of several different methods for semi-automatic generation of 3D objects from LIDAR data, Orthophotos, building footprints and data from various public registers. At the moment the system is only prepared for generation of static physical elements such as buildings, but later the system will be able to visualize traditional geoinformation such as socio-economic attribute values on "top" of the Virtual Environment. The buildings are generated as objects based on representation in the 2D technical/topographical map, the LIDAR data and information about each building from the national building and dwelling registry (BBR). After each entity is generated as an object it is saved in a custom built object database. This database is the heart of the system and several specific issues regarding the development of it will be discussed. At the front end, a 3D viewer based on a Java-driven scene graph is the core of the graphical user interface. The considerations behind a representational model for the objects will also be presented and finally some discussions about potential viewing platforms.

1. Introduction

For several years it has been possible to create very detailed models of urban areas as virtual environments for realtime visualization. This development has taken place while computer graphics hardware and software (CAD-systems) have been updated and have become more accessible in terms of better usability and a better price/performance relation. The number of polygons visualized in models is much higher today than it was just 5 years ago. But even though the technology can do much more today, the concepts and basics of urban modeling are still hanging on to the same design philosophy. Geometry and information are stored in separate databases. This means difficulties when it is necessary to update and refine each of the systems. The major part of the urban models are optimized for fast querying in relation to the demands put on the model from the perspective of the graphical user interface. Geoinformation can normally only be accessed through a database-link (such as ODBC) or through hyper linking of the map.

2. 3D Geovisualization

As stated in recent literature, the traditional cartography and other forms of visualization of the geographic world in one or another form are different kinds of geovisualization (MacEachren and Kraak, 2001). 3D geovisualization is a special form of geovisualization, where focus is on the creation and modeling of 3D spaces to represent objects as well as attribute information about these objects in virtual environments.

Three-dimensionality is not a new phenomenon. It has been possible for many years to create virtual 3D spaces that capture the spectator within a continuum, and provides him/her with a feeling of "being there" (Graft, 1943; Oettermann, 1997). But there still very little material is published about the effects and the most appropriate ways to use 3D geovisualization as interface to geoinformation in general (Granum and Musaeus, 2002; Fuhrmann and MacEachren, 1999; Raper, 1989; Raper, 2001). What can be gained from 3D and where are the challenges in relation to development of these virtual environments, if they really were supposed to become an alternative to or even outperform traditional mapping and GIS? This paper will especially focus on the differences between the traditional and existing systems for 3D mapping and modeling and the proposed solution from the 3DGI group at Aalborg University. These differences will here become clear in relation to the overall system architecture and design.

Urban planning and architecture are both application areas, where the use of 3D has proved to be a very strong medium (Batty et al., 2001; Zlatanova, 2000; Bodum, 2002; Bosselmann, 1998; Danahy and Wright, 1988; Fisher and Unwin, 2002; Whyte, 2002). In many of these cases, 3D models have provided the spatial framework for planning proposals where it has been specifically important to evaluate the form and the spatial concepts for the architecture and the planning. This is the most direct way to map and visualize in 3D. These 3D mappings and visualizations have also proven very valuable after realization in relation to building and property management. This is especially a fact in the more complex and high-rise urban environments. Other 3D modeling initiatives within the urban context can be seen in the field of facility management and urban management such as in the 3D modeling of a cadastre (Stoter and Zlatanova, 2003).

Another very successful implementation of 3D into geovisualization is within the field of large-scale terrain databases. These terrain visualizations can have several different aims, such as fly simulation and pilot training, but in recent years they have also been used more and more as interface for a fly-through of virtual landscapes. The terrain databases have in many cases also been combined with different GIS solutions so that they provide a possible visualization platform for other geoinformation.

The aim for the team at 3DGI in Aalborg as described in this paper will be to combine the different types of 3D geovisualization into one single approach heading towards the creation of a unified 3D geoinformation data structure, with the use of one common object-oriented database that is capable of handling real-time visualizations.

2.1 Technological progress

The main reason and driving force for the advance of 3D within geovisualization is first of all the impressive technological progress that has been seen in the last 5-10 years. Especially the gaming and entertainment industry has had a very strong influence on the development of computer graphics hardware in this period. That gives the industry and the developers a possibility to distribute their software that is used for running the 3D geovisualization to the spot where the users are. 3D gaming as well as 3D geovisualization has become available for mass market.

Along with the development of software for 3D modeling and rendering there has been a remarkable increase in the speed of the Internet. This means that it in a foreseeable future will become possible to distribute (or stream) the 3D objects. This will make the dynamics and interaction of the content even more useful and thereby emphasize the importance of these subjects in research. When the "hole" through becomes big enough to distribute high quality models with a reasonable representation and rendering, we will most certainly meet the constraints from the interfaces and from the limited interaction available.

2.2 Integrated database concept

The ultimate challenge for 3D mapping has been to develop an integrated system where the 3D objects including attributes and behaviours are kept in a database, which at the same time would be accessible for different kinds of queries and visualizations. The system that is described in this paper, will aim at both to fulfilling these demands and at the same time be the source for a real-time VR simulation of urban and rural areas.

The development of the data model for 3D mapping is an important part of the research program that has the objective to utilize the third dimension within the geoinformation society in a way so that it becomes more common to use 3D simulations and visualizations in both the public administration (as a tool) and in the many different media that we use to plan, navigate, seek information, organize and learn from in our lives. Within some years it should be just as natural to use 3D maps as we feel it is today when we make use of traditional 2D maps.

2.2.1 Separate 3D and GIS solutions (type 1)

There are many possible ways out of this. In this section three different data-models will be presented and a reason why it has been decided to use the most demanding data-model in the 3DGI project will be described later. The models can be classified in three different groups of solutions. The first type of data-model is a divided solution with two separate systems with a dedicated link between the systems. One for the traditional geoinformation (generic GIS) where all the 2D maps and respective attributes are kept in a proprietary system and the 3D modeling of the city is done in a CAD-related environment. The systems might use the same coordinate system and are therefore capable of doing spatial queries both ways.

This means you will be able to navigate in one map and still follow the spatial movement in the other. The solution is not sufficient when it comes to real GI queries since the data structures in the two geometric representations are very different.

2.2.2 Combined systems (type 2)

This type of system is the most widespread solution at this time. Here it is possible to query and visualize the two different data-types within each other's environments. The link between the two systems is built as a geo-referencing application. The systems are normally capable of exchanging data between 2D and 3D. By mapping the features over a dedicated surface representing the terrain, the 2D data is visualized in the 3D models. These features can even be elevated through the use of height attributes in the 2D database. There are several commercial systems that have been developed over this data-model.
1.2.3 Object-oriented solution (type 3)

The ultimate solution for a 3D data model is developed over an object-oriented database. The difference from traditional systems is very obvious. Traditional geoinformation databases are developed over the relational data-model. The relational database has its strength with simple geometric forms because of the tabular structure for the organization of data. As long as the database only has polygons, lines and points and more simple forms of topology, it is no problem to keep the data in the relational data-model, but as soon as we go from 2D to 3D we multiply the complexity with a very high number. At the same time the amount of data is exploding because we need to see more detailed visualizations and we need textures at the same time. This requires a lot from the database. It is both necessary to find an efficient way to create a spatial index for the database, so that objects can be found and retrieved for visualization purposes and at the same time we need the database to be very fast for real-time visualization purposes. Furthermore, there are issues such as Level-Of-Detail (LOD) and orientation of the individual objects to consider. The considerations are numerous and the only way to get through this will be by working closely together with others that have the same goals to reach.

The object-oriented database requires a very detailed object-classification to work correctly. It is not decided how this classification should be generated, but since there are other projects working on 3D mapping, it would be obvious to participate in this work and develop this as a collaborative effort. The goal is not to invent our own object-classification, but to join and encourage the use of open standards not only for database but also for the visualization and later for the distribution of these maps to everyone.

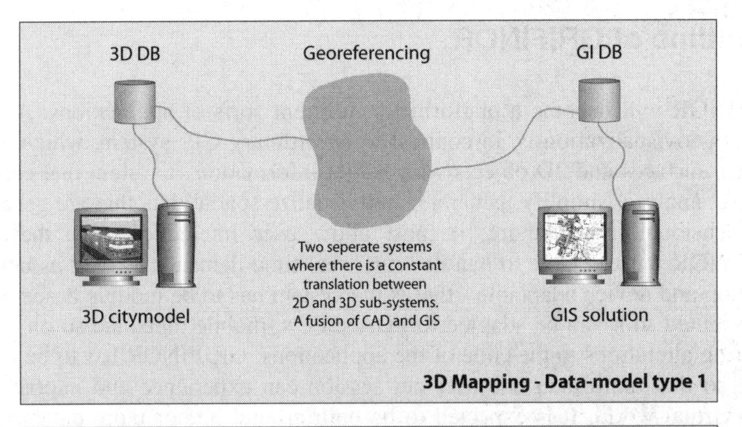

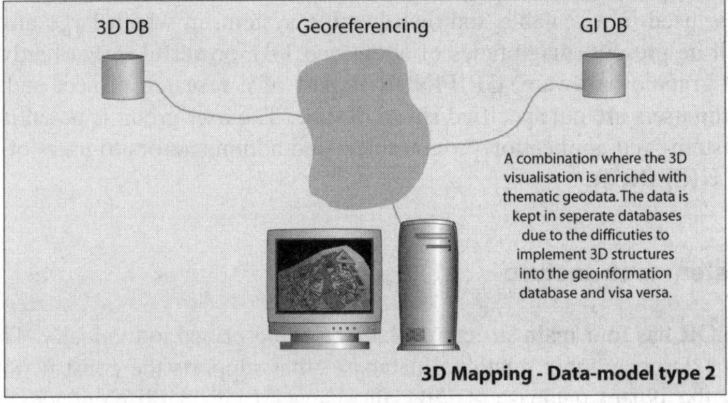

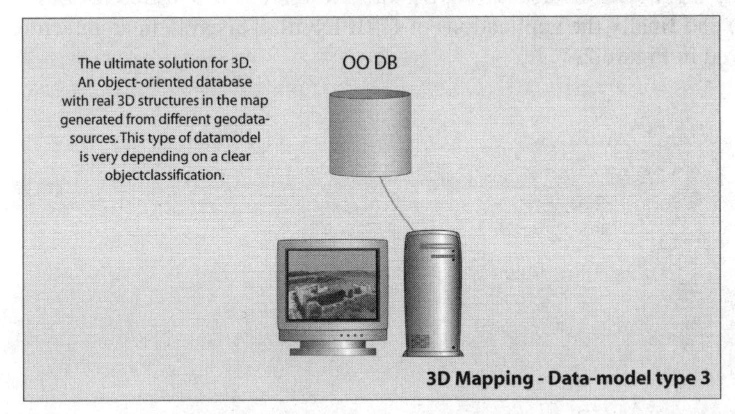

Fig. 1. The different 3D mapping solutions can be generalized into three different data-models. Most common solutions are the type 1 and 2 solutions. Type 3 is still an issue for basic research and development. There are no commercial type 3 solutions available yet.

3. Outline of GRIFINOR

GRIFINOR will become a platform for different sorts of applications. A system for 3D geovisualization is, in contrast to an ordinary GIS system, which at most handles surfaces and 2D objects with height information, a system that can store, retrieve, analyze, simplify, generate, and visualize spatial data that are generically 3 dimensional. Furthermore, it must allow user interaction with these data. GRIFINOR must be able to handle "soft" real-time demands as well as being application and device adaptable - that is the system has to be module based and object oriented so it can be adapted to PDA's, PC's, mobile units and so on, without requiring alterations to the code of the applications. GRIFINOR has to be collaborative so that more than one user per session can experience and interact in the same virtual world. It is expected to be built around one or more database technologies, used in a scalable and distributable system, in which large amounts of data will be present (magnitudes of about one TB), powerful server hardware and fast 3D graphic hardware. GRIFINOR is part of a research project and for that reason the users are not specified ahead of time. The user group is potentially vast, from system- and application programmers and administrators to users of applications in GRIFINOR.

3.1 System architecture

GRIFINOR has four main structures that can be described individually. This is the GeoDB (2D geographic relational database) that supports the construction of 3D objects, the object database (ODB), the viewer (with 3 different viewing platforms) and finally the applications of GRIFINOR. This system architecture can be observed in Figure 2.

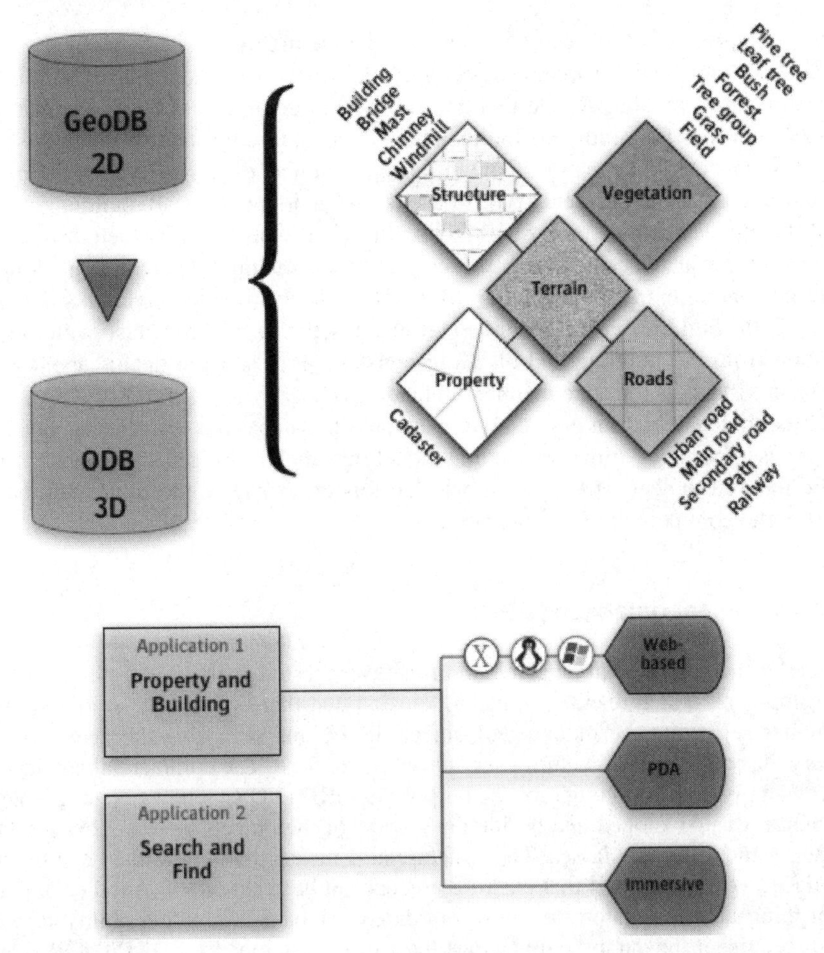

Fig. 2. The system architecture for the 3DGI object-oriented system called GRIFINOR for 3D Geovisualization

3.2 Semi-automatic construction of objects

The first part of the process is a conversion of traditional 2D geoinformation from the GeoDB to 3D objects that can be stored in the ODB. This will be done in separate steps for each of the object types. It will be too comprehensive to go through all the different types of entities. The final 3D model will consist of several very different object classes. Those will among others be: terrain, buildings, property, roads and vegetation. To give an idea about how e.g. buildings will be generated, a short description will be revealed here.

The information for the reconstruction of each building comes from several different sources. First of all, it is necessary to have a footprint of the building with very high precision. This can be provided from the highly detailed technical maps (1:1.000) from the municipalities. Secondly, information about heights of buildings is extracted from LIDAR data that are available in urban areas. Then information about each specific building is found in the national building and dwelling register. Here is registered information about the specific number of stories, type of building (residential, factory, shed etc.) type of use (housing or commercial), building material (for the best simulation of building texture) and type of roof. Much of this information's can be valuable for the correct reconstruction. The roofs are found through feature-extraction from the LIDAR data and even each individual unit (e.g. flats) in the building can be generated through a separate workflow also with information from the building and dwelling register. More specific details about this process will be documented in other publications (Overby et al., 2004).

It is obviously not an easy task to generate the database as 3D objects, since it involves a lot of very time-demanding modeling which is not general but very specific for each object type. It is a hope that this procedure in the future will be a part of the commercial map production.

3.3 The object database (ODB)

Some of the commercial database solutions were tested in the initial phase of the development, which resulted in the conclusion that there were very serious shortcomings regarding the indexing of objects in 3D and regarding the necessity to query the database in something very close to real time. The commercial databases were simply not fast enough for the purpose of GRIFINOR. A new object-oriented database was developed and beside the storage of the objects there is now a database for the different classes. The structure also means that a new class can be introduced very easily and that existing objects can be reclassified. Another important thing about a customized object-oriented database is also the ability to test and reengineer the database up against the viewer developed for GRIFINOR. This involves also the communication between server and client.

Among the many new facilities in the database is also the geographical index that enables the client to query at a global scale. This is one of the things that makes GRIFINOR a true geographical system. In fact, it is possible to store the world in the same database and therefore navigate around the globe and only query the same database (Kolar, 2004 (a,b)) .

3.4 3D viewer

The viewer should be both independent of certain hardware and very flexible in use. To fulfil these demands it was decided to use Java for the development of the viewer. At the moment only a very early implementation of the viewer has been developed. This viewer is built around the scene-graph called Xith3D (Yazel, 2003). The viewer will also offer a suite of navigational tools for interaction with the 3D virtual environment. The viewer will be developed in three flavours. There will be a web-client, which will work on at least three different platforms. The three different clients will be a Windows client, a Macintosh client and a Linux client. There will be a client for Personal Digital Assistants (PDA's) and other mobile hardware units, which will work through a wireless communication. Finally, there will be a client developed for more immersive display types such as panorama.

3.5 Applications

Although GRIFINOR will be able to support a large array of different applications, it has been decided to focus on two specific applications in the later phase of the development of GRIFINOR. The first one will be about buildings and property. The purpose of this application will be to show objects that represent sub classes of a building, and thereby also show how the property can be divided in 3 dimensional parts (Stoter et al., 2004; Stoter, 2004). The second application will show the ability of GRIFINOR to do a search and find. It could be through a street address or through another indirect geographical reference. Both applications will be an important part of a future digital administration, where traditional GIS will be replaced with 3 dimensional objects and the ability to use them for 3D geovisualization in a system such as GRIFINOR.

4. Conclusion

It is our belief that the future geographic interface for geospatial information used in the coming digital administration will not be built around a 2D flat concept of the world, such as the well-known paper map, but around a conceptual virtual 3D environment, where the semantics are very much like the real world. When you want to known more about a specific location near you, it will only take you a short virtual trip through the model to go there and just ask the question in the system. Also the administration will gain from this change, since the interface can be used as a common virtual space for meetings, for introduction of new planning initiatives etc. But the road towards this ideal situation is not straight or without holes. It will take time and much more research before it will be possible to implement these systems in a broad scale.

Acknowledgements

This initiative under the Centre for 3D GeoInformation is funded by:

- European Regional Development Fund (ERDF)
- Aalborg University, Denmark
- Kort & Matrikelstyrelsen (Danish National Survey and Cadastre)
- COWI A/S, Denmark (formerly known as Kampsax)
- Informi GIS, Denmark

References

Batty, M., Chapman, D., Evans, S., Haklay, M., Kueppers, S., Shiode, N., Smith, A. and Torrens, P.M. (2001). Visualizing the City: Communicating Urban Design to Planners and Decision Makers, in *Planning Support Systems*, (R. K. Brail and R. E. Klosterman, Eds.), Redlands: ESRI Press.

Bodum, L. (2002). 3D Mapping for Urban and Regional Planning, presented at URISA Annual Conference 02, Chicago.

Bosselmann, P. (1998). *Representation of places : reality and realism in city design*. Berkeley: University of California Press.

Danahy, J.W. and Wright, R. (1988). Exploring Design Through 3-dimensional simulations, *Landscape Architecture*, vol. 78, pp. 64-71.

Fisher, P. and Unwin, D. (2002). Virtual Reality in Geography, Taylor & Francis, pp. 404.

Fuhrmann, S. and MacEachren, A. (1999) Navigating Desktop GeoVirtual Environments, presented at IEEE Information Visualization Symposium, San Francisco.

Granum, E. and Musaeus, P. (2002). Constructing Virtual Environments for Visual Explorers, in *Virtual Space - Spatiality in Virtual Inhabited 3D Worlds*, (L. Qvortrup, Ed.), London: Springer-Verlag, 112-138.

Graf, U. (1943). Das Raum-Modell bei Stereoskopischen Verfahren in der Kartographie, *Petermanns Geographische Mitteilungen*, vol. 89, pp. 65-69.

Kolar, J. (2004a). Global indexing of 3d vector geographic features, presented at XXth ISPRS Congress, Istanbul.

Kolar, J. (2004b).Representation of geographic terrain surface using global indexing, presented at Geoinformatics 2004, Gävle, Sweden.

MacEachren, A.M. and Kraak, M.-J. (2001). Research Challenges in Geovisualization, *Cartography and Geographic Information Science*, vol. 28, pp. 3-12.

Oettermann, S. (1997) *The panorama : history of a mass medium*. New York: Zone Books.

Overby, J., Bodum, L, Kjems, E. and Ilsøe, P.M. (2004). Automatic 3D building reconstruction from airborne laser scanning and cadastral data using Hough transformation presented at XXth ISPRS Congress, Istanbul.

Raper, J. (1989). GIS - Three Dimensional Applications in Geographic Information Systems. London: Taylor & Francis.

Raper, J. (2001). Multidimensional geographical information science. London: Taylor & Francis.

Stoter, J. and Zlatanova, S. (2003) Visualisation and editing of 3D objects organised in a DBMS, presented at EuroSDR comm. 5 workshop on Visualisation and Rendering, ITC, Netherlands.

Stoter, J. (2004). 3D Cadastre, Technical University, Delft, Delft, Ph.D.

Stoter, J., Sørensen, E.M., and Bodum, L. (2004). 3D Registration of Real Property in Denmark, presented at FIG Working Week 2004, Athens.

Whyte, J. (2002). *Virtual Reality and the Built Environment*. Oxford: Architectural Press.

Yazel, D. (2003). Xith3D Overview, http://xith.org/tutes/xith3d.pdf.

Zlatanova, S. (2000). 3D GIS for Urban Development, ITC and Graz University of Technology, Enschede and Graz.

Complex Artificial Environments – ESLab's Experience

Juval Portugali and the ESLab team*

Abstract. The Environmental Simulation Laboratory (ESLab) is one of several laboratories, research centers and planning and design organizations that have emerged in the last two decades with a configuration that focuses on complex artificial environments in general, and on cities and their dynamics, in particular. The specific experience gained at ESLab is employed in this paper to discuss the various theoretical, methodological, social and ethical issues associated with the above emerging bodies.

1. Introduction

The aim of this paper is to show how the previous, somewhat abstract, discussion regarding the scope of complex artificial environments (Portugali, this volume), takes a concrete form in the structure and activities of ESLab – the Environmental Simulation Laboratory founded at Tel Aviv University in January 2001. As indicated above, the paper describes the research products of the entire ESLab team*.

Four years ago, the theoretical foundations of ESLab were *Self-Organization and the City* (Portugali, 1999) and the notion of *SIRN* – Synergetic Inter-Representation Networks (Haken and Portugali, 1996; Portugali, 2002). Based on these two sources, the basic principles of ESLab are as follows:

- The perception of the city as a *dual self-organizing system,* in the sense that the city as a whole and each of its parts is a complex self-organizing system.
- The use of agent base (AB) and cellular automata (CA) models as the main simulators of urban dynamics.
- The view that the current 2D AB/CA urban simulation models need to be further developed into 3D simulation models.
- Urban theory and modeling must be supplemented by a cognitive approach to cities and their dynamics.
- Our perception of planning must change: It should take into account the notions of entropic versus self-organized planning discussed previously (Portugali, this volume), and as a consequence be reformulated as a bottom-up democratic process.

---

* The ESLab team includes, in addition to the author, TAU faculty members Itzhak Benenson and Itzhak Omer, research associates Rivka Fabrikant, Hernan Kasakin and Karel Martens, and the following research students and programers: Tal Agmon, Slava Birfur, Ran Goldblatt, Erez Hatna, Vlad Kharbash, Talia Margalit, Hani Munk-Vitelson, Guy Nizry, Udi Or, Amir Porat, Asaf Roz, Karin Talmor and Michael Winograd. Their specific projects are referred to throughout the text and in the bibliography.

These principles show up in ESLab's software infrastructure, as well as in the various studies conducted within its frame.

Infrastructure-wise, ESLab can be described as a general purpose support system (GPSS) that can be employed in several ways, in particular as a research support system (RSS), a planning support system (PSS) and a community support system (CSS). The description of the above systems now follows.

2. The ESLab GPSS (General Purpose Support System)

The ESLab's GPSS is a 3-part system composed of two virtual reality (VR) simulators – MultiGen and Skyline, a family of urban simulation models and geographical information systems (GIS). Figure 1 illustrates this system.

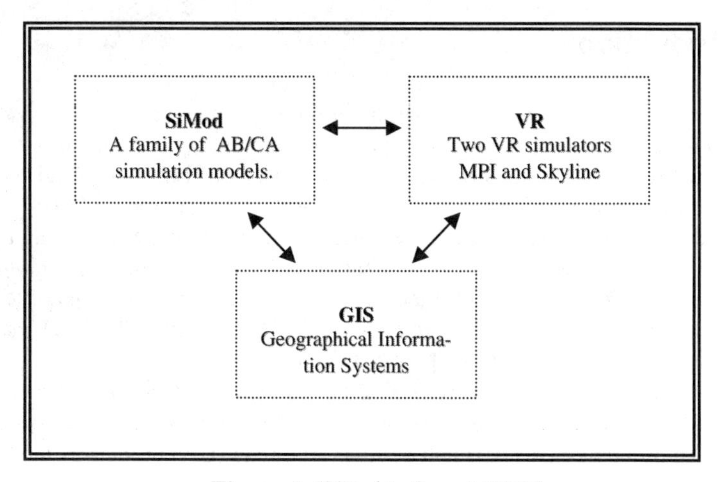

Figure 1. ESLab's 3-part GPSS

2.1 The two VR simulators

2.1.1 MultiGen paradigm

The *MultiGen paradigm* (MPI) is an architecturally oriented VR simulator. It enables building large scale virtual environments, such as cities and whole regions, with a high level of architectural realism. Its elementary parts are the *Creator*, which together with the creator terrain studio/generator (CTS), enables building real-time 3D content for use in visual simulation; the *SiteBuilder 3D*, which allows transforming GIS 2D map data into a 3D visual and the *Vega Prime*, which enables simulation of movement in environments such as cities. Employing the MultiGen, we have already built VR Tel Aviv (Figure 2) and several other virtual environments that will be specified below.

Figure 2. A snapshot from VR Tel Aviv

2.1.2. Skyline

Our second simulator is the *Skyline*. The main advantage of Skyline is that, unlike the MultiGen, it can be viewed on the Internet.[1] The three elementary parts of the Skyline are the *TerraBuilder*, which allows users to create, edit and maintain 3D databases; the *TerraExplorer, which* allows users to simulate movement through 3D Terra environments, and the *TerraGate*, a network data server technology designed to stream 3D geographic data in real time. As will be specified below, using Skyline we can "fly" over the entire country of Israel, as well as over the city of Tel Aviv-Yaffo, Tel Aviv University campus and UNESCO World Heritage sites in Israel.

A recent Skyline application is a project called *The SeaCity Interface,* which was on display during September–October 2004 in the Israeli Pavilion of the Biennale for architecture that took place in Venice.[2] The SeaCity interface is a system composed of three PC computers representing 1924 Jaffa and Tel-Aviv, 1949 Tel Aviv-Jaffa and 2004 Tel Aviv-Yaffo. The visitors/users can control the medium and speed of the movement, look around, stop, move again and listen to an audio at certain focal points along the way (Figures 3, 4).

---

[1] Click "Virtual Flights" on ESLab's website http://eslab.tau.ac.il/

[2] Prepared by the author, Asaf Roz and Karin Talmor, with the support of the entire ESLab team.

Figure 3. The SeaCity Interface: Snapshots from VR flight in 1924 Jaffa and Tel Aviv (bottom), 1949 Tel Aviv-Jaffa (middle) and 2004 Tel Aviv-Yaffo (top).

Figure 4. The SeaCity interface as placed in the Israeli pavilion, the Biennale for architecture, Venice, September-October 2004.

2.1.3 Building a virtual environment

The general process of building a virtual environment is basically the same in both Skyline and MPI (Figure 5). The starting step is a DTM layer of the area, on top of which is superimposed, firstly, an air photograph (orthophoto) and secondly, GIS layers (dxf files) referring to height of buildings, shape, facades, symbols, labels and so on. In Skyline, a layer of links, labels and symbols can be conveniently added.

Figure 5. Stages in the building of a virtual environment.

A major advantage of Skyline is, as noted, the possibility to post it on the Internet – making it an ideal tool for Internet-supported public participation systems, (see below for examples). On the other hand, the architecturally elaborated MPI, with its integrative link to VR helmet-mounted display (HMD) and similar devices, is an effective tool for producing genuine VR environments. The advantage of the

ESLab system is its ability to integrate both VR simulators. For example, in the above Skyline application *SeaCity* (p. 97, Figs. 3,4) that was on display in the Biennale for architecture 2004 in Venice, some of the detailed architectural elements of Tel Aviv's cityscape were first built using the MPI Creator and then were shifted to Skyline.

2.2 The family of urban simulation models

Our point of departure here were the models developed as part of a research project we conducted during the 1990s and published in the book *Self-Organization and the City* (Portugali, 1999). Within this project, we developed what we termed free agents on cellular space (FACS) and free agents on real space (FARS) models. A typical FACS model is built of two layers: a CA layer representing the infrastructure of the city – its buildings, road network etc., and an AB layer that is superimposed on the CA layer and mimics the behavior of the various agents operating in the city. In a FARS model, the abstract cellular space that represents the infrastructure of the city is replaced by the real space of the city as derived from GIS. The FACS and FARS approaches were recently combined and further elaborated within GAS – a framework of geographic automata systems (Benenson and Torrens, 2004).On the basis of the FACS, FARS and GAS models, we are now developing four new models: an object-based environment for urban simulations (OBEUS), which is a model enabling non-experts to build their own task-specific models by means of the interface it provides; CogCity, which suggests a cognitive approach to urban modeling; 3DCity, which is a 3D urban simulation model, and ParkingCity – an AB parking model.

A detailed description of OBEUS appears in a separate paper (Benenson et al, this volume). Here, suffice is to say that OBEUS demands the modeler to define the rules of objects' "behavior," leaving the rest of the work on model development to the system. By this, OBEUS drastically reduces the time of model development and makes the model transferable, transparent and understandable to the other developers. In the following section we introduce ESLab's other three models: CogCity, 3DCity and ParkingCity.

2.2.1. CogCity (Cognitive City)

This is a new type of environmental simulation model, currently in the final stages of development. Its essence is to model the dynamic of cities and other artificial environments, not only from an economic, social, political, or cultural starting point (as is common in this domain) but from humans' cognitive capabilities, as revealed by studies in cognitive science and cognitive geography. A paper entitled "Toward a cognitive approach to urban dynamics" (Portugali, 2004) elaborates on the theoretical rationale of this approach. It shows, first, that a cognitive approach allows a more realistic treatment of the cognitive dimensions that typify behavior and decision-making in all human domains (economic, political, etc.) – especially so in the context of the high levels of nonlinearity, uncertainty and unpredictability that characterize the city as a self-organizing system. In the latter, the somewhat simplistic economic or social behavioral models that underlie agents' behavior in the majority of current urban simulation models collapse, and in their stead one has to employ models revealed in the study of human behavior. Examples for such models are Tversky and Kahnman's (1981) decision heuristics and their adaptation-extension to the context of cities (Haken and Portugali, 1999).

Second, the classical urban models, such as the rank-size rule, central place theory, or von Thünen's type land-use models, which refer to the hierarchy of cities or to their core-periphery structure, are commonly assumed to be the result of economic or social forces. Portugali's (2004) paper shows that hierarchy and core-periphery are basic perceptual-behavioral models with the implication that the hierarchical or core-periphery structure of cities can be derived directly from first principles of cognition.

A third finding of the above study is that the very capacity of humans to construct cognitive maps implies that agents never come to the city *tabula rasa*. Rather they come with some "image of a city" in mind. The latter might be a conceptual cognitive map of the category *city* or a specific cognitive map of the certain city to which they come (*c-* and *s-cognitive* maps respectively). From this it follows that in many cognitive tasks and actions urban agents take top-down decisions and not bottom-up, as is usual is urban simulation models.

CogCity allows for this kind of spatial behavior to take place. An example of how such a model would look is given in the appendix of Portugali's (2004) paper. It starts, as is common in AB/CA urban simulation models, when new agents come to a city. In the above example, they are landowners or developers who are intending to buy an empty cell in the city and construct on it a building of a certain height and use. Each of them comes with a certain c-cognitive map "in mind" that refers to the overall structure of the city and with an aim to buy a parcel of land and construct on it a building of a certain height (4 to 8 floors) and use (residential, commercial, industrial, and so on).

Examining the city from its c-cognitive map, the agent constructs an s-cognitive map of the city, which then serves as a basis for it's subsequent spatial decision-making process: a decision about the area of the city appropriate for the intention in mind; a decision about the appropriate cell in it and a decision about the height and function of the building on it. Figure 6 shows a typical outcome. The fact that the outcome is that of an ordinary city indicates the possibility that the structure of the city might be the outcome of basic cognitive capabilities that are at work prior to, or in synergy with, economic, social or political considerations.

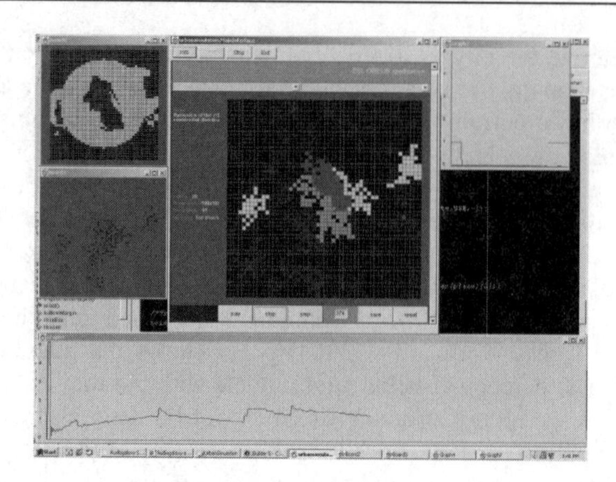

Figure 6. A typical outcome from CogCity.

The above properties of CogCity are three facets of the more fundamental nature of this model, which is that CogCity is another step in our attempt to simulate cities not only as complex self-organizing systems, but as *dual self-organizing systems*. As noted previously (Portugali, this volume), similarly to E-languages, cities are dual self-organizing systems in the sense that their parts – the agents – are themselves complex systems. As further noted, the majority of students and modelers of the city as a complex self-organizing system tend to overlook this property. In CogCity, these two self-organization processes – one that takes place at the local level of individual agents and one that takes place at the global level of the whole city – are treated as two facets of a single, dual process of self-organization.

2.2.2 3DCity

This is a 3-dimensional urban simulation model currently under development. As noted in a previous article (Portugali, this volume), the rationale behind the development of such a model is that several of the most significant urban phenomena, at all domains and scales, are directly related to the third dimension of the city. And yet, the vast majority of the models simulating the dynamics of cities and other artificial environments are essentially 2-dimensional, representing their environments as a flat surface, that is, as a map. 3D models are available, but they usually refer not to the complex dynamics of cities but, rather to their final static form. The MultiGen and Skyline software now available at ESLab make the task of building genuine 3D urban simulation models feasible.

We developed 3DCity as a superposition of two sub-models. The first describes the dynamics of land- and property-owner agents. Their behavior and decisions are only "2.5-dimensional" in the sense that their decisions regarding the third dimension are implemented on top of a 2D surface. That is, they buy land and decide the type and the height of a building to build on it. This sub-model can thus be seen also as a 3D visualization of an otherwise 2D dynamics. Figure 7 presents four snapshots from a dynamic model scenario that takes its input matrix

Figure 7. Four snapshots from the dynamics of 3DCity.

from CogCity and visualizes it in 3D by means of MultiGen.[3] The outcome of this model is a 3D urban structure, into which the second submodel adds the householders that attempt to buy or rent apartments, shops and offices in the various floors. These are the genuine 3D agents that take decision by reference to other agents or properties below, above, and around them. The development of the first sub-model is about to be completed.

2.2.3 ParkingCity

This is the first AB simulation model developed in ESLab that refers to transportation and movement in the city (Birfur, in progress). Generally, the model simulates the relation between car drivers, a given road network with its specific infrastructure of parking facilities, and parking policies. The model's agents mimic car drivers that can be programmed to exhibit a variety of behaviorial and decision making patterns. The road network and parking facilities are taken from urban GIS layers, and as a consequence explicitly represent the real world situation. Thus, given a certain road network, with a given infrastructure of parking facilities, the model can simulate scenarios resulting from different parking policies and drivers' behavioral patterns (Figure 8).

[3] This computer application was developed by Ran Sarig. The dynamics of this model can be observed on our website: http://www.eslab.tau.ac.il

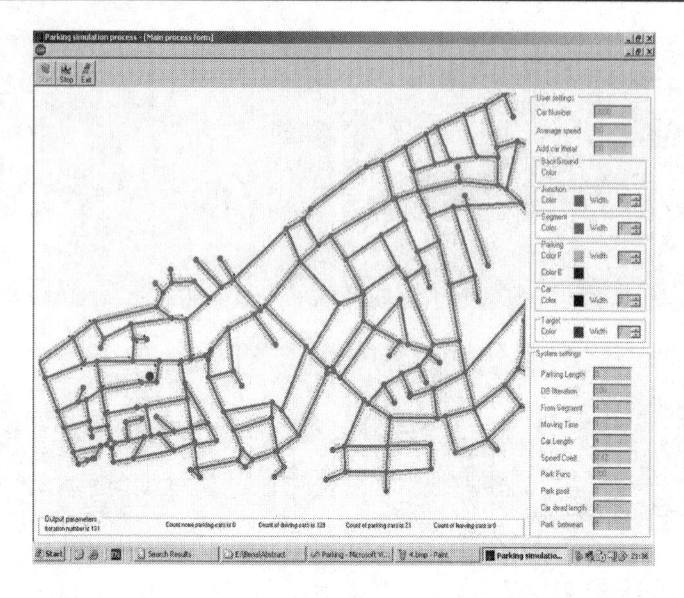

Figure 8. A snapshot from a dynamic scenario of ParkingCity.

2.3 Geographical Information Systems

Standard GISs such as *MapInfo* and *ArcInfo* provide the foundations for ESLab's GIS. The availability of the two afore-mentioned VR simulators immediately allows for several new GIS applications. Figure 9 shows a snapshots from a real-time simulated movement in the city of Tel Aviv, implemented via the MPI simulator. The different colors of the buildings indicate varying properties referring to the various uses in each building, to ethnic or socio-economic characteristics of the buildings' inhabitants and so on (the differences between the various floors can also be added). The user is allowed to move in this city in real time, to choose the means of movement – walking, driving, flying – as well as to shift from one means of movement to the other during the simulation. A nice feature in this simulation (prepared by Benenson and Peretz) is that as one flies up and looks at the city from a greater distance, the visualization shifts from building base to larger block base.

Figure 9. 3D GIS of VR Tel Aviv implemented via MPI.

Figure 10 shows a GIS application that makes use of Skyline. This project that was prepared for Tel Aviv University, and installed in its website, allows every Internet user in the world to visit VR Tel Aviv University.[4] The visitor can choose between a "guided tour" or fly around the campus alone. The GIS part of the simulation shows up in two ways: via the labels – clicking on them gives information about the various uses of each building – and by coloring buildings or whole areas. Other Skyline-derived GIS applications are described in detail below and illustrated in Figures 14 and 15.

Figure 10. A Skyline application of 3D GIS at VR Tel Aviv University.

[4] http://eslab.tau.ac.il/flyover.html

3. Some products of ESLab's RSS (Research Support System)

Employing the above GPSS as a RSS entailed several studies that further our understanding of "ordinary" artificial environments, while exploring the properties of virtual- or V-environments.

3.1 InfoCity

This project is a follow-up of Haken and Portugali's (2003) recent paper "The face of the city is its information." The paper's point of departure is Haken's (1988/2000) distinction between *semantic information* and *Shannonian information*. The first refers to the common usage of the term information, that is, to the meaning conveyed by a message, while the second to the notion of information as defined by Shannon's information theory (Shannon and Weaver, 1949). As is well recorded (Haken, ibid.), Shannon has defined information as a pure quantity and with meaning exorcised.

Commencing from the above distinction, Haken and Portugali (ibid) show the following: firstly, that different elements in the city afford different levels of Shannonian information; second, that the information afforded by the various urban elements can be measured using Shannon's information *bits*; third, that the very ability to do so depends on qualitative cognitive processes (e.g. categorization) that entail semantic information and fourth, that the above three properties are part of humans' cognitive capabilities. These findings entail potential implications to a variety of issues that concern complex artificial environments. A partial list includes the following domains and issues:

- **Legibility.** The notion was first introduced by Lynch (1960) in his *The Image of the City*. The above definition of information allows one to quantify, for example, the overall legibility of a city and of specific urban elements in it.
- **Transportation.** One can now measure the amount of information conveyed by a city's road network in general, and every single street in it in particular (see below).
- **Planning.** The above measures of information can provide an evaluation tool that can indicate, for instance, the impact of a newly proposed project (a building, a road, a park and so on) on its environment – Will it increase/decrease legibility in the city?
- **Navigation.** Given a navigation task, one can now use the above measures of information to define the most significant, that is, informative, elements along the desired route.

- **VR environments.** A VR environment such as a V-city, consumes a lot of computer memory space. As a consequence, when building such an environment, the immediate dilemma is: how much realism to add to the simulation. Too much will make the system very slow; too little, not legible. ESLab's solution is to build one or a few generic elements (a typical building, landscape, etc. – see Figures 2, 4 above) and to detail the minimum number of elements needed in order to make the city legible. We are now working on a method that will define the quantity of marginal information gained by detailing another element and its marginal information utility.

The aim of the InfoCity project is to realize the above potential. As a first step towards this aim we have extended the notion of semantic information by adding to it a variant termed *pragmatic information* (Portugali, 2004a). While semantic information refers to meaning in general, pragmatic information refers to the action afforded by various urban elements. For example, all roads in Tel Aviv (Figure 11) afford the action of driving. Of them, a large number of relatively short roads afford short moves, a small number of the roads also afford moving in between focal points in the city, while only a few roads allow driving short distances, moving between focal points and crossing the city from one side to the other.

Figure 11. Different quantities of information bits afforded by the streets of Tel Aviv.

As shown in the above study, the quantity of information bits conveyed by each of the short roads is very small; the quantity conveyed by the medium-sized roads is much higher, while the quantity of information conveyed by the cross-city roads is the highest. It is further suggested that despite the fact that in Tel Aviv (as in any other city) each road is unique and differs from all other roads by its specific location, landscape and name, people will not be able to memorize most of the short roads but will have no difficulty memorizing each of the medium-sized and the cross-city roads. This was recently confirmed by an empirical study conducted by Omer et al, (this volume).

A second example concerns two forms of legibility. Figure 12 shows the most legible building in a certain part of Tel Aviv. This building affords the highest quantity of information in that part of the city. It can thus be termed landmark in Lynch's (1960) sense. The study upon which Figure 12 is based has identified also the buildings affording the highest quantity of information as they unfold before a person walking in that street. A specifically interesting finding of this study is that the same buildings afford different quantities of information to a person when walking up and down the very same road, thus illustrating the action-dependent property of pragmatic information.

Figure 12. Buildings affording the highest quantity of information bits as they unfold before a person walking in the street.

3.2 Cognition in real and virtual environments

Nine studies can be mentioned here. The first, by Omer et al (this volume), suggests a method to increase the legibility of virtual environments. The second, by Porat (2005), explores navigation and environmental learning in 2- and 3-dimensional urban environments. The third, by Goldblatt (2005), explores the similarities and differences between real and virtual environments with respect to environmental learning and navigation. The fourth, by Winograd (in progress), studies the relations between information, navigation and emotional effects. The fifth, by Munk-Vitelson (in progress), deals with infants' spatial exploratory behavior. The sixth, by Kasakin (in press), focuses on cognitive aspects of urban design. The seventh, by Margalit (in progress) focuses on the role of discourse in shaping the face of cities. The eighth, by Hetna (in progress), studies the impact of agents' behavioral patterns on the overall dynamics of cities. The ninth by Or (2005), empirically evaluates classical models of residential dynamics and segregation.

Preliminary findings indicate, firstly, that people have difficulty perceiving, learning and memorizing 3D environments, and that they tend to overcome this difficulty by perceptually cutting the 3D space into 2D slices (though they can still integrate spatial knowledge across levels - Porat, ibid). Secondly, that there are major differences between learning in real environments as opposed to in VR environments (Goldblatt, ibid). Third, that exploratory behavior, meaning, the internal drive to explore a new environment, is typical of human infants and, by implication, of humans in general (Munk-Vitelson, ibid). Fourth, that analogies and metaphors play a central role in SIRN processes of design (Kasakin, ibid). Fifth, that discourse between professionals (architects, planners, etc.) and public discourse as it takes place in the media, play a central role in SIRN processes that shapes the face of cities (Margalit, ibid). Sixth, that agents' tolerance towards their neighbors is a dominant variable in the decision-making process of location (Or, ibid). This finding is based on high-resolution data from nine Israeli cities. Seventh, that agents' location decisions in cities are typified by the property of *bounded rationality* (Hatna, ibid). In developing his agent base model, Hetna makes use of Or's finding regarding neighbor's tolerance and thus, in his model, householder-agents of varying tolerance coefficients relocate in the city.

4. Some products of ESLab's PSS (Planning Support System)

4.1 PlanCity

PlanCity can be seen as a PSS. Such systems are commonly composed of environmental simulation models, GIS and visualization devices (Brail and Klosterman 2001, Brail, this volume). In this respect, the ESLab three-part system as described above provides the building blocks for the ESLab's PSS. To the above three blocks we add the cut, paste, plan (CPP) application currently under development at ESLab (Roz, in progress). CPP allows the user to cut a section of the environment, replace it with another and examine and visualize the implications (Figure 13).

Figure 13. Steps in operating the CPP application: Cutting a portion of an open square in Tel Aviv (left), pasting a building in its stead (right); studying the impact of the new building on its environment (middle).

All this is only part of the story, however. As noted earlier, our understanding of planning is related to the city as a dual self-organized system and to SIRN as the engine of its dynamics. This understanding led us to identify two groups of planners that are engaged in urban dynamics and the urban planning process: official-professional planners and "latent planners" – that is, every agent that is active in the city.

The idea of latent planners and their role in the city is the rationale behind the notion of *democratic planning* that is central to ESLab's activities.[5] The suggestion is that since every urban agent is a planner on a certain level, it is essential both to supply the planners with updated planning information (planning authorities often tend to refrain from supplying such information), and to be tuned to bottom-up planning ideas and solutions. The story of the balconies of Tel Aviv described previously (Portugali, this volume) is an example of such a bottom-up planning process.

PlanCity is designed for both types of planners. When directed to professional planners it can act as an ordinary PSS. When directed to latent planners it can be regarded as a means to supply real-time on-line planning information to all latent planners-agents operating in a city.

In a way our suggestion is an alternative to the conventional notion of public participation in planning. Our view of the usual notion of public participation is

[5] The inauguration of ESLab in early 2001 was associated with a conference about "The democratization of planning in Israel" (See "about us" in the lab's website http://www.eslab.tau.ac.il).

rather critical; we see it as a paternalistic approach by which professional planners in a top-down manner allow the public to express its view about planning issues (Portugali, 1999; Alfasi, 2003). The projects *AccessCity* and *CommunCity* are examples of how planning information can become public domain. They are described in the following section.

5. Some products of the ESLab's CSS (Community Support System)

5.1 AccessCity

The aim of this model is to develop measures of accessibility to focal objects in the city, such as public open spaces, schools, transportation nodes, community centers, and the like. Then, using our 2D- and 3D-visualization methodologies, to make this information available to every individual in the community. An example can be seen on our website: Levels of accessibility to public open spaces were calculated for every building in the area of Jaffa and, using our GIS-Skyline model, were posted on the Internet (Nizry, in progress). Each agent in the city can now see the accessibility of any building to public open space (Figure 14). The notion of accessibility is closely linked to issues of social justice and the artificial environment. Martens' (in progress) research on transport and justice deals directly with this issue, with special focus on transportation.

Figure 14. The accessibility of buildings in Jaffa to public open space – a Skyline application.

5.2 CommunCity

This project provides public access to information regarding plans prepared by professional and official planning bodies. A case in point is an experimental website that concerns the "Shefech Hayarkon" (Hebrew for Yarkon river estuary) project in the north of Tel Aviv.[6] This website provides the relevant information about the area and the project (history, current situation, land use maps etc.); it enables users a VR flight over the area, and to observe it in 3D as it is now, as it will be when alternative A is implemented, as it will be when alternative B is implemented and so on. It further provides the users with a GIS covering various themes and data, a *forum* in which they can ask questions and discuss the project among themselves and an on-line questionnaire enabling them to comment and give feedback on the various alternatives. All these in a high level of virtual realism that was made possible by the ESLab system described above. Figure 15 illustrates the main components of CommunCity.

As noted, the aim of both AccessCity and CommunCity is to use advanced communication technology to make planning information public domain. However, we are fully aware that such technology is not available to everyone in the community, especially not to those who might need it most. Though it is likely that in the future more people will have access to such technology, one has to take into consideration that this access will always be limited. In other words, technology alone will not democratize planning. Rather, technology must be seen as being one among several means toward this aim, including the action and initiatives of the civil society and changes that must be made in the very structure of the global, top-down planning process.

With respect to civil society, it can be said, firstly, that while the chances of sophisticated planning information technology becoming accessible to the public at large are rather slim, there is no doubt that they are already, and in the future will be even more, accessible to NGOs and the many other organizations that compose civil society. Second, that one aspect of the second urban revolution noted previously (Portugali, this volume), with its link to globalization and the decline of the welfare nation-state, is the rising power of civil society, which takes over many of the roles of past welfare states and as such already plays an important role in the democratization of planning.

With respect to changes in the structure and process of planning, it must be noted that ESLab is currently engaged in a project sponsored by the Israeli Ministry of Housing and Construction called "A New Structure to the Israeli Planning System." The first stage of the project was to build a new model for the procedural and legislative planning process that will be democratic in the sense noted above; it will allow a free flow of top-down and bottom-up planning ideas, initiatives and actions. The second stage is to adapt this model to the Israeli planning reality and suggest it as an alternative to the current planning law.

[6] http://www.eslab.tau.ac.il/yarkon.htm – in preparation by Talmor (in progress).

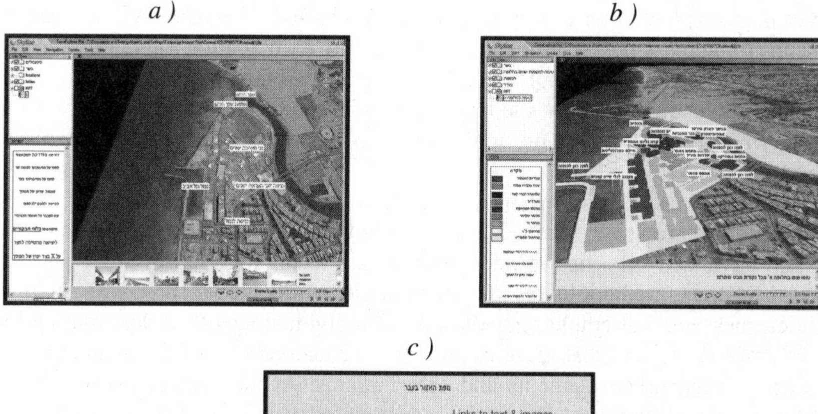

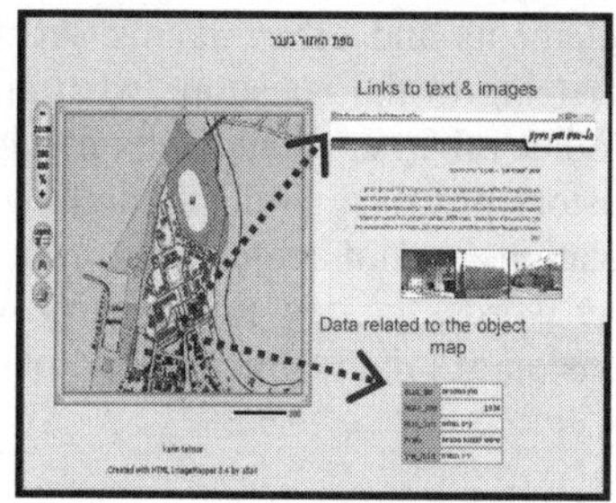

Figure 15. The main components of CommunCity: A VR flight over the planning area as it is (a) and over the area according to alternative A (b). An example of an interactive GIS (c).

Concluding notes

The structure of ESLab as described above is typical of several laboratories and research centers built during the last two decades. In fact, several of the contributors to this book are representatives of such centers: Batty from CASA, London; Pumain and Sanders from UMR Géographie-cités, Paris; Timmerman and Arentze from DDSS, Eindhoven; Bodum from the Centre for 3D GeoInformation, Aalborg; Semboloni from CSCD, Florence and Kwartler from ESC New York. Typical to such laboratories and centers is the use of state-of-the-art technology, which, while creating an exciting potential also raises concerns and doubts. These have been discussed at some length in pervious papers and include: the need to deepen our understanding regarding natural versus artificial environments (Haken, this volume), the increasing significance of visualization (Batty et al, this volume), the problematics of shifting ideas from the natural to the human domain (Pumain, this volume) and the tension between the technological ambition to achieve high realism and the need for explanation and understanding that require information reduction (Portugali, this volume).

In closing this paper it is appropriate to add to the above list two issues that were only partly discussed so far. The first is the issue of social responsibility.

State-of-the-art technology is by its nature a double-edged sword: While it can be a means to increase democracy and genuine public participation in planning, it can also become just another tool for only the strong sections of society. Much depends on the choices made at the level of individuals and research groups: what to study, what kind of tools and applications to develop, what research grants to submit and so on.

The second issue concerns what one might call "the deeper message" of the new media we are using – namely, complexity theory, agent base simulation models and advanced information and visualization technologies. Beyond being advanced theories, methodologies and technologies, these media carry with them an implicit message that emphasizes the role of individual agents in determining the structure of cities, and that requires an explicit consideration of the qualitative relations between urban elements and urban agents. At the core of every AB/CA model is a parameter referring to the question "what are the relations between a cell, a building, or any other urban element, to its neighboring cells, buildings or elements?" But as the history of urban modeling and planning indicates, the question of "what are" the relations cannot be detached from the issue of "what should be" the relations. In other words, thc ncw mcdia carry with them a need to explicate the qualitative dimension of urban dynamics and planning.

References

Alfasi, N. (2003). Is public participation making urban planning more democratic? The Israeli experience. *Planning Theory and Practice* 4, 2, 185-202.

Benenson, I. and Torrens, P.M. (2004). *Geosimulation: Automata-based modeling of urban phenomena.* Wiley, England

Birfir, S. (in progress). building model of parking human behavior. Ph.D thesis, Tel Aviv University.

Brail, R.K. and Klosterman, R.E. (Eds.) (2001). *Planning Support Systems*, ESRI Press, Redlands California.

Goldblatt, R. (2005). Learning an unfamiliar environment: Learning a virtual environment versus learning the same real one. M.A thesis, Tel Aviv University.

Haken, H. (1988/2000). *Information and Self-organization.* Springer, Heidelberg.

Haken, H. (1996). *Principles of Brain Functioning.* Heidelberg: Springer.

Haken, H., and Portugali, J. (1996). Synergetics, Inter-representation networks and cognitive maps. Pp. 45-67 in *The construction of cognitive maps*, edited by J Portugali. Dordrecht: Kluwer academic publishers.

Haken, H. and Portugali, J. (2003). The face of the city is its information. *Journal of Environmental Psychology* 23 385-408.

Hatna, E. (in progress). Modeling urban structure dynamics as an outcome of individual choice behavior. Ph.D thesis, Tel Aviv University.

Kasakin (in press). Design aided by visual display: a cognitive approach. *The Journal of Architectural Research.*

Lynch K. 1960. *The Image of the City.* MIT press, Cambridge Mass.

Martens, K. (in progress). Beyond equity in transport: a review of the justice and transport field.

Munk-Vitelson, H. (in progress). Spatial behavior of prewalking infants. Ph.D thesis, Tel Aviv university.

Nizry, G. (in progress). Evaluating accessibility to urban public gardens in an individual level resolution. M.A thesis, Tel Aviv University.

Or, E.(2005). High-Resolution Residential Heterogeneity in Israeli Cities. M.A thesis, Tel Aviv University.

Porat, A. (2005). wayfinding in open and closed urban spaces. Ph.D thesis, Tel Aviv University.

Portugali, J. (1996). Inter-representation networks and cognitive maps. Pp. 11-43 in *The construction of cognitive maps*, edited by J. Portugali. Dordrecht: Kluwer academic publishers.

Portugali, J. (1999). *Self-Organization and the City*. Heidelberg: Springer.

Portugali, J. (2002). The seven basic propositions of SIRN (Synergetic Inter-Representation Networks). *Nonlinear Phenomena in Complex Systems* 5(4) 428-444.

Portugali, J. (2004). Toward a cognitive approach to urban dynamics, *Environment and Planning B, Planning and Design* 31, 589-613.

Portugali, J. (2004a). Shannonian, semantic and pragmatic Geoinformation. Geoinformatics 2004 – proceedings of the 12th international conference, Universsity of Gavle, 15-21.

Roz, A. (in progress). Integrated planning and modeling planning support systems. M.A. thesis, Tel Aviv University.

Shannon, C. E. and Weaver W. (1949). *The Mathematical Theory of Communication*. Urbana, IL, University of Illinois Press.

Talmor, K. (in progress). A Web-based Public Participation System in the Planning Process: Developing an e-participation Prototype. M.A theisi, Technion, Haifa.

Tversky, A. and Kahaneman, D. (1981). The framing of decision and psychology of choice. *Science* 211.

Winograd, M. (in progress). The relations between information, navigation and emotional effects. M.A. thesis, Tel Aviv university.

Part three: Urban simulation models

Geosimulation and its Application to Urban Growth Modeling

Paul M. Torrens

Abstract. Automata-based models have enjoyed widespread application to urban simulation in recent years. Cellular automata (CA) and multi-agent systems (MAS) have been particularly popular. However, CA and MAS are often confused. In many instances, CA are paraphrased as agent-based models and simply re-interpreted as MAS. This is interesting from a geographical standpoint, because the two may be distinguished by their spatial attributes. First, they differ in terms of their mobility: CA cannot „move", but MAS are mobile entities. Second, in terms of interaction, CA transmit information by diffusion over neighborhoods; MAS transmit information by themselves, moving between locations that can be at any distance from an agent's current position. These different views on the basic geography of the system can have important implications for urban simulations developed using the tools. It may result in different space-time dynamics between model runs and may have important consequences for the use of the models as applied tools. In this chapter, a patently spatial framework for urban simulation with automata Tools is described: Geographic Automata Systems (GAS). The applicability of the GAS approach will be demonstrated with reference to practical implementations, showing how the framework can be used to develop intuitive models of urban dynamics.

1. Introduction

The practice of model-design, model-building, and the application of models in the geographical sciences is in the midst of a transformation. Recent shifts in the art and activity of spatial simulation may be considered as the end-result of a decade or so of research and development, currently gathering critical momentum. This is manifest, most vividly, in the emergence of a new class of models, and a new generation of applications, an approach that some authors have begun to refer to as geosimulation (Benenson and Torrens 2004a; Benenson and Torrens 2004d).

In this chapter, we will explore the concept of geosimulation, in the context of its use in building urban models. We will introduce a new methodology for constructing geosimulation models, focused on the idea of spatial automata devices—what we call Geographic Automata Systems. We will also demonstrate the use of these techniques for urban applications, referring to the development of simulations of urban growth.

In section 2 geosimulation is discussed as a new approach to simulation, which is defined more concretely in section 3. Automata are introduced in section 4 as the favored modeling tool for geosimulation work. In section 5, it is argued that there is strong need for a patently spatial set of geosimulation tools (elsewhere in this volume, Benenson and colleagues describe software for this very purpose). Our work in this area is introduced in section 6, with reference to Geographic Automata Systems. The problem of modeling urban growth—sprawl in particular—is discussed in section 7. Naturally, we argue for a geosimulation approach and we describe an urban growth model based on geosimulation ideas, and built as a Geographic Automata System, in section 8. The practical use of this model to explore growth phenomena is described in section 9, before concluding remarks appear in section 10.

2. Geosimulation as a new trend in spatial simulation

There is a distinguished lineage to the development of spatial simulation methodology, and geosimulation represents what we might consider as the new wave of a long line of spatial simulation developments. The *idea* behind geosimulation is best considered in terms of the distinction between that approach to modeling and what may have come before it. The distinction between older and newer is not discrete; very little is in the world of geography.

More conventional spatial simulation is perhaps aptly considered as dealing with the exchange of entities and activities between relatively coarsely-considered units of space, and describing those exchanges in relatively aggregate terms. That is a naïve characterization, but it is only intended to serve a comparative use. By means of contrast, we can consider newer-style approaches—our notion of geosimulation—as extending, substituting, and supplanting conventional models. The geosimulation approach is more likely to be characteristic of models that handle massive quantities of geographic entities, each represented at an atomic (individual and independent) scale of consideration. Exchanges of and between these entities is mediated by the connections that exist between elementary components of geographical systems, considered dynamically and interactively. Our ability to simulate geographical phenomena has advanced to the point where entity-level behaviors can be translated, directly, into artificial computational environments—code. That code can be used to generate and play with incredibly life-like geographic systems—spaces, phenomena, entities—in completely artificial simulated environments; *in silico,* as Steven Levy might refer to them.

Of course, the distinction between possessing the ability to do something, and actually doing it, is rather important in simulation contexts. Spatial models are hungry things and developers must feed them data, methodology, and tools, before they can get them to perform any tricks. We think our Geographic Automata Systems can help.

Abstracting from geography for a moment, geosimulation could be thought of in terms of broader trends in general simulation. We might draw analogies between geosimulation and parallel developments in the social and physical sciences: bottom-up modeling as an alternative or extension of top-down simulation (Epstein, 1999); open and transparent simulation in lieu of black-box modeling (Wiener, 1961); notions of phenomena as complex adaptive systems (Johnson, 2001), etc. The contribution of geographers to these developments is significant. Geographers are building new tools within a larger simulation community (Dibble and Feldman, 2004), some open source in nature (Clarke and Gaydos, 1998), and others to be shared as software (Benenson et al., 2004; Semboloni et al., 2004).

The contribution of geographer's ideas to a growing debate about real-world systems should not be under-estimated; space is beginning to feature prominently in cross-disciplinary theory-building and testing in this context (Gimblett, 2002). Rather than poaching methodologies, tools, and ideas from other fields; geography is beginning to have a significantly reciprocal *influence* across a wide range of fields on the outskirts of its interests. To a certain extent, geosimulation is a catalyst for this activity.

3. Defining geosimulation

We have devoted lots of dead trees to specifying geosimulation as a modeling approach, and readers that are particularly interested in that material might wish to read some of that work (Benenson and Torrens, 2004a; Benenson and Torrens, 2004c; Torrens, 2004). Put succinctly, geosimulation might be defined with reference to its explicit attention to space and geography, both methodologically and in terms of its intellectual foundations.

First, we can consider issues of representation in geosimulation models. Whereas more conventional spatial modeling handles representation of geographic units in a relatively aggregate fashion, geosimulation-style models are more judicious in their representation of geography. The traditional consideration of average and spatially-modifiable geographical units or (statistically) mean individuals is replaced in geosimulation; units are regarded, instead, as spatially non-modifiable entities, with individual descriptions and independent functionality. Where aggregates are considered, they are more than likely formulated generatively, built from the bottom up by assembling individual entities for the purposes of accomplishing an aggregate task or amassing an aggregate structure.

Second, the treatment of behavior in geosimulation models is important. Under the geosimulation approach, simulated entities are often individual; likewise, they are commonly independent and autonomous in their behavior. From a synoptic perspective, the behavioral focus is often on disaggregate interactions in a systems setting. The independence is significant; attention turns to the specification of individual-level behaviors, and immediately this casts the developer's attention to issues such as cognition, motivation, mobility, etc. Independence has further implications for considering space-time dynamics; we will discuss this in more detail shortly. The move toward autonomy in behavior simulation is also noteworthy; entity behavior is not necessarily treated as homogenous across the system being considered: coffee in the city, but cocoa in the suburbs. Moreover, and borrowing from complexity studies, *collective* behavior is often modeled as a by-product of spatial interaction; communities emerge as a function of neighbor interaction, for example, with interaction defined in terms of a range of behaviors from perception to budgeting (Torrens, 2001).

Third, geosimulation is markedly distinct in its treatment of time and dynamics, particularly so when compared against more conventional techniques that are popularly employed in spatial simulation. Under the geosimulation approach, models are commonly designed as event-driven, rather than time-driven. Time in such simulations moves within discrete packets of change, based on the internal clocks of simulated components. When put together to form a system, update of these clocks may be synchronous or asynchronous; the methodology is relatively flexible in this regard.

4. Automata as the favored geosimulation methodology

Methodologically, geosimulation research and development has been dominated by automata-based approaches to model-building. Cellular automata (CA), and their sibling multi-agent systems (MAS), are particularly popular (O'Sullivan and Torrens, 2000; Torrens, 2002a, 2003, 2004; Torrens and O'Sullivan 2001). We would like to argue that an approach based on spatial-specific processing devices—Geographic Automata—is perhaps more appropriate for geographic research. Let us examine automata before we begin that discussion.

At its heart, an automaton is a processing mechanism (whether tangible, or mathematical). It is a discrete entity endowed with some structural variables (states) and capable of receiving similar information as input from the outside world. A given automaton's states change over time (transition) according to a set of rules; these rules evaluate the internal state of the automaton at a point in time and the information input to the automaton at the same time, to determine the automaton's state in a subsequent point in time (Figure 1a). Changes to automata states operate in a discrete temporal domain. Alan Turing's hypothetical computing device is a classic example.

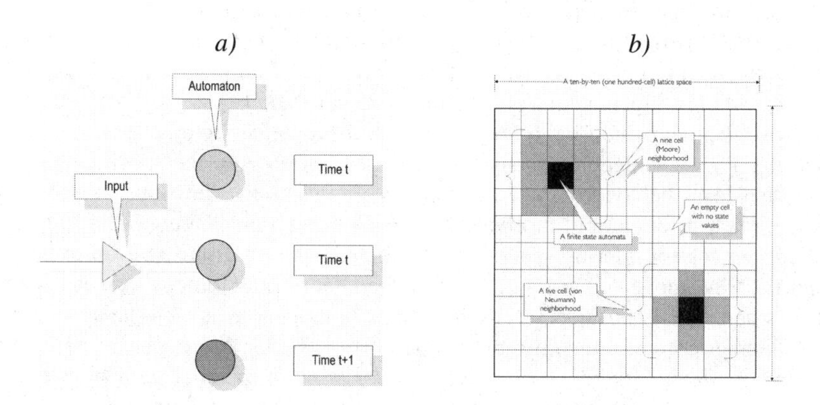

Fig. 1. (a) An automaton changes state (color) between two time steps, based on input. (b) Varying cellular automata neighborhood configurations in a 2D cell-space.

Cellular automata are an extension of the automata concept, in which the space in which an automaton resides becomes important to the specification. CA are an arrangement of connected individual automata, arranged to form a partitioned space. Think of the pixels on a TV screen and you have the general idea. Each unit in the partition is an automaton as described previously. However, input is now drawn from other automata in a localized area around a given automaton—the neighborhood (Figure 1b).

Multi-agent systems are also automata, although the term has become quite generic in recent years. MAS are generally endowed with some agency; various state variables and transition rules are introduced to lend agent-automata life-like qualities, for example for use in Artificial Intelligence research (Ferber, 1999). In social science, agency is generally expressed with reference to decision-making and choice behavior (Kohler and Gumerman, 2001), cooperation and conflict (Epstein and Axtell, 1996), economic reasoning (Luna and Stefansson, 2000), etc. In other fields, agency is used to mimic insect (Bonabeau et al., 1999) and animal behavior (Meyer and Guillot, 1994), and to specify Internet bots (Leonard, 1997) and Web-crawlers (Pallman, 1999). Agents pop up in all sorts of places, from Lord of the Rings films to Xbox games. In general social science contexts, agents are usually non-spatial in nature. This is not so in geographical contexts, where agency relates quite closely to the mobility of agents in a simulation. This is one of the distinguishing factors between CA and MAS in geographic research; agents may be designed with the ability to move within a simulated space—and carry their state information and rules of interaction with them as they do so. CA, by contrast, are static in their lattice space; they may diffuse information to neighbors, but they cannot alter their position. The distinction is important, at least to geographers. We would like to argue, in fact, that the distinction is so important as to warrant a whole new class of automata—what we call Geographic Automata.

5. Why we need geographic automata

Considering my own needs as a model developer, I generally need my model to support some key components of the systems that I wish to simulate. Likely, my needs are quite similar to those of the reader. Of course, I deal mostly with human geographical systems, so my wish-list is understandably biased in that direction. Entities in the model generally need to be distinguishable in terms of the space in which they are situated (or bounded). A flexible expression of the spatial relationships between entities is desirable. Generally, entities in my simulations have some form of mobility, and so I need to be able to track them as they wander around a simulated space, or as their occupation of that space alters. Similarly, I need to be able to describe their behavior, whether spatial or non-spatial.

The trouble with the automata approaches that I outlined in the previous section is that none of them is capable of supporting this sort of functionality in a cohesive manner. CA are handicapped by their inability to simulate true movement. The velocity of information transmission could be approximated in some fashion, but a cell will never be able to uproot and jump around its lattice. MAS can do all of this, but much of the current MAS methodology underestimates the importance of space and movement behavior. As desirable as mobility methodology might be to a geographer such as myself, MAS have other shortcomings; agents cannot be fields, for example.

Geography is all about the behavior and distribution of things in space, and different things in different spaces. When we throw time into that soup, we have to consider all of this in the context of space-time dynamics. Why not fall back on our geosimulation approach, infusing spatial properties into the aforementioned automata tools? We might, for example, adopt a fully automata-based view of geographic systems, fabricating systems, from the bottom up and using building blocks fashioned as geographic automata. Relying on notions such as emergence (Johnson, 2001) or network theory (Watts, 2003), we might relate these building blocks to each other as a Geographic Automata System.

6. Geographic Automata Systems

Based on our understanding of geographical systems, we can extend the automata idea with space-specific functionality to account for the general needs of geographical modelers. We have actually explored the suitability of the following approach for general urban modeling, at least, and it seems sufficient for most needs that we can consider (Torrens and Benenson, 2005). The usefulness of the approach for modeling urban growth will be demonstrated shortly, but first let us define what we mean by a Geographic Automata System.

A Geographic Automata System retains all of the basic functionality of automata, CA, and MAS:

- States
- State transition rules
- Neighborhoods

To this mixture, we add some peculiarly geographic functionality:

- A typology, describing automata types
- Neighborhood transition rules
- Dynamic location conventions
- Movement rules

The result is a unique class of automata, somewhere beyond CA and MAS, with a dash of Geographic Information Science. The former set of functionalities outlined above now hinge on the latter set—state, state transition rules, and neighborhoods are formulated on the basis of automata type, neighborhood transition rules, location conventions, and movement rules. As ever, all of these components are dynamic with respect to time. In a simulation context, exploration with the Geographic Automata Systems then invokes qualitative and/or quantitative investigation of the influence of these components on system behavior; the ability of the framework to support representation of geographical systems; and specification of the spatio-temporal behavior of geographic entities in an artificial Geographic Automata System simulation environment. The Geographic Automata Systems idea expressed here is a framework for modeling, but it can be used to build models and software for spatial simulation. Thus far we have begun to construct minimal, but wholly sufficient, simulations based on the framework (Benenson and Torrens, 2004b). In addition, Benenson and colleagues have used the framework as the basis for a library of urban simulation tools (Benenson et al., 2004). Before describing one such modeling example in more detail, let us explain the concept in fuller detail.

In the framework, we allow for *typologies* of Geographic Automata. In our own work thus far, we distinguish between fixed and non-fixed Geographic Automata (GA). Fixed GA are used to model entities that do not change their location over time (although their spatial extent may change in size—shrink or grow, for example—or shape). In an urban context, such entities might be roads, land parcels, or parks. Considering the list of functionality outlined previously, fixed GA may succumb to the influence of state and neighborhood transition rules, but not those of motion. Non-fixed entities are those that have the ability to change their location in space and time. Again, in an urban context, we might think of pedestrian walkers, migrating renters, or subway trains. The full range of rules may be applied to non-fixed GA; location conventions and movement rules are of obvious importance, but neighborhood rules take on a curious form when entities are mobile with respect to other entities, and here the fixture of those GA becomes significant. The movement of a car relative to other cars as opposed to that relative to a traffic light is one example you might consider. The distinction is equally important when employing algorithms, mathematics, or databases in their representation.

GA may also possess *state descriptors* and *state transition rules*, as with automata, CA, and MAS. Once again, these may be dynamic with respect to time and space. Their formulation in Geographic Automata Systems differs, however. State transition depends on input from fixed neighborhoods in the context of automata and CA. So, cells appear to magically mutate within a lattice. This is fine for describing phenomena such as local-scale urban decline, but less so for other scenarios—migration is an obvious example. If we consider a model with fixed and non-fixed GA, however, state transition falls under the additional influence of the spatial behavior of other objects in the system, or even *within the cell.*

Indeed, the addition of functionality to enable, determine, and describe *movement* of GA within the system opens up all manner of possibilities for developing spatial models, and using them to test theories and simulate phenomena of interest. Movement, and its representation in a simulation context, is a popular research thread in fields peripheral to geography; the video game industry is one example that springs to mind, with an emphasis on movement choreography (Reynolds, 1999). It seems obvious that geographical simulations should accommodate that sort of functionality, as well as adding unique geographical theory and methodologies relating to mobility, search behavior, way-finding, etc. Movement rules have thus been added to the framework. Indeed, as we will demonstrate later, they become a key ingredient of our urban growth models.

The introduction of functionality to support movement necessitates inclusion of components to track the *location* of entities and objects in simulated spaces. Considering a typology of geographic entities that is based on fixture, location conventions should be amenable to supporting entities of fixed and non-fixed type, their stable and mobile locations, as well as facilitating evaluation of relationships between the two on the basis of those conventions. In the examples that we have developed thus far, we allow for two varieties of location convention. Direct location is specified rather obviously in terms of the current location of an entity in the system, and this might relate to a coordinate point, additional height information, centroids, a network location, or a set of vectors bounding a polygonal coverage. Location by indirect means is somewhat more complicated, and we have formulated this by means of pointers. An entity has its own location conventions, as well as an additional set of location primitives that express its location *relative* to other objects or entities. Two brothers might duplicate the same location when at home in their townhouse, but an indirect pointer will be employed to convey the existence of that relationship as they separate in space, for example when they go to different nightclubs on a Friday evening.

This brings us to the issue of *neighbors* and *neighborhood rules*. Neighborhoods feature in CA models, as described previously. Similarly, agents in MAS may have neighbors. In CA models, neighborhoods are usually static and symmetrical. They are not really suited to describing dynamic spatial relationships between objects and their variance in space and time. The neighbor concept in MAS is more flexible; flying or swimming Boids may have nearest neighbors for example (Reynolds, 1987). However, geographers' interest in neighborhoods is much broader, encompassing notions such as adjacency, connectivity, and proximity that are not always geometrical in form. A generalized framework for describing such concepts is needed. We allow for such functionality in the Geographic Automata System framework, by separating neighborhoods and neighborhood rules that govern the ways in which those neighborhoods might change in space in time, as Voronoi relations, social networks, leader-follower partnerships, etc. There may be instances in which these are non-spatial, or a mixture of spatial and non-spatial. Consider a household example. Neighbor relations could be expressed in terms of family ties: parent-to-child, sibling-to-sibling. As family members go about their business over the course of a day, the spatial relations between these people will change (or not; siblings may go to the same school), but the family tie remains. If there is a new birth in the family, the family tie will be altered, again non-spatially. As children grow and leave for university, they are de-coupled from the household space and interaction may take on a new form (email, phone). If the family fissions, both spatial and non-spatial relations may change yet again, with parents divorcing and moving apart. The neighbor and neighborhood rules need to be flexible to accommodate these sorts of behaviors.

Next, let us illustrate the use of these components in a unified manner, to build models of urban growth and to run simulations of suburban sprawl.

7. An application to simulating sprawl

We have built the modeling framework to be discussed using a geosimulation approach, and on the basis of a GAS foundation. The model is designed to simulate urban growth. Under simulated conditions, city-systems evolve from initial seed settlement sites, going on to urbanize through compaction, polynucleation, infill, inner-city densification and decline, and peripheral suburban sprawl. We have designed the model as an artificial laboratory to test ideas and hypotheses about sprawl in particular.

A lengthy discussion about sprawl is somewhat beyond the general remit of this text; the reader should consult Torrens & Alberti (2000) for more substantial treatment of the topic. Put succinctly, sprawl refers to a phenomenon that is particularly prevalent (and popularly studied) in the United States. Sprawl is suburban growth, first and foremost, understood to extend peripherally around cities in swaths of development that are much lower in density than the core area of the city in question; it is also much more scattered in its spatial distribution.

Sprawl has a number of empirical characteristics. It is also manifest with lots of softer, non-quantifiable, characteristics; consult the literature in architecture and urban design for examples (Calthorpe et al., 2001; Duany et al., 2000; Duany et al., 2001; Katz, 1993). A minimal set of descriptors of sprawl might well include measures of density (of population, employment, or some other activity); the functionality of urban space in the city-system; spatial distribution or structure of the urban extent; and dynamic attributes that would allow all of these things to change in space and time. Moreover, these characteristics may be important at varying scales of observation or consideration.

The potential causes of sprawl are numerous and hotly debated (Ewing, 1997; Gordon and Richardson, 1997a). In fact, evaluation of the veracity of debated factors might be one of the goals of a sprawl model. System growth is very important as a sprawl mechanism; it sets the metabolism of the city and sprawling cities are generally either fast-growing in absolute population totals, or fast-growing in the decentralization of that population to the city's periphery (even if it means leaving a donut hole in the center of the city). The distribution of that growth is particularly significant, and this is what distinguishes sprawl from general suburbanization in most instances. The distribution of sprawl is low in density, scattered in nature, rapid in its appetite for land, and is almost always manifest on the periphery of the main urban mass.

Space and geography, then, are absolutely essential to consideration of sprawl. But representation of the spatial components of sprawl necessitates a uniquely spatial modeling approach. Not surprisingly, we would like to argue that sprawl is an ideal test-bed for geosimulation and GAS; similarly, geosimulation and GAS offer much potential for generating and testing ideas relating to sprawl. Let us try to demonstrate that with a modeling example.

8. Model description

We have constructed an urban growth model, formulated on the basis of a Geographic Automata Systems engine, as mentioned, with components of the model formulated in that scheme: a typology of fixed and mobile geographic automata, each described in space and time by means of a set of state variables, state transition rules, geo-referencing conventions, movement rules (if mobile), neighbor conventions, and neighborhood rules.

The *typology* delineates two types of geographic automata—fixed and mobile. Fixed GA correspond to landscape and infrastructure elements. The simulated space is characterized as a landscape, until developed into urban infrastructure by the other—mobile—type of GA in the model. Mobile GA serve as the agents of change in the simulated system. They are designed to mimic developers and settlers, wandering the landscape with informed behaviors, converting it from non-urban to urban uses, and depositing population as they proceed. In this way, then, fixed GA act as a container for mobile GA; the landscape supports settlement through urbanization.

Fixed GA may also be designated as *gateways*. Essentially, gateways serve as an entry-point to incoming growth (population) to the simulated system. *A priori*, certain sites are designated as gateways in the simulated city, corresponding to the initial seed locations from which an urban system evolves. Gateways may also manifest over the course of a simulation run—during run-time—and this is used to denote the emergence of new centers of urbanization within the city-system as it evolves. Whereas the seed gateways are used as a proxy for exogenous in-migration to the city-system, run-time gateways are used to generate endogenous growth (and decline) within the city.

A set of *state variables* are used to introduce a minimal set of relevant characteristics of the evolving urban system: whether or not a GA is a gateway, the development condition of the simulated landscape, and the state of its settlement. Fixed GA units are endowed with variables to describe whether they are developable or not; this allows for space to be delineated as functional (suitable for urbanization) or not. Similarly, those units are described with an additional variable to indicate whether they have been developed. The end result of urbanization in the model is settlement of a fixed GA with some volume of population; a population count variable is thus introduced to denote the number of people residing on a given fixed GA unit. Because fixed GA are equal in size in the simulation, this may be interpreted as a population density state.

Of course, all of these variables are dynamic with respect to space and time. Change takes place in the model through general *state transition rules*, as well as the movement activity of mobile GA units. Onc of thc interesting features of employing a GAS-based approach to modeling urban growth is that much of the state transition functionality that appears in traditional CA-style urban growth models (Clarke and Gaydos, 1998; Engelen et al., 1995; White and Engelen, 2000; Xie, 1994; Yeh and Li, 2000) can actually be handled through movement rules, thus avoiding the sort of methodology that would have cell states mutate magically within a lattice, rather than initiating as the result of agent-based activity *within* them. Consequently, a handful of general state transition rules are employed. A dispersal function is used, to distribute population very locally between fixed developed GA units, within a small neighborhood radius. This neighborhood is specified in an eight-cell Moore configuration and the dispersal works by diffusing a percentage of population within that neighborhood to a target cell. This dispersal actually works in two directions: growth is diffused between neighbors, but so too is decline. If the average population in the neighborhood is below the value in a target cell, the total in the reference cell will decline accordingly. In this way, then, phenomena such as urban blight and gentrification are handled, albeit in a proxy manner.

A number of *georeferencing conventions* are used in the model, to situate important features in the model, such as seed gateways, and to track the movement of mobile GA as they propagate through the system. Georeferencing is performed by direct and indirect means. The actual coordinates of modeled entities within the simulated space are noted by fixed means, as (x,y) coordinates on a Cartesian plane with origin in the centroid of the simulated space. If a GA is fixed, these coordinates will not change; if it is mobile, the coordinates will be updated to reflect shifts in their position. Indirect georeferencing is employed with respect to mobile GA and the seed gateways from which they originate. As a mobile GA moves through the simulated landscape, its direct location will change, but it retains a *pointer* to the seed gateway from which it originated in the system. This designation remains with the GA as long as it is present within the system.

Movement rules constitute the real work-horse of the model. The rules are designed to mimic proposed drivers of sprawl, as discussed in the literature. Specifically, we use the *geographic* factors understood to be responsible for sprawl in American cities as inspiration for the formulation of the movement regimes. Thus, we have specified a variety of forms of movement for the modeled developer-settler GA. Growth enters the system either from exogenous or endogenous sources, and that growth is distributed spatially over the urbanizing landscape using mobile GA (Figure 2).

Compact development regimes are mimicked by *immediate* and *nearby* movement rules that see mobile GA develop fixed GA in small eight-cell and 24-cell neighborhoods of influence. The compact rules correspond to the sorts of development that might take place early in the evolution of a city-system, when space is considered with relatively less premium than may be the case at a later stage in the city's growth. The compact rules also mimic conventional New Urbanist ideas about denser forms of development (Katz, 1993). The compact movement rules lead to relatively small clusters of dense settlement.

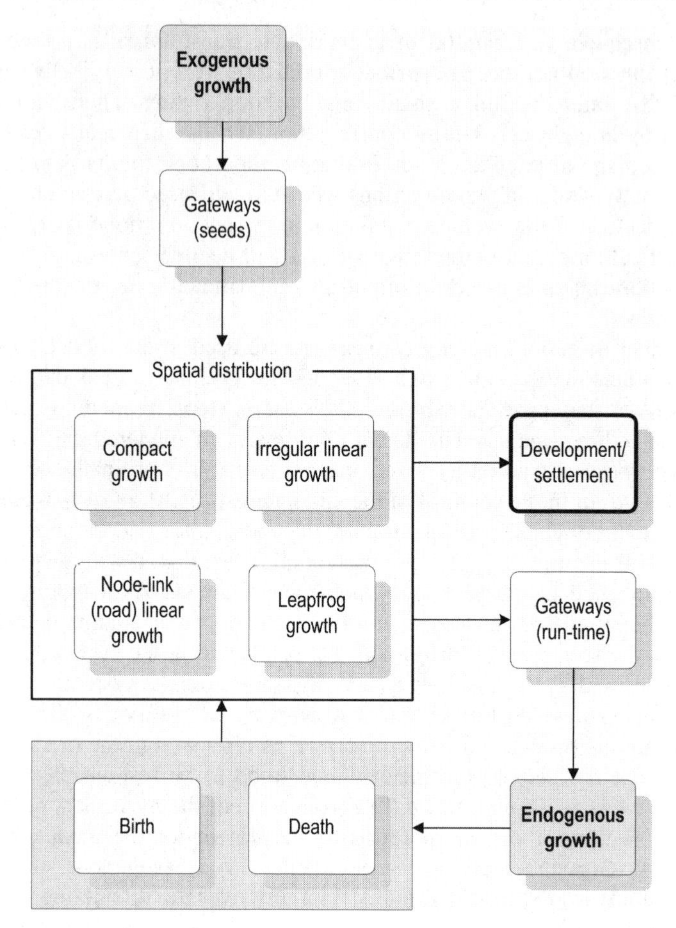

Fig. 2. . A conceptual diagram of the urban growth model engine.

More linear forms of development, such as that which might flourish around transport corridors or flanking arterial highways, are modeled using two other movement rules: an *irregular* and *road-like* movement regime. The irregular function is used to simulate the sorts of development features that might occur when development is constrained to an irregular linear process due to natural or political boundaries. This rule leads to a dendritic form of development, much like that generated by diffusion-limited aggregation (Batty et al., 1989) or random walks (Batty and Longley, 1994). The road-like movement rule is used to grow roads in a simulation. Rather than having roads pop-up out of the ether, the rule is formulated in a chained node and link manner, whereby a developing GA moves over the landscape, laying down nodes to be connected; these nodes are then linked by linear strips of development, thereby mimicking road-like urbanization. This is particularly useful in simulating ribbon-sprawl, i.e., sprawl that tends to feature in linear swaths buffering major roads on the periphery of cities.

Fragmentation is a major feature of conventional suburbanization; as mentioned previously the notion of scatter is crucial to sprawl. Scattered development on the periphery of cities, at lower-than-average densities of settlement, is now commonplace in many American cities. It is largely a by-product of speculative land development (Bahl, 1968), often in areas that were previously devoted to agricultural use. In a sense, the formation of edge cities (Garreau, 1992) is also a process of scattering, albeit on a larger scale of consideration. Scattered development is mimicked by means of a *leapfrog* movement rule, whereby a mobile developer GA can skip ahead of the main urban mass and settle sites outside of that periphery.

The movement rules may also be *combined*, so that a leapfrog movement may be followed by a road-like movement or a compact movement, or other such combinations.

Thus far, we have covered specification of the model as a Geographic Automata System, referring to the typology of entities in the model, the set of state variables used, state transition rules, georeferencing conventions, and movement rules. The final set of components that we need to describe relate to spatial relationships in the model: neighborhoods and neighborhood rules. As mentioned with respect to the compact movement rules, variable *neighborhoods* are employed in the model and these allow for different areas of influence for development and settlement to be introduced. In addition, action-at-a-distance is also supported, as in the case of the road-like and leapfrog movement rules. In this way, then, the influence of a given GA may extend beyond the neighborhood filter.

Neighborhood rules are employed in a rather simple fashion. The actual *shape* of a neighborhood does not vary as the simulation evolves, mobile GA just choose to employ varying *sizes* of neighborhood, and this choice is tied quite simply to the movement rule that they are ordered to employ in a given place at a given time.

This brings us to broader issues of time and dynamics, and their use in the model. *Time* is discrete in the model; it moves in packets or bundles of change. These bundles are characterized as a simulated year in the simulation we will present shortly. In this sense, time is event-driven. Each packet of change (a year) may involve hundreds of individual transitions and movements, and these movements will proceed based on their own internal clocks. A road-like movement rule will take up many cycles of the CPU's clock, whereas the in-migration of growth to a gateway may take up only a handful of cycles. The volume of activity that occupies a single temporal packet will understandably grow as the simulated city fills with more and more people—more and more agents of change. But, by encapsulating these dynamics within events time will appear to flow quite organically within the simulation. There is also some theoretical justification to this approach to dynamics in the model. Different processes have different cycles and we wished to accommodate that functionality. The mechanisms designed to diffuse growth and decline between fixed GA are slower (in event time) than those employed by movement rules; the lifecycle of development and settlement events is shorter than that of neighborhood transition through gentrification and decline.

Finally, the model works in a *constrained* fashion. The state variables for *developable* or *not-developable* conditions allow for the simulated space to be devised as functional (open to development) or non-functional (closed to development, for example, if the area is a large water mass). Automata models are known to be very sensitive to initial parameterization, and urban growth CA are quite sensitive to the specification of seed sites. Seed gateways are thus introduced with some *a priori* understanding of where the model developer would like the city-system to start growing, although other rival sites may emerge over the course of a simulation run. The general metabolism of a simulation is constrained by use of these seed gateways; at each time-step in a simulation run, a volume of growth is metered to the simulation. Although the actual spatial distribution of that growth is left to the model, the general rate of exogenously-derived growth can be constrained. Similarly, the rate of endogenous growth can be constrained. In some simulations, we have tailored the seed conditions and growth rates to known conditions and are able to generate realistic urban evolution and patterns of growth for real-world city-systems (Torrens, 2002b); moreover, quantitative and structural measurements confirm that the simulated conditions match those present on the ground, although that discussion is beyond the main thread of this chapter.

9. Simulating urban growth

We have just described how the model is formulated as a Geographic Automata System, and the functionality that framework affords is particularly useful in modeling urban growth. We will now demonstrate how the model can be used to build realistic simulations of urban growth regimes in artificial cities. These simulations allow for various ideas about the factors responsible for sprawl to be tested in an artificial and controlled computational environment. In particular, we can test the ways in which a city-system might evolve by conventional means (sprawl, in this case), and by smart growth mechanisms in which growth is managed in some sustainable form. This actually echoes a hot debate in the literature at the moment, pertaining to questions of whether sprawl or smart growth is desirable, feasible, cost-effective, and socially just (see Gordon and Richardson, 1997b). The purpose of this chapter is to introduce the idea of geosimulation, Geographic Automata Systems as a framework for geosimulation, and to demonstrate how they might be used to build urban growth models. The simulation experiments described here were designed to test specific ideas about conventional urban growth in American cities, and much of that discussion falls outside of the relevance of this chapter. Nonetheless, some brief description of the simulations may serve to emphasize the points that we are trying to make in this discussion. We will introduce two simulations based on the model described in the last section. One relates to sprawl; the other relates to smart growth.

The sprawl simulation is designed to mimic the general evolution of a contemporary city-system in the United States. Five initial gateways seeds are introduced into the model, with differing growth rates. One seed, in the center of the simulated space, is chosen to dominate *a priori*, and the assignment of growth to it is established to reflect this. 75% of the way through a simulation run, the supply of external growth to all sites, save this dominant city, is halted and growth in those areas proceeds by endogenous means alone. (Growth rates are thus treated heterogeneously across the city-system.) Theoretical justification for this is as follows. The dominant city is afforded a historical advantage from the outset of the simulation; it was an initial settlement site and has inertia, geographically, and path dependence in terms of system dynamics. As the city-system evolves, that site flourishes and gains a competitive advantage that 75% of the way through the simulation (when its urban extent reaches the hinterlands of competing sites) begins to draw growth away from its competitors. The city-system evolves to a familiar sprawling pattern, with road-influenced fingers of ribbon sprawl and a sea of low-density and fragmented urbanization on the periphery. Urban decline is evident in the core of the city (Figure 3).

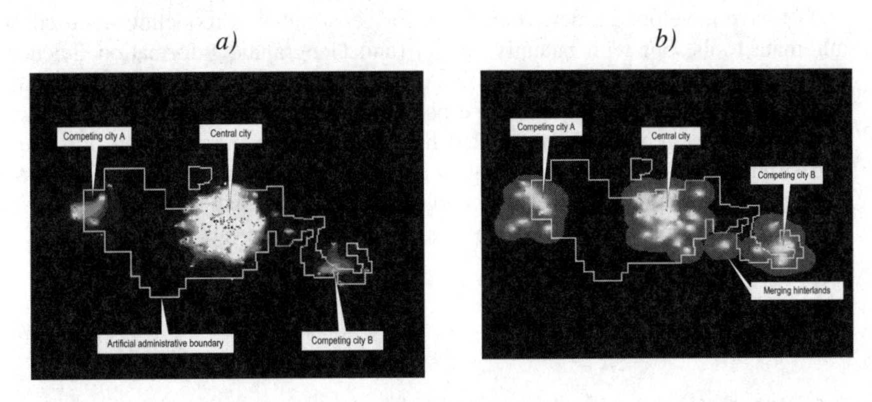

Fig. 3. The results of a sprawl-based (a) and a smart growth (b) simulation run (darker areas refer to low population density; lighter areas house high densities of population).

A second simulation is formulated on a smart growth regime, whereby development is encouraged in smaller compact clusters. A dominant city is established *a priori*, as before. However, initial peripheral sites gain a competitive advantage outside of the main urban mass, and instead of sprawling become relatively dense in their own right. When the supply of exogenous growth is cut, these settlements actually survive. The result is a polycentric spatial structure to the city-system. Sprawl still predominates on the periphery, but it is bound within edge cities that constitute well-established polycentric cores.

These simulations demonstrate the suitability of the geosimulation approach and the Geographic Automata Systems framework. In related work, we have also begun to use this model to explore ideas about the dynamic formation of sprawl as a generative and inherently spatial process. We have also considered the phenomenon of sprawl at a more micro-scale, and have built a GAS-based model of community dynamics within one single GA "cell" of these simulations (Torrens, 2001), looking at issues of residential mobility, community demographics, and socio-spatial segregation.

10. Conclusions

This chapter has introduced a few new research ideas in the field of urban simulation, hopefully not too generally for the reader to grasp fully. We have proposed geosimulation as a new and particularly useful approach to spatial simulation. Work in the area of geosimulation and urban analysis is really beginning to gather steam; Itzhak Benenson and I have edited a journal's double special issue devoted to the topic (Benenson and Torrens, 2004d). Much of the work discussed in those pages relies on automata tools. In many instances, geosimulation work might benefit from more explicitly spatial simulation tools; the need in geography work is obvious.

We have developed a new framework for geosimulation modeling, focused on automata tools, but with patently spatial (and Geographic Information Science) appeal and functionality—Geographic Automata Systems. We believe the framework to be very useful for modeling urban systems.

One such example is demonstrated here, focusing on urban growth modeling and simulation of scenarios relating to suburban sprawl in American cities. We hope that the framework will be found to be more extensible across a range of geographic examples, relating urban analysis and other areas of geography.

References

Bahl, R.W. (1968). A land speculation model: the role of the property tax as a constraint to urban sprawl. *Journal of Regional Science* 8 (2):199-208.

Batty, M., and Longley, P. (1994). *Fractal Cities*. London: Academic Press.

Batty, M., Longley, P.A. and Fotheringham, A.S. (1989). Urban growth and form: scaling, fractal geometry, and diffusion-limited-aggregation. *Environment and Planning A* 21:1447-1472.

Benenson, I., and Torrens, P.M. (2004a). *Geosimulation: Automata-Based Modeling of Urban Phenomena*. London: John Wiley & Sons.

Benenson, I. (2004b). A minimal prototype for integrating GIS and geographic simulation through Geographic Automata Systems. In *GeoDynamics*, edited by P. Atkinson, G. Foody, S. Darby and F. Wu. Florida: CRC Press.

Benenson, I., Aronovich, S., and Noam, S. (2004). Let's talk objects. *Computers, Environment and Urban Systems*, 29 (4): 425-471.

Benenson, I., and Torrens, P.M. (2004c). Geosimulation: object-based modeling of urban phenomena. *Computers, Environment and Urban Systems* 28 (1/2):1-8.

Benenson, I. & Torrens, P. M. (2004d). Special Issue: Geosimulation: object-based modeling of urban phenomena. *Computers, Environment and Urban Systems* 28:1-8.

Bonabeau, E., Dorigo, M., and Theraulaz, G. (1999). Swarm Intelligence: From Natural to Artificial Systems, Santa Fe Institue Studies in the Sciences of Complexity. New York: Oxford University Press.

Calthorpe, P., Fulton, W. and Fishman, R. (2001). *The Regional City: Planning for the End of Sprawl*. Washington, D.C.: Island Press.

Clarke, K.C., and Gaydos, L. (1998). Loose coupling a cellular automaton model and GIS: long-term growth prediction for San Francisco and Washington/Baltimore. *International Journal of Geographical Information Science* 12 (7):699-714.

Dibble, C., and Feldman, P.G. (2004). The GeoGraph 3D Computational Laboratory: Network and Terrain Landscapes for RePast. *Journal of Artificial Societies and Social Simulation* 7 (1).

Duany, A., Plater-Zyberk, E. and Speck, J. (2000). *Suburban Nation: The Rise of Sprawl and the Decline of the American Dream*. New York: North Point Press.

Duany, A., Speck, J. and Plater-Zyberk, E. (2001). *Smart Growth: New Urbanism in American Communities*. New York: McGraw-Hill.

Engelen, G., White, R. Uljee, I. and Drazan, P. (1995). Using cellular automata for integrated modeling of socio-environmental systems. *Environmental Monitoring and Assessment* 30:203-214.

Epstein, J.M. (1999). Agent-based computational models and generative social science. *Complexity* 4 (5):41-60.

Epstein, J.M., and Axtell, R. (1996). *Growing Artificial Societies from the Bottom Up*. Washington D.C.: Brookings Institution.

Ewing, R. (1997). Is Los Angeles-style sprawl desirable? *Journal of the American Planning Association* 63 (1):107-126.

Ferber, J. (1999). Multi-Agent Systems: An Introduction to Distributed Artificial Intelligence. Harlow (UK): Addison-Wesley.

Garreau, J. (1992). *Edge City: Life on the New Frontier*. New York: Anchor Books/Doubleday.

Gimblett, H.R. (ed.) (2002). Integrating Geographic Information Systems and Agent-Based Modeling Techniques for Simulating Social and Ecological Processes, Santa Fe Institute Studies in the Sciences of Complexity. Oxford: Oxford University Press.

Gordon, P. and Richardson, H.W. (1997a). Are compact cities a desirable planning goal? *Journal of the American Planning Association* 63 (1):95-106.

Gordon, P. (1997b). Where's the sprawl? *Journal of the American Planning Association* 63 (2):275-278.

Johnson, S. (2001). Emergence: The Connected Lives of Ants, Brains, Cities, and Software. London: Allen Lane, The Penguin Press.

Katz, P. (1993). The New Urbanism : Toward an Architecture of Community. New York: McGraw-Hill.

Kohler, T.A., and Gumerman, G. (2001). *Dynamics in Human and Primate Societies*. New York.: Oxford University Press.

Leonard, A. (1997). *Bots: The Origin of a New Species*. San Francisco: Hardwired.

Luna, F, and Stefansson, B. (eds.) (2000). Economic Simulation in Swarm: Agent-based Modeling and Object Oriented Programming. Dordrecht: Kluwer.

Meyer, J.A., and Guillot, A. (1994). From SAB90 to SAB94: four years of animat research. In *From Animals to Animats 3. Proceedings of the Third International Conference on Simulation of Adaptive Behavior*, edited by D. Cliff, P. Husbands, J.-A. Meyer and S. Wilson. Cambridge, MA: The MIT Press, 2-11.

O'Sullivan, D., and Torrens, P.M. (2000). Cellular models of urban systems. In *Theoretical and Practical Issues on Cellular Automata*, edited by S. Bandini and T. Worsch. London: Springer-Verlag, 108-117.

Pallman, D. (1999). Programming Bots, Spiders, and Intelligent Agents in Microsoft Visual C++, Microsoft Programming Series. Redmond, WA: Microsoft Press.

Reynolds, C. (1987). Flocks, herds, and schools: A distributed behavioral model. *Computer Graphics* 21 (4):25-34.

Reynolds, C. (1999). Steering behaviors for autonomous characters. Paper read at Game Developers Conference, at San Jose, CA.

Semboloni, F., Assfalg, J. Armeni, S., Gianassi, R. and Marsoni, F. (2004). CityDev, an interactive multi-agents urban model on the web. *Computers, Environment and Urban Systems* 28 (1/2):45-64.

Torrens, P.M. (2001). New tools for simulating housing choices. Program on Housing and Urban Policy Conference Paper Series. Berkeley, CA: University of California Institute of Business and Economic Research and Fisher Center for Real Estate and Urban Economics. http://urbanpolicy.berkeley.edu/pdf/torrens.pdf.

Torrens, P.M. (2002a). Cellular automata and multi-agent systems as planning support tools. In *Planning Support Systems in Practice*, edited by S. Geertman and J. Stillwell. London: Springer-Verlag, 205-222.

Torrens, P.M. (2002b). SprawlSim: modeling sprawling urban growth using automata-based models. In *Agent-Based Models of Land-Use/Land-Cover Change*, edited by D. C. Parker, T. Berger, S. M. Manson and W. J. McConnell. Louvain-la-Neuve, Belgium: LUCC International Project Office, 69-76.

Torrens, P.M. (2003). Automata-based models of urban systems. In *Advanced Spatial Analysis*, edited by P. A. Longley and M. Batty. Redlands, CA: ESRI Press, 61-81.

Torrens, P.M. (2004). Geosimulation approaches to traffic modeling. In P. Stopher, K. Button, K. Haynes & D. Hensher (Eds.). *Transport geography and spatial systems*. London: Pergamon, pp. 549-565.

Torrens, P.M., and Alberti, M. (2000). Measuring sprawl. CASA Working Paper. London: University College London, Centre for Advanced Spatial Analysis. http://www.casa.ucl.ac.uk/measuring_sprawl.pdf.

Torrens, P.M., and Benenson, I. (2005). Geographic automata systems. *International Journal of Geographical Information Science*, 19 (4), 385-412.

Torrens, P.M., and O'Sullivan, D. (2001). Cellular automata and urban simulation: where do we go from here? *Environment and Planning B* 28 (2):163-168.

Watts, D.J. (2003). *Six Degrees: The Science of a Connected Age*. New York: W.W. Norton & Company.

White, R., and Engelen, G. (2000). High-resolution integrated modeling of the spatial dynamics of urban and regional systems. *Computers, Environment and Urban Systems* 24:383-400.

Wiener, N. (1961). Cybernetics: or Control and Communication in the Animal and the Machine. Cambridge, MA: MIT Press.

Xie, Y. (1994). Analytical models and algorithims for cellular urban dynamics. Ph.D., Department of Geography, University of New York at Buffalo, Buffalo, NY.

Yeh, A. Gar-On, and Xia L. (2000). Simulation of compact cities based on the integration of cellular automata and GIS. In *Theoretical and Practical Issues on Cellular Automata*, edited by S. Bandini and T. Worsch. London: Springer-Verlag, 170-178.

Geographic Automata Systems and the OBEUS Software for Their Implementation

Itzhak Benenson, Slava Birfur, Vlad Kharbash

Abstract. The concept of Geographic Automata System (GAS) formalizes an object-based view of city structure and functioning; OBEUS software implements this view on the operational level. The paper presents the GAS paradigm and latest user-friendly version of OBEUS, the latter based on .NET technology and developed according to OODBMS logic. OBEUS boosts further development of GAS theory, especially regarding the treatment of time in models describing collectives of multiple interacting autonomous urban objects. We claim that all high-resolution urban Cellular Automata and Multi-Agent models of which we are aware can be described in GAS terms and represented as OBEUS applications. GAS and OBEUS can thus serve as a universal, transferable framework for object-based urban simulation.

1. From arbitrary spatial units to geographic objects

During the first two decades of their development (1960s-1980s), urban and regional modeling and simulation were almost exclusively based on a compartmental view of systems. Units of flat partitions of space into black-box *regions* were characterized by vectors of state variables representing regional characteristics such as land, population, jobs, transportation and services (Allen and Sanglier, 1979). Although providing a basic outline of urban dynamics, this framework has too many degrees of freedom to be amenable for theoretical inference and is too data-ponderous for practical purposes. Lee's list of regional model sins – Hyper-comprehensiveness (too many phenomena for one model) and Wrong-headedness (too many relationships), among others (Lee, 1973; Lee, 1994), remains valid. Successful simulations of real-world urban systems within regional frameworks are consequently quite rare (Batty, 2003).

Lee demanded that urban models be 'simple'; Cellular Automata (CA) and Multi-Agent System (MAS) model approaches, intensively developed during the last decade, comply with Lee's claim (Benenson and Torrens, 2004b). The idea behind CA and MAS is that high-resolution views of urban systems, which distinguish between real-world objects, simplify system description. CA and MAS models flourished during the 1990s (Batty, 1997; Clarke, Hoppen et al. 1997; Portugali, 2000; White and Engelen, 2000; Benenson, Omer et al., 2002). Compared to black-box regions, CA and MAS are intuitive and avoid many of the sins Lee cites. At the same time, this framework remains circumscribed within urban contexts. For instance, regular partition of space into cells engenders problems when representing road networks; a more general problem is the representation of spatial objects of different sizes and forms. Recently introduced Geographic Automata Systems (Benenson and Torrens, 2003; Benenson and Torrens, 2004a, 2004b; Torrens and Benenson, 2005) resolve these and other limitations.

Contrary to CA, the GAS framework does not demand division of space into small compartments in order to define the artificial skeleton that represents urban elements. Individual urban objects are directly represented in GAS by discrete *automata*. These automata can be of different nature, spatial extension and hierarchical rank; they can also represent spatially fixed as well as non-fixed objects, the locations of which change in time. Spatial relationships between automata can therefore be used to determine the structure of urban space. Although the object approach to representation of urban reality has been mentioned in the literature (Erickson and Lloyd-Jones, 1997; Semboloni, 2000; Galton, 2001), the idea itself remained inchoate up to date.

The recent boom in GIS data production and research provides strong empirical support for the GAS approach: Layers of urban GIS are nothing but collections of urban objects of the same kind. Spatial relationships between objects can be estimated within GIS based on adjacency, overlay, visibility, or accessibility. Moreover, the pattern recognition methods applied to objects of GIS layers can be utilized to grasp spatial emergence and self-organization.

To take the next step and portray *urban dynamics*, we have to 'animate' geographic features, that is, formulate the rules by which urban objects are created, relocated, changed, and destroyed. This step demands formulation of geographic *automata transition rules*; informally, these rules describe the *behavior* of urban objects. Benenson and Torrens (Benenson and Torrens, 2004a),who label the GAS view of urban model construction *Geosimulation,* claim that every high-resolution urban model developed to date can be reformulated in GAS terms.

The claim for GAS universality remains bombastic until a constructive proof can be provided. We propose that a software system capable of implementing the GAS framework can be considered as such a proof. This paper presents the most recent progress made in development of Object-Based Environment for Urban Simulation (OBEUS), a system that verily operationalizes the GAS concept.

We assume that the reader is familiar with the background of Relational DBMS, the Entity-Relational Data Model (ERM) (Howe, 1983), and the Object-Oriented (OO) programming paradigm (Booch, 1994). In the following, the GAS paradigm is presented in rather abstract form, with a very limited number of examples; other illustrations can be found in the literature on this topic (Benenson, Aronovich et al., 2004; Benenson and Torrens, 2005; Torrens and Benenson, 2005). OBEUS software accompanies *Geosimulation: Automata-Based Modeling of Urban Phenomena* (Benenson and Torrens, 2004b), a book that provides readers with a wealth of models that comply with the GAS paradigm.

2. Geographic Automata System (GAS): a short introduction

The GAS concept treats urban infrastructure and social objects as spatially located automata and Benenson and Torrens refer to the latter *Geographic Automata* (GA) (Benenson and Torrens, 2004a). To quickly recapitulate:

- An abstract automaton $\boldsymbol{A}$ is characterized by (vector) state $\boldsymbol{S}$, which changes in time according to state transition rules $\boldsymbol{T}$, depending on current input $\boldsymbol{I}$.
- Cellular Automata theory considers $\boldsymbol{A}$'s neighborhood $\boldsymbol{N(A)}$ and assumes that input $\boldsymbol{I}$ is defined by automata belonging to the $\boldsymbol{N(A)}$.
- Agent $\boldsymbol{A}$ of a Multi-Agent System (MAS) can relocate in space, that is, its representation should be capable of managing location and neighborhood changing over time.

The minimal set of transition rules $\boldsymbol{T}$ for geographic automata $\boldsymbol{G}$ should, thus, specify changes in:

- Non-spatial attributes of $\boldsymbol{G}$'s state.
- Location of $\boldsymbol{G}$
- Relationships of $\boldsymbol{G}$ with other geographic automata

To formalize these demands, Benenson and Torrens (Benenson and Torrens, 2004a) consider Geographic Automata System G to consist of automata of different *types* ($\boldsymbol{K}$). Automata of given type $\boldsymbol{k} \in \boldsymbol{K}$ are characterized by a non-spatial set of states - $\boldsymbol{S}^k$, *geo-referencing conventions* — $\boldsymbol{L}^k$ — that determine how automata are located in space , and their *relationships* to automata of the same and other types - $\boldsymbol{N}^k$ (we later omit the upper index $\boldsymbol{k}$). Geo-referencing conventions and relationships usually include automata of different types.

State transition rules $\boldsymbol{T_S}$ determine how automata non-spatial states change in time, *movement rules* $\boldsymbol{M_L}$ govern changes of location whereas *relationship transition rules* $\boldsymbol{R_N}$ specify how automata relationships change in time.

Altogether, a Geographic Automata System $\boldsymbol{G}$ may be defined by seven components:

$$\boldsymbol{G} \sim <\boldsymbol{K}, \boldsymbol{S}, \boldsymbol{T_S}, \boldsymbol{L}, \boldsymbol{M_L}, \boldsymbol{N}, \boldsymbol{R_N}> \quad (1)$$

Unitary geographic automaton $\boldsymbol{G}$ is characterized at t by state S_t, location L_t, and relationship N_t; the state, location and relationships at time t + 1 are determined by the transition rules $\boldsymbol{T_S}, \boldsymbol{M_L}, \boldsymbol{R_N}$ as follows:

$$\begin{aligned} \boldsymbol{T_S}&: (S_t, L_t, N_t) \rightarrow S_{t+1} \\ \boldsymbol{M_L}&: (S_t, L_t, N_t) \rightarrow L_{t+1} \\ \boldsymbol{R_N}&: (S_t, L_t, N_t) \rightarrow N_{t+1} \end{aligned} \quad (2)$$

Exploration with GAS $\boldsymbol{G}$ involves the qualitative and quantitative investigation of its spatial and temporal dynamics, given all the components defined above. In this way, GAS models offer *Geosimulation* framework for considering spatially enabled interactive behavior of the collective of geographic objects.

3. From a Geographic Automata System to software

To interpret GAS paradigm in software, we have to translate all the components of (1) into software objects and methods. What might be the specificity of the software required for GAS? We claim that this specificity rests on automata *relationships*. The priority of relationship information is the core of GAS interpretation as Object-Based Environment for Urban Simulation (OBEUS).

3.1 Automata of a given type $k \in K \rightarrow$ Instances of *population* class

Urban objects always utilize information above the individual level. Objects belonging to an OBEUS **population** class represent "containers" for this meta-data. For instance, populations of human residential agents can be restricted in their property operations by laws that are applicable to (shared by) all agents. Thus, these laws are characteristic of the population of householders as a whole.

3.2 Individual automata of type $k \rightarrow$ Class of *Objects* of a type k

At a conceptual level, OBEUS considers unitary **objects**, which are distinguished as either **fixed** and **non-fixed** (Benenson, Aronovich et al., 2005). Fixed objects — buildings, parks, road links, traffic lights, etc. — do not change their location once established whereas non-fixed objects — tenants, firms, pedestrians, vehicles, etc. — do. In the majority of situations, fixed objects can be referred to as *immobile* while non-fixed objects can be referred to as *mobile.* However, a more general notion capturing the above qualities includes examples of a 'landlord' kind: The latter can change their possessions and thus are non-fixed, although their mobility is irrelevant for descriptions of their market behavior. Fixed objects can also change: for example, a building can be reconstructed or destroyed.

The difference between fixed and non-fixed objects lies in the way that they are geo-referenced. Fixed objects are located in space **directly**, by means of a coordinate list, in a manner similar to vector GIS. The coordinate list contains the object's spatial representations: vertices, centroid, minimal bounding rectangle, and so on. Features of planar GIS layers or 3D CAD-models can represent urban objects in spatially explicit models. For theoretical models, the points of a regular grid usually suffice. The information of location of the fixed object is usually included into its state vector $\boldsymbol{S}$.

Change of location is characteristic of non-fixed objects. In OBEUS, the fundamental method of locating non-fixed urban objects is indirect, by **pointing** to one or several fixed objects. For example, tenants are located by pointing to their habitats.

The above approach to locating fixed and non-fixed objects is not at all absolute. Location by pointing is especially flexible, and can be used to locate anything from fixed apartments in a fixed house, or landlords, whose individual location may be unimportant while they have to point to their estates (Benenson, Aronovich et al., 2005). Looking ahead, note that geo-referencing by pointing is captured as relationship. In the above examples, the relationships are those of tenant-habitat, apartment-house and landlord-estate. In OBEUS we assume that all objects of a given type $\boldsymbol{k} \in \boldsymbol{K}$ are located in the same manner.

We are unaware of any examples of urban models, where objects cannot be easily classified according to the main OBEUS fixed – non-fixed dichotomy.

3.3 Relationships between automata → Class of *relationships*

OBEUS follows the Entity-Relationship Data Model (ERM) (Howe, 1983); it therefore considers relationships between entities explicitly, as software objects.

To illustrate the use of relationships, let us consider a hypothetical system that consists of objects of the two types - **landowners** and land **parcels**. The **landowner's** decision to sell or develop a land **parcel** might depend on properties of model objects of both types, say, on (a) **parcel** value, (b) **landowner** economic abilities, and on *relationships* between them, as (c) neighborhood potential, given by **parcel-parcel** relationships and (d) ownership characteristics, given by **landowner-parcel** relationships.

It is important to note that relationships have their own properties. A decision to develop might depend on parcel accessibility from neighboring parcels — a property of **parcel-parcel** relationships — or on the owner's rights regarding the given parcel – a property of **landowner-parcel** relationships.

The compelling dominance of the Entity-Relationship model in modern DBMS is the best proof available of the benefits reaped from the distinguishing between objects and relationships and the view of relationships as self-existing software objects. In OBEUS, we consider relationships between unitary entities of types **j** and **k** as a new class of objects — **jk_Relationship** — that encapsulates the relationship's properties. By doing so, we implement the Object-Oriented Database (OODB) view (Booch, 1994) of Geographic Automata Systems.

Relationships treated as separate software classes unify CA and explicit GIS-based land-use models, both of which are rooted in neighborhood relationships. For a regular square CA grid, von Neumann or Moore neighborhoods entail neighborhood relationship of 1:4 or 1:8 degrees. The popular 'constrained CA' is based on neighborhoods of radius 6 around the cell, entailing neighborhood relationship of 1:113 degree. Constrained CA accounts for a relationship's properties – viewed as weights — which reflect the influence of neighboring cells on the land use of given cell (White and Engelen, 2000). The only difference between a regular CA grid and an irregular land partition is the variation in degree of neighborhood relationship between the parcels (Flache and Hegselmann, 2001; Benenson, Omer et al., 2002). The distance between a cell and its neighbor, the length of the common boundary and other frequently used geographic characteristics are properties of the software objects belonging to the **ParcelParcel_Relationship** (or **CellCell_Relationship** in a more abstract formulation) relationship class.

According to the Relational DBMS theory, relationship objects and their properties are stored in DBMS tables. The more complex the definition of a relationship and its properties, the greater the computational advantage of using a relationship class and a table presentation over on the fly evaluations as to whether two objects are related, made during simulation runs. Retrieving all the parcels having a common boundary with a given one from the table representing a relationship (thereby automatically obtaining the length of the common boundary, which is the property of this relationship) demands much less time than implementing an algorithm that determines a given parcel's neighbors and then the length of the common boundary based on the GIS coverage of parcels.

3.4 Limitations of relationships in OBEUS

GAS is a dynamic system and the properties of the unitary geographic automata as well as their relationships change over time. The changes can cause inconsistencies: for instance, which of two 'automata' — the owner or tenant — has the right to cancel the rental contract (that is, destroy the relationship) between them? What order of actions will be taken should one partner want to end the relationship when the other does not? To avoid inconsistencies of this and other types, three essential limitations are imposed on the semantics of the relationships covered by OBEUS:

First, we assume that relationships between fixed objects are also fixed, that is, they do not change once established.

Second, no direct relationships between non-fixed objects are permitted. That is, non-fixed unitary objects can be related to fixed objects only; in particular, non-fixed objects can be located by pointing solely to fixed objects. Non-fixed objects can maintain relationships in a transitive fashion, when non-fixed object is related to another non-fixed object only if they are related to fixed objects which are themselves related.

This constraint implies, in particular, that direct **neighborhood relationships** can be defined for fixed objects only. A householder's neighbors, for example, are thus defined as inhabitants of the (fixed) houses neighboring to that householder's house.

Direct relationships between non-fixed objects are undoubtedly important in the real world. For example, an aggressive neighbor can force another neighbors to leave their apartments. However, locating tenants via tenant-house relationships and recognition neighboring tenants in a transitive way is sufficient for representing this interaction. To sense the flavor of models where this decomposition of relationship between non-fixed automata is insufficient and where direct relationships between non-fixed objects seem inevitable, we direct the reader to descriptions of the dynamics of flocks and fish schools (Reynolds, 1987; Vicsek, Czirok et al., 1995; Brogan and Hodgins, 1997; Toner and Tu, 1998; Levine and Rappel, 2001). Just to mention, these models could be implemented in OBEUS, if the continuous Euclidean space is represented in a discrete way.

Third, we encapsulate the methods that alter relationships between non-fixed and fixed objects in the classes of the non-fixed objects and assume a **leader-follower pattern** of relationship updating (Noble, 2000). According to this pattern, non-fixed objects have exclusive update rights whereas for the fixed objects relationships are read-only. Some simple examples — e.g., a tenant automaton destroying its relationships with the apartment automata it occupies while establishing a new relationship with the next apartment it moves to, or the car automata establishing a relationship with the parking spot it occupies — give the impression that the leader-follower pattern is a self-evident condition. The leader-follower pattern does impose constraints of the model style — for example, the case where a landlord increases the rent, forcing the tenant to vacate the residence should be formally represented in OBEUS by means of tree elementary steps: updating Landlord-Estate relationship property 'payment demand', executing query that provides Estate rent, and the destruction of the Tenant-Estate relationship.

In OBEUS, the leader and follower are defined separately for each relationship. We as yet have no proof that all or the majority of real-world urban situations can be captured by the *leader-follower* pattern although we are unaware of any urban model where more complex patterns (circular or multiple causation, for example) are necessary.

Relationships between fixed objects should be initialized when those objects are initiated; afterwards, they are retrieved only. Relationships between non-fixed and fixed objects are initiated or eliminated according to requests from the former because they are always the leaders.

The leader-follower pattern provides a fundamental solution for the consistency problem; at the same time, as previously mentioned, real-world relationships between non-fixed objects can be defined transitively. In the transitive mode, OBEUS automatically enables retrieval of relationships between non-fixed software objects related to the related non-fixed object but limits that retrieval to the read-only mode. In more complex cases, relationships between non-fixed objects can be retrieved by formulating assessment rules (see below).

3.5 Location and movement rules are nothing but relationship transition rules

As noted, location agreements $\boldsymbol{L}$ of the GAS definition (1) become components of object state $\boldsymbol{S}$ in cases of direct location (fixed objects) or are represented by the relationship $\boldsymbol{N}$ in cases of indirect location (fixed and non-fixed objects). Movement rules are consequently either state transition rules or relationship transition rules, that is, they can be considered as belonging to either T_S or R_N sets. As a result, GAS can be formally restricted to

$$G \sim \langle K, S, T_S, N, R_N \rangle \qquad (3)$$

We ignore here the discussion of which of the two forms, (1) or (3), better reflects our understanding of the urban system and its dynamics. As for OBEUS software, it is based on the representation (3).

3.6 State and relationship transition rules T_S and $R_N \rightarrow$ Automata behavior and assessment rules

Geographic objects 'behave.' In terms of GAS, geographic automata change states and relationships with other automata. Interpretation of the behavior of real-world objects in terms of the transition rules applied to software objects is the essence of any model; in OBEUS, this interpretation is 'coded' as methods of the software classes that represent automata of different types.

In what follows, we distinguish between **assessment rules**, aimed at estimating the parameters objects react to, and **automation rules,** which describe the behavioral act itself. We assume that the number of assessment rules is limitless but that only one (arbitrarily complex) automation rule can be active during any single simulation run.

4. Beyond-GAS features of OBEUS

Two basic components of OBEUS go beyond features deliberated by GAS. The first one pertains self-organization, especially its spatial aspects, while the second pertains to synchronization of events, necessary for animating the model objects. It is well known that these components are interrelated. For example, the well-known patterns produced by the 'Game of Life' do not self-organize when an *asynchronous* updating scheme is employed (Schonfisch and de Roos, 1999).

4.1 Patterns in OBEUS

GAS definition does not contain any remnant of *collective phenomena*, a cornerstone of contemporary views of the city. Ensembles of urban entities that inherit properties from their unitary components while displaying properties of their own (Portugali, 2000) such as, say, residential neighborhoods, repeatedly emerge and disintegrate in the urban space over time. In the case of GAS, collective phenomena, if exhibited, are, by definition, the outcomes of rules of automata behavior; one goal of modeling urban systems with GAS, like any other systems research, is to identify those rules that entail self-organization and to characterize the patterns produced. For example, the modeler can be interested in recognizing and characterizing areas populated predominantly by individuals belonging to specific socio-economic groups (e.g., poor, rich, immigrants) or areas of high density of parcels for sale, to facilitate acquisition of a construction permit. Another criteria may be building age proximity, an idea utilized in deltatron CA (Clarke, 1997; Candau, Rasmussen et al., 2000).

From the perspective of the object-based approach, OBEUS considers these self-organizing ensembles of *unitary* urban entities – *patterns* - as autonomously existing objects.

A pivotal question asked by complex systems theory with respect to self-organization is to what extent the pattern's properties are determined by the properties of the assembled objects. Whatever the answer, the simulation environment should define its position regarding the ensuing patterns. This strategy is adopted in OBEUS for the specific case of 'foreseeable' self-organization. Stated differently, OBEUS demands an *a priori* formulation of the criteria defining when an object is a candidate for membership in a pattern as well as a detection method for determining whether the pattern exists as conceptualized.

4.2 Population time versus unitary objects time

In OBEUS, we distinguish between observer time and unitary object time. The student of GAS is interested in observing collective processes that occur at population and above levels in **population time**, which is measured in **iterations**. **Automata** have their own **time**, which is measured in **ticks**, depending on object type. Population characteristics are updated, aggregate criteria are tested, and the system state is stored in the course of iterations. Alternatively, objects change properties, locations, and relationships in ticks. Ticks and iterations can be variously interpreted and are model-specific although specified in part by the synchronization scheme applied.

4.3 Synchronization of events in OBEUS

4.3.1 Objects of the same type

Two householders, ignorant of each other, try to occupy the same apartment. Who will finally settle there? Management of the events that occur simultaneously yet influence one another raises synchronization problems, a major issue intensively discussed in CA, MAS, temporal GIS and other discrete-time systems (Berec, 2002).

Developments in Object-Oriented Databases indicate that there is no general, conceptual solution to the synchronization problem. Instead, existing methodologies provide software development patterns for specific classes of problems

(Zeigler, Praehofer et al., 2000). OBEUS architecture utilizes two of those solutions when contending with synchronization problems.

Synchronous mode: In this mode, the properties of urban entities are assumed to change simultaneously. The calling order of the objects has no influence on the model's outcome.

Note that the logic of synchronous updating can push conflicts further down the time path. If two mutually avoiding agents occupy adjacent locations and simultaneously leave them at a given time-step, nothing in the synchronous mode can prevent occupation of these locations by yet another pair of avoiding agents.

Asynchronous mode: In this mode, objects change in turn, with each observing the urban reality left by the previous object. Hence, conflicts between objects can be resolved; in such cases, the updating order may influence the results. OBEUS demands that when applying the asynchronous mode, the modeler defines an order of object actions in advance and supplies templates for temporal updating: for automata of a given class, random sequence and sequence in order of one of the automata characteristics are currently being implemented.

4.3.2 Objects of different types and relationships

The order of updating automata belonging to different populations demands additional specification. This order is currently assumed to be hierarchical and established by the user. Population entities belonging to the lowest level of the hierarchy are updated first, followed by entities of the upper level, etc.

An important feature of OBEUS is the *asynchronous and immediate updating of relationships* irrespective of the mode chosen for automata updating. To illustrate, Tenant-House relationships are always updated immediately after tenants leave/occupy a house.

5. User's view of OBEUS

In OBEUS, the user defines the model components via Model Tree window, the synchronization scheme via Model Flow window, and formulates automata behavior rules with C# compiler. Model output is displayed in Map and Graph windows and stored as DBMS tables.

5.1 Building a model tree

By means of a model tree, the user defines classes of unitary automata, relationships and their properties. When a new type of automata is defined, the population of this type of automata is constructed automatically (Figure 1).

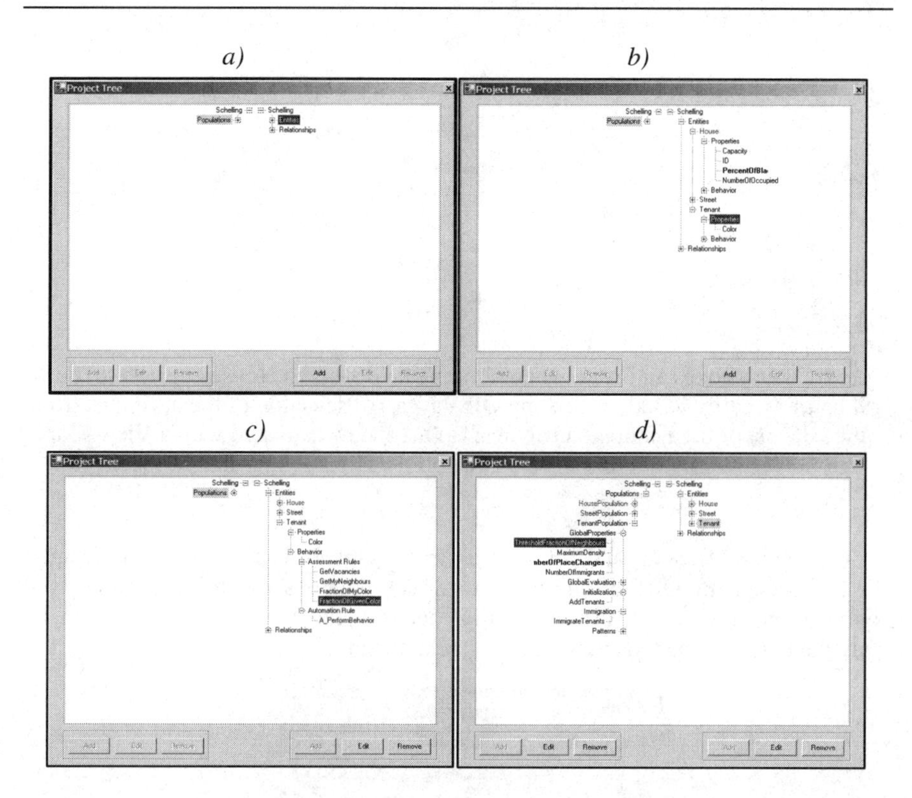

Fig. 1. Stages of development of the generalized Schelling residential dynamics model: (a) Initial state of the model tree; (b) Entities and entities' properties defined; (c) Assessment and behavioral rules defined; (d) Global parameters, initialization and immigration rules defined

Automata and their properties can be defined from scratch and constructed according to templates (e.g., a cell grid) or acquired from a GIS layer (Figure 2). New classes of automata are introduced in the right-hand branch of the scheme and automatically cause construction of corresponding population classes in the left-hand branch (Figure 1). For fixed spatial entities a map view is available.

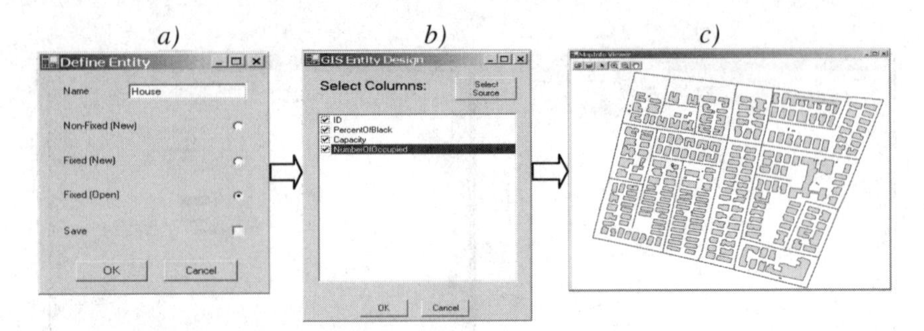

Fig. 2. Dialog boxes activated when defining a GIS-based House entities: (a) definition of an entity based on existing GIS-layer; (b) selection of the attributes from the GIS list of the features' attributes; (c) map view activated with a View Map option of the right button mouse click on the House entity in a Model Tree.

The choice of leader and follower is necessarily for each relationship (Figure 3). According to OBEUS limitations, direct relationships between non-fixed automata are prohibited; in relationships between fixed and non-fixed automata, only the latter can be thus the leader in a relationship.

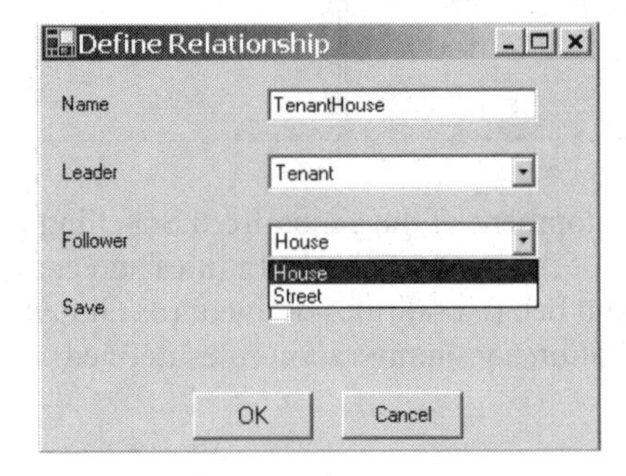

Fig. 3. An example of a relationship for generalized Schelling model; Tenant entity does not appear as a possible Follower in the pull-down list.

5.2 Defining behavioral rules

The core of model design lies in defining automata behavior. As previously mentioned, the information necessary for behavior definition is collected via **Assessment Rules**, after which an **Automation Rule** is applied in order to alter automata state. According to the OBEUS concept, any limitations on this stage will constrain the modelers' ability to directly interpret their understanding of the investigated system. Assessment and behavior rules are consequently formulated as classes' methods via a C# compiler. We base currently on Borland C# Builder. When a C# compiler is activated, all the objects and previously defined assessment rules are available at prompts when constructing new models (Figure 4). Each act of compilation initiates the updating of the model tree.

a)

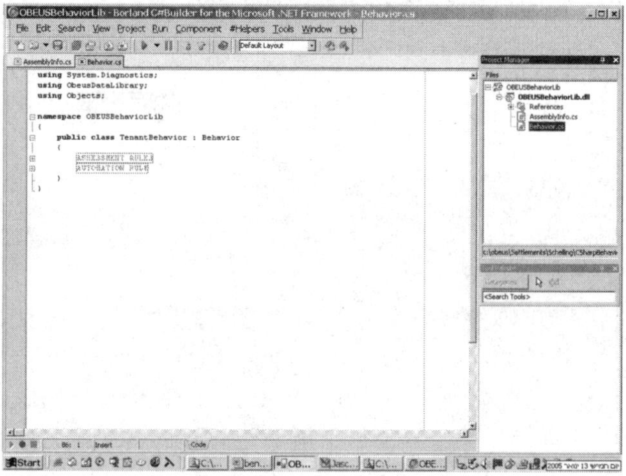

OBEUSBehaviorLib - Borland C#Builder for the Microsoft .NET Framework - Behavior.cs

File Edit Search View Project Run Component #Helpers Tools Window Help

Default Layout

AssemblyInfo.cs Behavior.cs

```
using System.Diagnostics;
using ObeusDataLibrary;
using Objects;

namespace OBEUSBehaviorLib
{
    public class TenantBehavior : Behavior
    {
        ASSESSMENT RULES
        AUTOMATION RULE
    }
}
```

Project Manager

Files

OBEUSBehaviorLib

OBEUSBehaviorLib.dll

References

AssemblyInfo.cs

Behavior.cs

c:\obeus\Settlements\Schelling\CSharpBehav

<Search Tools>

86: 1 Insert Code

Start C:\... ben... OB... Jasc... C:\... OBE... 2005 ינואר 13 יום חמיש

b)

OBEUSBehaviorLib - Borland C#Builder for the Microsoft .NET Framework - Behavior.cs

File Edit Search View Project Run Component #Helpers Tools Window Help

Default Layout

AssemblyInfo.cs Behavior.cs

```
using ObeusDataLibrary;
using Objects;

namespace OBEUSBehaviorLib
{
    public class TenantBehavior : Behavior
    {
        #region ASSESSMENT RULES
        private ArrayList GetMyNeighbours(Tenant man)
        {
            //A tenant might have several houses he is related to. That is why
            //I use an array and not a simple variable for neighboring houses
            House[] myNeighbouringHouses = man.GetRelatedHouses();
            ArrayList myNeighbours = new ArrayList();
            Tenant[] tenants = man.GetRelatedTenantsThruHouses_Order1();
            if(tenants != null)
            {
                foreach(Tenant tenant in tenants)
                {
                    myNeighbours.Add(tenant);
                }
            }
            return myNeighbours;
        }
            private double FractionOfMyColor(Tenant man) ...
            private double FractionOfGivenColor(House house, bool givenColor) ...

        private ArrayList GetVacancies(House house) ...
        #endregion ASSESSMENT RULES
        #region AUTOMATION RULE
        Random rnd = new Random();
        public void A_PerformBehavior(Entity entity) ...
        #endregion AUTOMATION RULE
    }
}
```

Files

OBEUSBehaviorLib

OBEUSBehaviorLib.dll

References

AssemblyInfo.cs

Behavior.cs

c:\obeus\Settlements\Schelling\CSharpBehav

<Search Tools>

15: 56 Insert Modified Code

Start C:... be... O... Ja... C:... O... 18:58

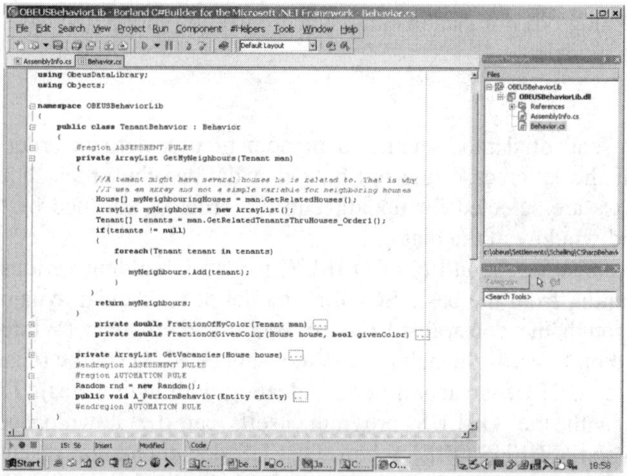

Fig. 4. Coding Assessment and Behavioral rules: (a) Initial state of the Borland C# compiler window; (b) Developed behavioral rules of the generalized Schelling model (one of them fully shown)

5.3 Building a synchronization chart

At the final stage of the model construction, the synchronization mode that determines the temporal sequence for applying behavioral rules should be established via the synchronization window (Figure 5). Recall that in OBEUS, relationships are always updated asynchronously.

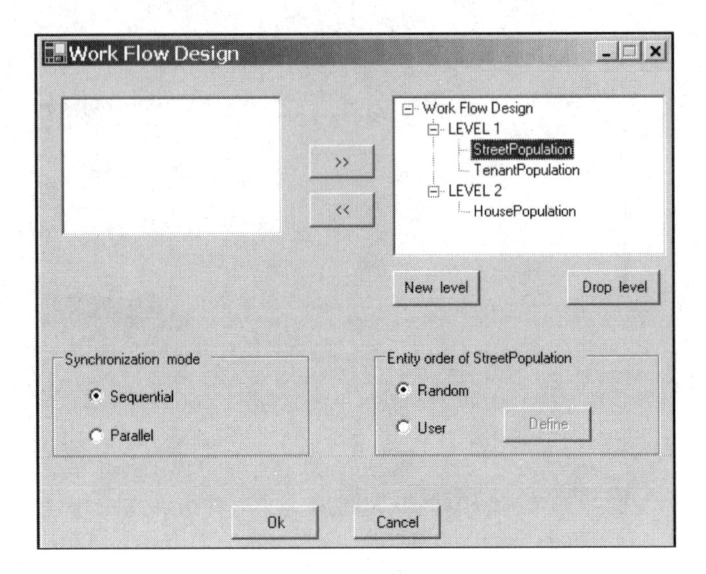

Fig. 5. Dialog box for establishing synchronization mode (the case of generalized Schelling model)

In the sequential mode, either a random or user-defined order should be selected; in the latter case, the attribute that defines the order should be chosen. Populations are selected for updating in the order established by the user in the right-hand window of the dialog.

To demonstrate the utility of OBEUS, we are rebuilding various popular models, beginning from the basic Schelling model of residential dynamics (Schelling, 1971) through the constrained CA of White and Engelen (White and Engelen, 1997) and on towards the deltatron model suggested by Clarke (Clarke, 1997) (the last utilizes OBEUS's capabilities for dealing with aggregates). These examples, together with the OBEUS program itself, can be downloaded from http://eslab.tau.ac.il/OBEUS/OBEUS.htm. OBEUS documentation discusses its advantages over other popular software for CA-MAS simulations, such as RePast and NetLogo (RePast, 2003, Tobias and Hofman, 2004).

6. What we get and will get with OBEUS?

Several decades ago, dynamic regional modeling was born from a common reference point: the system of differential or difference equations that describe changes in regions' aggregate state variables over time. This meant that one could easily distinguish between regional models by comparing their analytical presentations, initial conditions and parameter values. With time, the regional framework was abandoned in favor of high-resolution simulations. This evolution, however, revived the problem of a sharable or common reference, which is compulsory if we want to advance urban modeling beyond art to engineering. We argue that the GAS concept together with an OBEUS environment provide this mutual reference.

We argue that GAS serves as the common reference for object-based urban models, while OBEUS acts as the framework for those models' operational implementation. Just as with differential or difference equations in the case of the regional model, model development is a two-step process: formulation of the model in general terms at the first step and formal implementation at the second (Figure 6).

Working with OBEUS, the modeler should, as an initial step, formalize her understanding of the world in terms of GAS, that is, to define the types of objects to be included in the model, the relationships to be accounted for and how, if ever, self-organizing patterns will be recognized.

As a subsequent step, all these definitions should be implemented in OBEUS, a stage where behavioral rules and time-synchronization schemes are finalized and implemented. Note that the view of the model as dynamic database facilitates changes in model structure, which encourages the reader to rebuild the model just because this is easy and convenient with OBEUS.

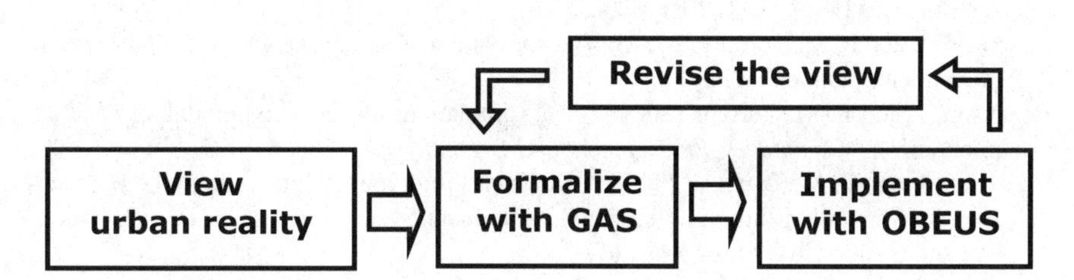

Fig. 6. The process of GAS model construction and implementation

An essential part of the model implementation in OBEUS is based on the C# or other .NET language that demand an expertise not necessarily displayed by urban modelers. Yet, it remains unclear whether the flexibility necessary for formulating automata behavior can be achieved with a higher-level language. The capacity of many existing computer languages regarding urban objects' relocation 'behavior' (Schumacher, 2001) remains to be tested although the idea of a **language for formulating automata behavior** is becoming popular (Galton, 2003).

To conclude, the concept of GAS and the OBEUS software are very recent developments, and thorough investigation of their limitations yet should to be done. Our view is that such an investigation should be merely operational – each published urban high-resolution Cellular Automata or Multi-Agent model should be taken 'as is' and tested whether it can be reformulated in GAS terms and further represented as OBEUS applications. The test of this kind is a topic of a separate paper; shortly, we didn't fail with quite a number of models we tested in this line till now. We claim that GAS and OBEUS can thus serve as a universal, transferable conceptual and operational framework for object-based urban simulation modeling.

References

Allen, P. M. and M. Sanglier (1979). "A dynamic model of growth in a central place system." *Geographical Analysis* 11(3): 256-272.

Batty, M. (1997). "Cellular automata and urban form: A primer." *Journal of the American Planning Association* 63(2): 266-274.

Batty, M. (2003). New Developments in Urban Modeling: Simulation, Representation, and Visualization. *Integrated Land Use and Environmental Models*. S. Guhathakurta. Berlin, Springer: 13-44.

Benenson, I., S. Aronovich, et al. (2005). "Let's Talk Objects: Generic Methodology for Urban High-Resolution Simulation." *Computers, Environment and Urban Systems*, 29: 425-453.

Benenson, I., I. Omer, et al. (2002). "Entity-based modeling of urban residential dynamics - the case of Yaffo, Tel-Aviv." *Environment and Planning B: Planning & Design*. 29: 491-512.

Benenson, I. and P. M. Torrens (2003). *Geographic Automata Systems: a new paradigm for integrating GIS and geographic simulation*. Association Geographic Information Laboratories Europe (AGILE), Lyons.

Benenson, I. and P. M. Torrens (2004a). Geosimulation: automata-based modeling of urban phenomena. London, Wiley.

Benenson, I. and P. M. Torrens (2004b). "Geosimulation: object-based modeling of urban phenomena." *Computers, Environment and Urban Systems* 28(1/2): 1-8.

Benenson, I. and P. M. Torrens (2005). A Minimal Prototype for Integrating GIS and Geographic Simulation through Geographic Automata Systems. *Geodynamics*. P. Aitkinson and F. Wu, CRC Press: 347-367.

Berec, L. (2002). "Techniques of spatially explicit individual-based models: construction, simulation, and mean-field analysis." *Ecological Modeling* 150: 55-81.

Booch, G. (1994). Object-oriented analysis and design with applications. Menlo Park, Ca, Addison-Wesley.

Brogan, D. C. and J. K. Hodgins (1997). "Group Behaviors for Systems with Significant Dynamics." *Autonomous Robots* 4: 137-153.

Candau, J. T., S. Rasmussen, et al. (2000). *A coupled cellular automata model for land use/land cover change dynamics*. 4th International Conference on Integrating GIS and Environmental Modeling (GIS/EM4): Problems, Prospects and Research Needs, Banff, Alberta, Canada.

Clarke, K. C. (1997). Land Transition Modeling With Deltatrons, http://www.geog.ucsb.edu/~kclarke/Papers/deltatron.html. 2002.

Clarke, K. C., S. Hoppen, et al. (1997). "A self-modifying cellular automata model of historical urbanization in the San Francisco Bay area." *Environment and Planning B: Planning and Design* 24(2): 247-261.

Erickson, B. and T. Lloyd-Jones (1997). "Experiments with settlement aggregation models." *Environment and Planning B: Planning & Design* 24(6): 903-928.

Flache, A. and R. Hegselmann (2001). "Do Irregular Grids make a Difference? Relaxing the Spatial Regularity Assumption in Cellular Models of Social Dynamics." *Journal of Artificial Societies and Social Simulation* 4(4): http://www.soc.surrey.ac.uk/JASSS/4/4/6.html.

Galton, A. (2001). "Space, Time, and the Representation of Geographical Reality." *Topoi* 20: 173-187.

Galton, A. (2003). "A generalized topological view of motion in discrete space." *Theoretical Computer Science* 305: 111-134.

Howe, D. R. (1983). *Data analysis for data base design*. London, Edward Arnold.

Lee, D. B. (1973). "A Requiem for Large Scale Modeling." *Journal of the American Institute of Planners* 39(3): 163-178.

Lee, D. B. (1994). "Retrospective on Large-Scale Urban Models." Journal of the American Planning Association 60(1): 35-40.

Levine, H. and W. J. Rappel (2001). "Self organization in systems of self-propelled particles." *Physical Review E* 63: 208-211.

Noble, J. (2000). Chapter 6. Basic Relationship Patterns. *Pattern Languages of Program Design Vol. 4*. N. Harrison, B. Foote and H. Rohnert, Addison-Wesley: 73-89.

Portugali, J. (2000). *Self-Organization and the City*. Berlin, Springer.

RePast (2003). University of Chicago, *RePast 2.0*. (Software), Chicago: Social Science Research Computing Program.

Reynolds, C. (1987). "Flocks, birds, and schools: a distributed behavioral model." *Computer Graphics* 21: 25-34.

Schelling, T. C. (1971). "Dynamic models of segregation." *Journal of Mathematical Sociology* 1: 143-186.

Schonfisch, B. and A. M. de Roos (1999). "Synchronous and asynchronous updating in cellular automata." *Biosystems* 51: 123-143.

Schumacher, M. (2001). Objective coordination in multi-agent system engineering. Berlin, Springer.

Semboloni, F. (2000). "The growth of an urban cluster into a dynamic self-modifying spatial pattern." *Environment and Planning B: Planning & Design* 27(4): 549-564.

Tobias, R. and C. Hofmann (2004). "Evaluation of free Java-libraries for social-scientific agent based simulation ".Journal of Artificial Societies and Social Simulation :(1)7 http://jasss.soc.surrey.ac.uk/7/1/6.html

Toner, J. and Y. Tu (1998). "Flocks, herds, and schools: A quantitative theory of flocking." *Physical Review E* 58: 4828-4858.

Torrens, P. M. and I. Benenson (2005). "Geographic Automata Systems." *International Journal of Geographic Information Science,* 19(4), 385-412.

Vicsek, T., A. Czirok, et al. (1995). "Novel type of phase transitions in a system of self-driven particles." *Physical Review Letters* 75: 1226-1229.

White, R. and G. Engelen (1997). "Cellular automata as the basis of integrated dynamic regional modeling." *Environment and Planning B: Planning & Design* 24(2): 235-246.

White, R. and G. Engelen (2000). "High-resolution integrated modeling of the spatial dynamics of urban and regional systems." *Computers, Environment and Urban Systems* 24(5): 383-400.

Zeigler, B. P., H. Praehofer, et al. (2000). Theory of modeling and simulation:integrating discrete event and continuous complex dynamic systems. New York, Academic Press.

The CityDev Project: An Interactive Multi-agent Urban Model on the Web

Ferdinando Semboloni

Abstract. In this paper I present a multi-agent simulation model of the development of a city. The model, CityDev, is based on agents, goods and markets. Each agent (family, industrial firm, developer, etc.) produces goods by using other goods, and trades the goods in the markets. Each good has a price, and the monetary aspects are included in the simulation. When agents produce goods and interact in the markets, the urban fabric is built and transformed. The computer model (simulator) runs on a 3-D spatial pattern organized in cubic cells. In the present paper the model is described and results are shown.

1. Introduction

The simulation of the urban dynamic has been approached by using different methods: from gravitational model to cellular automata and multi-agent systems (MAS) (Batty and Jiang, 1999; Portugali, 1999). While MAS method reproduces the behavior of real actors, cellular automata simulation has to translate the phenomena under study in its specific language. Even if the advantages of the cellular automata approach mainly rely in its simplified representation of the reality, MAS seem the best candidate for the simulation in social sciences and hence in urban dynamic.

In this paper is presented a multi-agent model of urban dynamic. This multi-agent model, CityDev (Semboloni et al., 2004), is an urban multi-agent simulation which includes the whole aspects of the urban system and is conceived for the interaction with human user. Even if this last aspect is an important characteristic of the model, this paper focuses on the dynamics of the city as an economic system. For this reason the urban dynamic is simulated in its real and monetary aspects. The spatial aspect comes out from the transportation costs and from the need for each agent to utilize a building. CityDev is based on agents, goods and markets. Each agent produces goods by using other goods and trades the produced goods in the markets. Agents are usually located in the physical space. In this simulation the evolution of the urban fabric results from the agents' interaction. A grid of 100x100 squared cells, each 100x100 meters sized, is the spatial pattern of the simulation. Built 3-D cells can be superposed in case of a multi-floors building. In other words buildings of the city are considered as composed of indivisible 3-D cells (Semboloni, 2000).

In addition to the agent interactions, the model simulates the dynamic of the general aspects of the city. These concern the generation and death of agents, the re-valuation of the land rent, and the simulation of traffic congestion. The simulation runs by steps. Each of these steps is supposed to represent one year of the real life of the city. Each step has four phases: trading in the markets, including selling and buying, production, and simulation of general aspects.

In essence the model is based on the economic base theory: an outside demand of industrial product stimulates the production of industry. Families are generated in order to work in industrial firms. Because families demand final consumption goods, commercial firms are generated, which in turn demand workers and so on (Figure 1). Because each agent demands a building were to live or work, developers are generated which build the urban fabric.

In the following sections first the core structure of the model – agents' strategies, markets, and interaction among agents – is shown, second, the general aspects of the simulation are described, and third the results of simulations are presented.

2. Agents' strategies and goods

Agents are the subjects, the actors of the play, while goods are the objects, the basic elements which are utilized, produced and traded by agents in the markets. Agents are divided in consumers and producers. Families (a group of inhabitants living in the same 3-D cell) belong to the first group, while industrial firms, commercial firms, private service firms, public services, and developers belong to the second group. Goods include: land, labor, buildings (housing, commercial, and industrial), exported goods, imported goods, consumption goods and services.

Each agent produces a good by using other goods as input. The goods produced by an agent are not utilized by the same agent. For this reason agents trade goods they have produced. Markets are the virtual places where these exchanges occur. Different markets exist for different types of goods available in the simulation: land market, buildings market, labor market, export goods market, consumption goods market, private service market, and public services market. In markets agents offer the goods they have produced, set a price for each good and sell it to the first buyer agreeing with this price, or to the highest bidder. Markets are in fact distinguished in two groups: that in which the consumer simply gets the desired good and pays the price established by the producer, and markets in which consumer bids for the desired good which, in turn, is sold to the highest bidder. The consumption goods market and the private services market belong to the first group, while the buildings market belongs to the second group. In the buildings market (housing, commercial and industrial) a buyer first chooses the desired building and second bids for this building. Bids for the same building are collected and the building is assigned to the highest bidder.

All these actions concerning selling and buying require choices among alternatives, i.e. decisions which maximize a variable. This variable differs if an agent is a consumer or a producer. Consumers maximize the quantity of goods, under the constraint of an established budget, while producers maximize the earnings under the constraint of a production function. Each agent is characterized by incoming and outgoing flows of goods, as it is shown in table 1, see (Semboloni et al., 2004). In this table, columns show the production function of each agent, whereas rows show the origin and destination of goods, otherwise stated the in and out fluxes in the market. For instance, the row of consumption goods indicates that these goods are the input for families, and the output for commercial firms. In other words, commercial firms supply the consumption goods in the market and the families buy these goods.

Goods	Agents						Outside
	Family	Ind. firm	Comm. firm	Priv. services firm	Publ. service	Developer	
Land plot	-	-	-	-	-	In	-
Labor	Out	In	In	In	In	In	-
Building	In	In	In	In	In	Out	-
Export goods	-	Out	-	-	-	-	In
Consumption goods	In	-	Out	-	-	-	-
Public services	In	-	-	-	Out	-	-
Private services	In	In	In	Out	-	-	-
Raw materials for industry	-	In	-	-	-	-	Out
Raw materials for building	-	-	-	-	-	In	Out
Wholesale goods	-	-	In	-	-	-	Out

Table 1. Relationships among agents and goods. "In" means that a good is utilized as an input by the agent; "Out" that a good is produced and supplied by the agent.

As table 1 shows, all agents, except developers, need a building where to live or work in. A building is a 3-D cell having an a surface of 100x100 meters, which can be utilized by an only agent. In turn, 3-D buildings can be superposed thus realizing a mix of activities and housing over the same ground cell. The 100x100 surface is considered as a gross measure including floorspace, roads, and open unbuilt space. In additions the floorspace available for agent depends on the average height of the cell, which is normally set to 3 meters. A variation of this value means a corresponding proportional variation of the floorspace of the building. A building is further characterized by a quality. This quality is set to 1 when the building is carried out and is decreased at each step. The number of inhabitants grouped together in a family is related to the size of the 3-D cell. In fact, a family is considered as a set comprising one hundred people (about 25 real average families). Hence, each inhabitant is supposed to occupy an average of 100 squared meters for housing, including all the surface utilized for secondary roads, private gardens etc. Among these people, forty are considered being active workers. Consequently, an industrial firm employing two families should be compared with a real firm having about 80 employees. This simplification has been introduced to ease the programming and to reduce the computing time required by the simulation, which depends strictly, on the number of agents.

Agents are divided in two basic groups: consumers and producers. Families are the only consumer agents while producer are further subdivided in: export sector (industries), service sector (commercial and private services firms, and public services) and developers, which are the builders of the city. The relations among agents are shown in Figure 1, see (Semboloni et al., 2004) while agents' strategies are analyzed in depth hereafter.

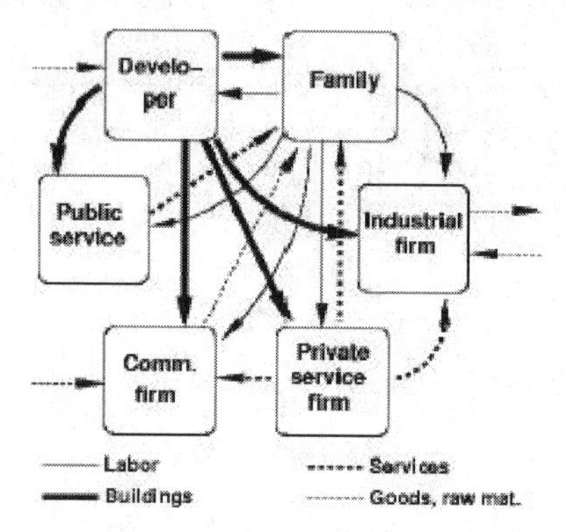

Fig. 1. The network of interactions among agents.

Families are the only final consumers. Their task is to consume goods and to supply labor. The decision areas for families are: to find a job, an house, to buy consumption goods, and to utilize public services. First a family has to earn for paying the rent and for buying goods. For this reason, a family gets the best payed job offered in the labor market. From the job a family gets a salary and eventually a part of the profit of the firm where a family works.

Under the constraint of an established budget a family spends a share (say 50 percent) of it for consumption goods, private services and the rest for housing and transportation costs. Consumption goods are a class including all the goods consumed by a family. Among these goods, some are more frequently bought, other less. As in literature is accepted, the goods less frequently bought, are sold in large commercial centers, which in turn sell also all the other goods. In order to consider this aspect, a family first chooses at random the size of the commercial center where to buy consumption goods. Second a family chooses the cheaper – transportation cost included – good offered in commercial centers having a size greater than that chosen. In turn, in the housing market a family first chooses an house among that having a cost – rent plus transportation cost to working place – lesser than the established budget. Further, among the houses satisfying this condition, a family chooses that having the greatest surface, the better quality and the minimum distance from the workplace. The amount of bid is established subtracting by the established budget the estimated transportation costs. However a family can change house as well as job. Roughly a family considers this eventuality with a probability proportional to the period from which job started or house was rent. In addition, because the rent of an house can change during the simulation, a family is constrained to look for a new house if the rent plus transportation costs overcomes the established budget. In this last case a family chooses from the housing market in the usual way. In the previous case a family compares the house where it is currently living with the best house supplied in the market.

2.2. Producers

Producers are distinguished by activity sector: industrial, commercial and private services firms, public services and developers. However, these last are quite different from the other producers. In fact a developer, doesn't need a building where to produce. In addition the good produced by the developer, i.e. the building, is the only capital good produced in the simulated economic system. Finally, a developer utilizes as raw material the land which is a special good. In fact it is not produced, doesn't perish, and cannot be transported. For all these reasons the developers are treated in a following separate section. Now, considering all the producers, except developers, they have common strategies related to the quantity produced, to the price of the produced goods, and to the location. The cost of production usually includes fixed costs (rent plus salaries), variable costs (raw materials) and a congestion cost proportional to the quantity produced over an established threshold. By using the marginal costs function the producer establishes the supply curve, which relates price and quantity. The number of produced goods is related to the variation of the quantity sold in the previous steps, and the price is established applying the supply curve. In order to choose the location, i.e. to rent a building, a producer follows the example of the existing similar producers. In fact, an agent with a limited forecasting capability, follows the decision of similar agents if the result of this decision is positive, i.e. if these agents' activities are profitable. In other words, first he chooses the building for which the amount of profits of similar surrounding agents is the greatest, and second he bids for the chosen building an amount equal to difference of the average sales, minus the average production cost and minus the average profit of the surrounding similar agents. As consumers, producers can change the location with a probability proportional to the period from which the building has been rent. In this case a producer compares the advantage of the existing location with that of the best building offered in the market.

While the previous strategies are common to all the producer, except developers, producer agents have different task in relation to their activity sector. Industrial firms produce export goods that are consumed outside the city. The main decision areas of an industrial firm are: where to produce, the price of the goods produced, and the quantity of production. Commercial firms sell consumption goods to the families, while private service firms provide private service which are exploited by families, industrial and commercial firms. Their main decision areas comprise: where to locate, the establishment of the price and of the quantity of production. Public services provide families with such facilities as schools, hospitals, and public administration. Public services receive public funds in relationship to the number of families in the city. The main discretionary power of a public service concerns its location. This decision is connected to a spatial analysis of the relation of supply and demand of public services. In other words, a public service decides to locate in the worst serviced area. In fact, the strategy of public services is oriented to the maximization of the social utility, and a public services locates where the difference: demand of public services minus offer of public services is highest.

2.3. Developers and the building process

Developers construct buildings in which families, private firms and public services live or work. Hence, they have an important role in the establishment of the shape of the urban fabric and are the key actors in the building process which comprises the following steps. First the developer buys one or more land plots in the land market. Second the developer decides which type of buildings to build and where. Hereafter, the developer produces the established type of building and offers it in the buildings market. Third agents bid for the desired building in the concerned buildings market. In addition abandoned buildings can be supplied in the market of the buildings to be rehabilitated and eventually bought by a developer for rehabilitation.

A land plot can be sold if it is suitable for the urban development, which usually happens next to already built ares or near roads. In other words, if it is bordered by a built cell or if it is bordered by a road and is located at a distance from a built cell not longer than 3 hundred meters. Among all the cells candidates to be sold, only a part chosen at random is offered in the land marked. The price is set according to a basic land price, related to the agricultural land. This basic price is increased in relation to the quantity of built cells admitted by the urban plan, and it is doubled if the land plot is next to already built cells. Finally the rent depending on the transportation costs in added. This share is calculated as transportation costs of the marginal cell, minus the transportations costs of the cell in question. By using this method, the average rent increases with the increasing of the spatial extension of the urban cluster.

A developer chooses to buy the land plot for which the difference: average price of the surrounding buildings minus the price of the land plot is the highest. In turn, the decision where to build and which type of building (housing, commercial or industrial) to build is more complex, because two aspects are involved: the demand and offer of each type of building and the local convenience in relation to the land plots owned by the developer. In essence a developer chooses to build the cell and the type of building with a probability proportional to the expected bid multiplied by the expected demand. Once carried out, the building is supplied in the buildings market. This buildings market has three separate sections: housing, industrial buildings (sheds), and commercial or service buildings (stores and offices). While only industrial firms are allowed to bid both for industrial buildings, commercial and private service firms may bid both for commercial buildings and for housing. Because the quality of a building is decreasing during time, a building if abandoned by the renter, can be rehabilitated. For this reason the building is offered in the market for the buildings to be rehabilitated. This happens when the difference between the potential rent and its quality overcomes an established threshold. Developers can buy an existent building from this market, rehabilitate it and then offer it in the concerned market.

3. General aspects

The model simulates aspects which are considered as outside to the agent's interactions. These concern the establishment of the demand of export goods, the generation and death of agents, the revaluation of the land rent as well as the decrease of the quality of buildings, and the simulation of traffic congestion. The demand for export goods is established by using a logistic growth function. According to the economic base theory, this demand is the key quantity for the growth of the model.

New agents are generated if in the previous step the total amount of extraprofits make possible the existence of other similar firms, such as for industrial, commercial, and private service firms, or the demand for the output produced by the agents has been greater than the supply, such as for developers and public services. In turn, an agent is eliminated if his budget is negative. Because the city grows at each step, the land rent is reevaluated and this additional value is transferred to the building rent. In turn the buildings' quality is decreased during the simulation. Because families and firms have a limited budget the increase of rent gives rise to the relocation as well as to the upgrading of buildings.

Finally, commuting fluxes are assigned to the roads network and distance is recalculated in order to simulate the congestion.

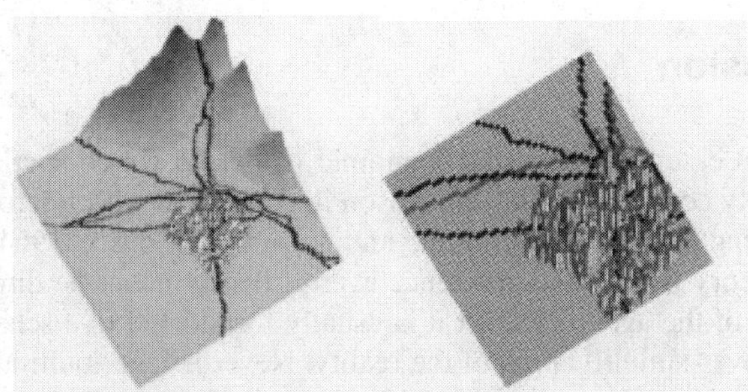

Fig. 2. The urban cluster after 150 iterations. Right: a zoom in the central area.

4. Results

The CityDev project is currently under development. A brief sketch of the results obtained is here shown. It includes a map of a simulation after 150 steps (Figure 2), the variation of agents during the simulation (Figure 3), and the variation of land rent as a function of the distance from the city center (Figure 4).

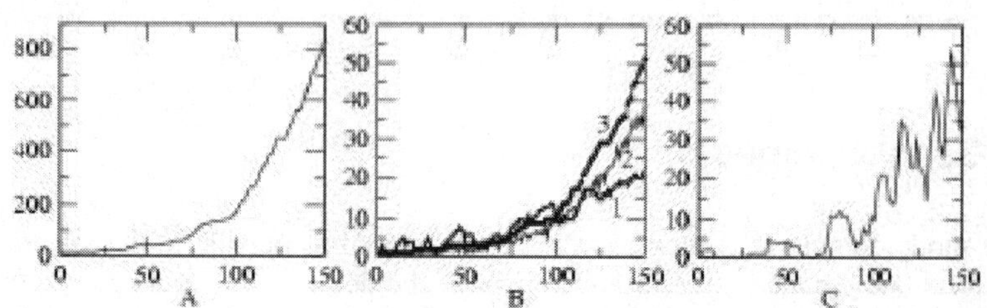

Fig. 3. The variation of agents number during 150 iterations. Graph A, Families. Graph B, 1: Commercial firms, 2: Private services, 3: Public services. C, Developers.

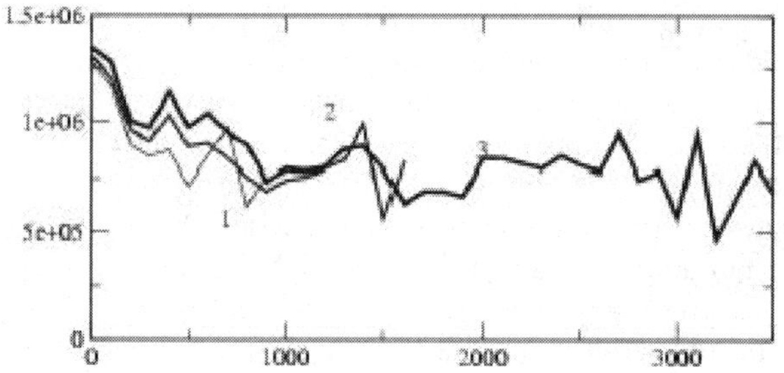

Fig. 4. The rent structure after 50 (graph 1), 100 (graph 2) and 150 (graph 3) iterations. X axis, distance from the city center in meters, Y axis, price of a land plot in Euros

5. Discussion

CityDev is in essence a dynamic economic model, in which spatial aspects have been explicitly considered. For this reason it is indebted with micro-economic theory concerning market equilibrium, and location (Alonso, 1964). This micro-economic theory is useful as reference even if hardly it can be directly applied to the behavior of the agents. In fact it is usually formulated in a schematic way and is based on over-simplification of the reality. Nevertheless, multi-agent models allow to extract from the reality the rules of the model, i.e. the behavior of agents, by a direct inquire of the real actors of the process. This is an important aspect which allows the direct calibration of parameters such as the amount of a wage or the rent of an house.

6. Conclusion

The multi-agent simulation is an useful tool for the simulation of urban dynamic, in connection with an interactive environment with human users. Agents allow the complete simulation of the urban dynamic, in which the building of the urban fabric is only one of the many facets of the process. Even if microeconomic theory does help in building the agents behavior, the possibility to model agents behavior on the basis of the corresponding real humans actors is a positive factor for the design of the model.

Acknowledgments

The CityDev project has been previously funded by the University of Florence under a program or the innovation in teaching. In this first phase the project has been developed in collaboration with Jurgen Assfalg, Saverio Armeni, Roberto Gianassi, and Francesco Marsoni (Faculty of Engendering of the University of Florence). Currently the project is funded by the Italian Ministry of the University. Valerio Melani (Faculty of Engineering of the University of Florence) collaborates to the project.

References

Alonso, W. (1964). *Location and Land Use*. Harvard University Press, Cambridge, Mass.

Batty, M. and Jiang, B. (1999). Multi-agent simulation: New approaches to exploring space-time dynamics within GIS. *CASA Working Papers*, 10:1–25.

Portugali, J. (1999). *Self-Organization and the City*. Springer-Verlag, Berlin.

Semboloni, F., Assfalg, J., Armeni, S., Gianassi, R., and Marsoni F. (2004). CityDev, an interactive multi-agents model on the web. Computers environment and urban system, 4:45–64.

Semboloni, F. (2000). The dynamic of an urban cellular automata model in a 3-d spatial pattern. In XXI National Conference Aisre: Regional and Urban Growth in a Global Market, Palermo.

Modeling Multi-scale Processes in a Cellular Automata Framework

Roger White

Abstract. When modeling land use changes in large regions or countries it has been necessary to combine two or more models operating at different scales. Typically location and relocation of population and economic activity is handled by a spatial interaction based model defined on statistical units like census tracts or counties, and the output of this model then drives a CA based model of land use, constraining cell totals in each of the regions. This approach works relatively well when the statistical areas are numerous and functionally coherent (e.g. urban centred regions). But when the areas are few, and worse, polycentric, results are very poor. An alternative approach is to attribute the activities to the cells of the corresponding land uses, and then to treat the dynamics at both scales using a single CA. In order to do this, the CA is defined with a variable size grid, so that the neighbourhood of each cell includes the entire modelled area, but the number of cells in the neighbourhood is relatively small, since the cells in each successive ring of cells in the neighbourhood is nine times as large as the cells in the preceding ring. And since each cell neighbourhood includes the entire modelled area, spatial processes at all scales are included in the cellular transition rules. The theoretical strength of the approach and the practical advantages in many applications are clear. First tests in applications previously modelled with a single scale CA linked to a spatial interaction based regional model indicate that the approach eliminates several problems inherent in the conventional approach; for example boundary effects, where urban growth cannot cross regional boundaries, disappear, and in large regions growth is distributed more realistically.

1. Introduction

When modeling land use changes in large regions or countries it has been necessary to combine two or more models operating at different scales. Typically the process of location and relocation of population and economic activity is handled by a spatial interaction based model defined on statistical units like census tracts or counties, and the output of this model then drives a CA based model of land use, constraining cell totals in each of the regions. This approach works relatively well when the statistical areas are numerous and functionally coherent (e.g. urban centred regions). But when the areas are few, and worse, polycentric, results are very poor. An alternative approach is to attribute the activities to the cells of the corresponding land uses, and then to treat the dynamics at both scales using a single CA. In order to do this, the CA is defined with a variable size grid (Andersson *et al.*, 2002a, 2002b), so that the neighbourhood of each cell includes the entire modelled area, but the number of cells in the neighbourhood is relatively small, since the cells in each successive ring of cells in the neighbourhood are nine times as large as the cells in the preceding ring. And since each cell neighbourhood includes the entire modelled area, spatial processes at all scales can be included in the cellular transition rules. A number of difficulties in implementation must be overcome, but the theoretical strength of the approach and the practical advantages in many applications are clear. The approach is outlined here.

2. Regionalized constrained CA—the problems

A regionalized CA land use model typically involves three coupled levels: (1) a global level, where constraints for the entire modelled area are supplied, usually in the form of a growth scenario, (2) a regional level, in which the modelled area is subdivided into regions such as provinces, counties, or census tracts and a spatial interaction based macro-model relocates activity among the regions at each iteration to yield a regionalized growth dynamics, and (3) a cellular level, where the land use dynamics is modelled with the CA, with cell demands for each land use type in each region being calculated from the levels of the corresponding activity predicted for the region by the regional model.

While this approach has proven to work relatively well in applications to the Dublin area and the Netherlands (Engelen *et. al.*, 2003) it is nevertheless subject to several problems and limitations. First, large regions are in effect modelled as points. Thus distances between regions are rather arbitrary, especially if there are several concentrations of activity within a single region. In the case of the application to the Netherlands, the 40 economic regions used in the macro-model are for the most part relatively compact urban centred regions, and so this problem is not serious. But in the case of the Dublin area application, in which there are only nine counties, many containing several towns, both the inter-county distance measures and the self distances, i.e. the distances from each region to itself, are highly arbitrary. Furthermore, in these interaction based models, the self distances are the most powerful parameters in the model. It is clearly undesirable that the most powerful parameters should have essentially arbitrary values.

A second problem is that in order to improve the performance of the regionalized CA models, information from the cellular level, such as average density of activity on cells actually occupied by the activity, is provided to the regional model. This increases the complexity of the model, and thus the number of parameters to be calibrated.

A third problem is that the output of the regional model is used to determine *regional* cell demands. Thus regional boundaries become visible at the cellular level. In effect the boundaries act as invisible barriers to the growth of spatial structure at the cellular level, so land use patterns do not spread naturally from one region to another. For example, in the Dublin application, the macro-model generates a demand for additional residential land use cells in Counties Meath and Kildare just to the west of Dublin, but there is nothing at the CA level that makes the parts of these counties that are adjacent to the Dublin agglomeration, and thus ripe to receive overspill, more attractive to residential location than other parts of the counties (Fig. 1a).

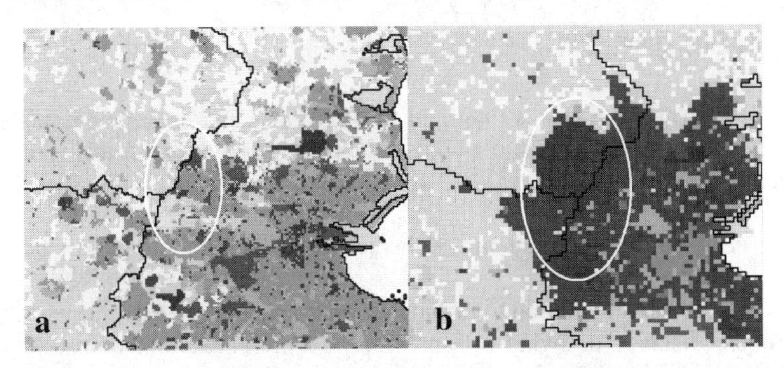

Fig. 1. The boundary effect in two simulations of the Dublin area. (a) Output of the regionalized CA model showing simulated land use for the year 2025: urban development does not cross the county boundary from Dublin to Meath. (b) Output of the variable grid model, simulated land use for 2025: urban development crosses county boundaries.

Finally, the boundary problem is made worse by the inclusion of a term representing diseconomies of agglomeration in the macro-model equations. The term necessarily applies to the entire region, even though some parts of the region may not be experiencing diseconomies. For example, the diseconomies term reduces the tendency of activity to relocate into Dublin county, thus in effect redirecting some growth to surrounding counties. But parts of Dublin county, especially in the north, are far from the urbanizing core, and thus not experiencing the diseconomies of congestion and land price inflation (Fig. 2). Nevertheless, the inhibition of growth due to the diseconomies term applies equally to these areas.

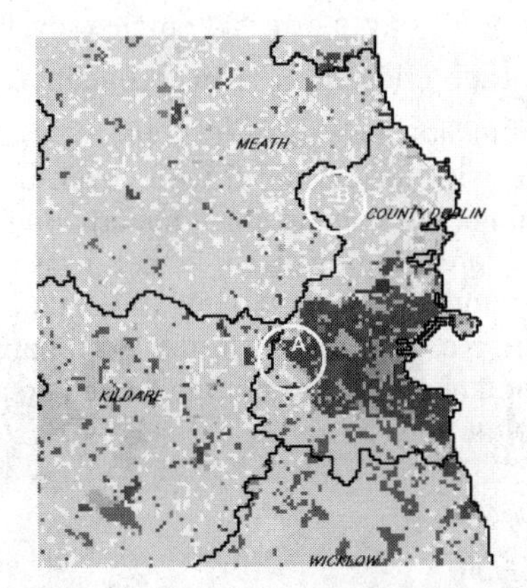

Fig. 2. Land use in Dublin County. Diseconomies of agglomeration are likely to be high on the fringe of the developed area (A), but low in areas far from the urbanizing zone, as in the north-western part of the County (B).

3. A Variable grid CA

The solution to these problems is to eliminate the regions and the regional model by effectively reducing the regional model to the CA model. In brief, this is done as follows:

- Allocate each cell a certain amount of the activity corresponding to its land use.
- Expand the cell neighbourhood to include the entire area modelled.
- Capture the distance decay effects previously represented in the macro-model in the neighbourhood weighting functions. Each cell is now effectively its own region, competing with all others for activity.
- Finally, aggregate the cellular activity levels within any desired set of regions to produce regional activity levels.

3.1. Definition of the cell neighbourhood

Since the land use raster of a modelled region may contain on the order of a million cells, in order for the CA to remain computationally tractable, the number of cells in the neighbourhood must be drastically reduced. This is done by implementing a variable size, nested grid structure in the neighbourhood. In the classic CA approach, the state of a cell depends on the states of the cells in its neighbourhood. Typically the neighbourhood is small, where small might mean anything from the 4-cell von Neumann neighbourhood to the 196-cell neighbourhood used in the model of the Dublin region. Here we refer to the fundamental grid of the land use raster as the level-zero or l_0 grid, and grids of progressively higher levels of aggregation are defined recursively so that each grid cell of level l contains 3^2 cells of the level l_1 grid, or $(3^2)^l$ cells of the fundamental, l_0 , grid (Fig. 3). Because of the exponential increase in cell size with increasing distance from the centre of the neighbourhood, the maximum number of cells in a neighbourhood is relatively small. For example, the Dublin area regionalized model uses a neighbourhood with a radius of 8 cells (representing 1.6km), giving 197 cells in the neighbourhood In contrast, with the variable grid approach, the maximum number of cells required in a cell neighbourhood to cover the entire modelled area (consisting of 101,232 grid cells of 200m resolution) is 57.

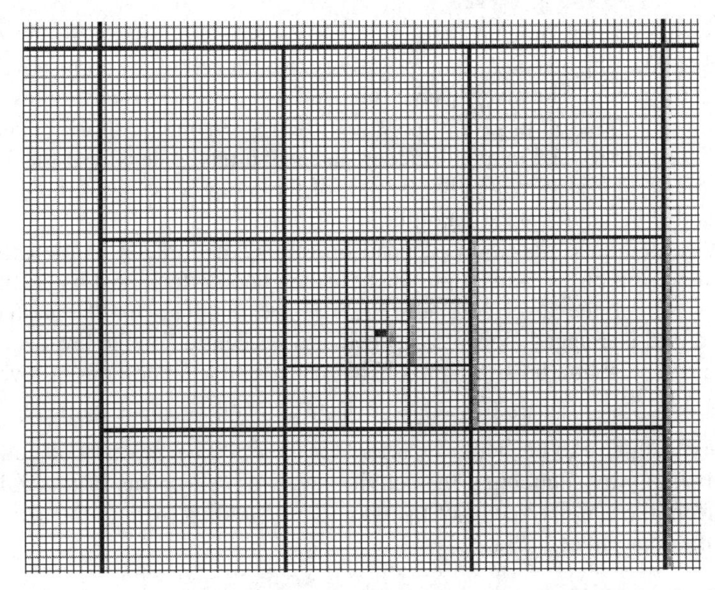

Fig. 3. The variable grid template and update procedure for a zero level cell. When the level zero cell for which the potential is being calculated shifts from the black to the red cell, the neighbourhood cell template shifts with it, thus moving one level zero cell to the right. To update the activity vectors of the l >0 cells, values in the level zero cells in the leftmost column of each higher level cell (examples are shown in green) are subtracted from the cell total and added to the total(s) of the cell(s) to the left.

3.2. Cell states and activities

Each level zero cell is assigned a state according to its dominant land use. This land use may correspond to one of the activities being modelled at the macro scale (e.g. residential land use corresponds to population), or it may be a land use or land cover for which no macro scale dynamics are included in the model (e.g. agriculture or water). In the former case, an activity level corresponding to the land use is also associated with the cell (e.g. the actual population attributed to that residential cell). Grid cells for levels $l > 0$ are characterized by activity vectors $\boldsymbol{a}(\boldsymbol{a}, \boldsymbol{a'})$, $\boldsymbol{a} = a_1 \ldots a_n$, $\boldsymbol{a'} = a_{n+1} \ldots a_N$, rather than states, since in general they will include multiple land uses and the corresponding activities. For each modelled activity a_i, the activity level consists of the sum of the activity levels of the m level zero cells contained in the higher level cell. For other activities and land uses, the activity level is simply the number of level zero cells of the given land use contained in the cell. These latter activities and land uses could, however, be given a measure representing, say, quality of the land cover (e.g., how nice is the park?).

At the initialization of a simulation, activity data that is typically available by administrative or statistical regions must be distributed over the level zero cells in some way. In the current implementation of the variable grid approach this is done by simply allocating an equal share of activity to each cell of the corresponding land use in the region. However, where time and data permit, a more sophisticated algorithm may be used to give a more realistic representation of the actual distribution of the activity and to avoid sudden discontinuities of density at the regional boundaries. This step may either be part of the pre-processing of the data or may be included in the initialization of the simulation.

As in all CA, cell state transitions depend on the neighbourhood of the cell. Thus at each time step of the simulation, the neighbourhood effect must be calculated for each cell of the level zero grid. Since the template for the nested neighbourhood grid is dragged along cell by cell, always centred on the level zero cell for which the neighbourhood effect is being calculated, the column of level zero cells that constitutes the trailing edge of one $l > 0$ cell becomes the leading edge of the next cell, and so the activity vectors for the $l > 0$ neighbourhood cells are calculated simply by subtracting the activity levels or cell counts of this column of cells from the activity vector of the one cell and adding them to the activity vector of the adjacent cell (Fig. 3).

3.3. Calculating cell state transitions and activity allocation

Cell state transitions and activity location are essentially—but not quite—the same operation. The transition of a level zero cell from one state to another represents the removal of one activity and the implantation of another. However, if the new activity is one for which the dynamics are being modelled, then the amount of activity located on the cell will depend on the desirability of the cell for that activity. Level zero cells are changed to the state and allocated the activity for which they have the highest potential until all the activity is allocated.

The potential V of a cell *i* for activity a is calculated as

$$V_{ia} = rZ_{ia}X_{ia}S_{ia}E_{ia}N_{ia} \tag{1}$$

where r = a random perturbation term

Z_{ia} = the land use zoning status of the cell for the activity

X_{ia} = the accessibility of the cell to the transportation network

S_{ia} = the intrinsic suitability of the cell for the activity

E_{ia} = a term representing diseconomies of agglomeration

N_{ia} = the neighbourhood effect

The three terms Z_{ia}, X_{ia}, and S_{ia} capture very local effects. They have been described elsewhere (e.g. White and Engelen, 2000) and will not be discussed further here. The neighbourhood effect and the diseconomies of agglomeration term, on the other hand, capture both local and long distance effects. These terms integrate the dynamics of the two scales and permit the reduction of the two linked models of the regionalized CA to a single CA model. These terms thus warrant further discussion.

The neighbourhood effect is calculated as

$$N_{ia} = \Sigma_j \Sigma_a w_{adj} m_{ja} \tag{2}$$

where m_{ja} = the amount of activity a in cell j

w_{ad} = the weight the activity is given as a function of the distance of cell j from cell i.

Since in general cell j is a higher order cell, $m_{ja} = \sum_k m_{ka}$, where k indexes all level zero cells contained in j. The weighting term captures the distance decay effect, with $w_{ad} = f(d_{ij})$. Beyond a distance in cell space representing approximately 1 km, the calibrated weighting functions typically approximate a traditional negative exponential or inverse power function of distance. At shorter distances they may maintain this form, but in some cases the configuration is quite different, with the curve becoming negative, for example, reflecting a short range repulsion effect between the pair of activities. The neighbourhood effects as calculated here are thus very similar to the potentials calculated using gravity equations in the macro model of the regionalized CA.

The diseconomies of agglomeration term captures the negative consequences of the positive neighbourhood effect. Conventional regionalized models in which the net migration of activity is a positive function of potential predict the ultimate collapse of all activity into the dominant region unless a countervailing term representing diseconomies of agglomeration (e.g. congestion costs, land costs) is introduced. And the larger (and fewer) the regions in the modelled area, the worse the problem. In the cell based approach described here, this problem is not as severe. Nevertheless, the problem does not entirely disappear, because in general, a certain amount of the growth of large metropolitan areas occurs not on the fringes but in regional centres that lie beyond the main urbanized area. This phenomenon is in large part due to an attempt to escape the diseconomies of agglomeration (e.g. congestion, high land prices) present in the main centre. In the model, these satellite cities represent local peaks in the potential surface, but the peaks are frequently not high enough to attract as large a proportion of regional growth as is observed in reality. The problem is solved by including the term representing diseconomies of agglomeration, E_{ia}, in the potential function. Thus,

$$E_{ik} = \frac{1}{1 + e^{\lambda_k \left(\frac{V_{i,pop}}{V_{crit}} - 1 \right)}} \tag{3}$$

$$V_{crit} = \varepsilon < V_{init} >$$

where

E_{ia} = relative level of economies or diseconomies of agglomeration

$V_{i,pop} = \sum_j w_{pop,dj} m_{j,pop}$, the population potential, i.e. the population component of the neighbourhood effect

V_{crit} = critical level of population potential at which diseconomies appear

$<V_{init}>$ = mean value of initial population potentials

λ_a = relative importance of economies or diseconomies for activity a

ε = parameter expressing critical level relative to initial average population potential

The parameter λ_a accommodates the fact that some activities like financial services show very little sensitivity to diseconomies of agglomeration, whereas others, for example manufacturing, are highly sensitive.

Actual cell state transitions and the consequent activity allocation depend on eq. (1). In conventional regionalized spatial interaction based models of location, activity migrates from region *i* to region j on the basis of the distance-discounted mass or level of activity at j relative to the total distance discounted activity level in all regions—i.e. relative to the potential at *i*. In the approach developed here there are no regions, and the comparisons are made directly between potentials calculated for each cell in eq. (1). As in the previously developed constrained CA land use model (White and Engelen, 2000), for each cell the potentials for the various activities are ranked, and the cells are then ranked on the basis of their highest potential. Starting with the highest ranked cell, cells are changed to (or remain in) the state for which they have the highest potential, and a certain amount of the corresponding activity is allocated to the cell. A running total of activity allocated is maintained, and when all of an activity has been allocated to cells, no further cells are changed to that state.

The amount of activity A_{ki} allocated to a cell is a positive function of the relative neighbourhood effect for that activity on that cell:

$$A_{ki} = \langle A_k \rangle \left(\frac{N_{ki}}{\langle N_k \rangle} \right)^{\lambda_k}, \qquad \lambda_k \leq 1$$

A cell with an above average neighbourhood effect will thus receive an above average amount of activity. This reflects the tendency of land in better locations to be developed more intensively. At the end of each iteration, activity levels are summed to provide totals for each activity in each of the original statistical or administrative regions as part of the model output, since users are typically interested in information at this level of aggregation.

3.4. Calculating distances

Calculating the neighbourhood effects from the weighting functions requires that the distances from the reference cell to each cell of its neighbourhood be known. In the current experimental implementation of the variable grid model Euclidean distances rather than distances through the transport network are used. This expedient was adopted because establishing distances through the transportation network to cells in the neighbourhood is perhaps the most difficult problem to solve in the variable grid approach, in the sense that compromises must be made between accuracy and computational efficiency in order to keep run times reasonable. In region based approaches, nodes on the transport network are selected to represent the regions, and a matrix of minimum path (or minimum time path) distances among these nodes is calculated. As long as the network does not change, the distance matrix need be calculated only once. In the cell based approach, on the other hand, while the neighbourhood of each cell contains a relatively small number of "regions" –i.e. the various sized cells that comprise the neighbourhood, the location of these cells is different for every level zero cell, so potentially the nodes representing the cells in the neighbourhood change each time the neighbourhood template is moved to the next level zero cell on the grid. Network distances will be introduced in future versions of the model.

4. Application to the Dublin region

A first, simplified, version of the variable grid CA model is applied to the Dublin region data set. As mentioned above, the current implementation calculates distances within the cell neighbourhood along Euclidian lines rather than along the transportation network. The full network is, however, used to calculate the cellular accessibility values, X_{ia}. Another simplification is that the four residential land use categories have been combined into a single class. Finally, in the applications shown here, the l_0 cell size is increased to 500m from the 200m resolution used in the regionalized CA applications. The larger cell size is simply an expedient to permit a very rapid exploration of the properties of the new variable grid model.

An example of a ten year simulation is shown in Fig. 4. The 1990 map (Fig. 4a) shows the actual land use in the initial year, and the 2000 map (Fig. 4b), the simulated land use. In general the model seems to perform well, giving reasonable predictions of land use patterns and responding as expected to changes in parameter values, especially those defined by the neighbourhood weighting functions. The cellular transition potential map for residential land use is shown in Fig. 5. It is not yet possible to make definitive tests of the performance of this model in comparison to that of the regionalized CA model, since the current implementation of the new model is, as mentioned, a simplified one. Furthermore, the new model has not yet been as exhaustively calibrated as the regionalized model. However, we are currently developing automatic calibration routines

(Straatman *et al.*, 2004), and when these are fully functional it will be possible to use the same routine to calibrate both models, and thus to compare the performance of equivalent calibrations of the two models.

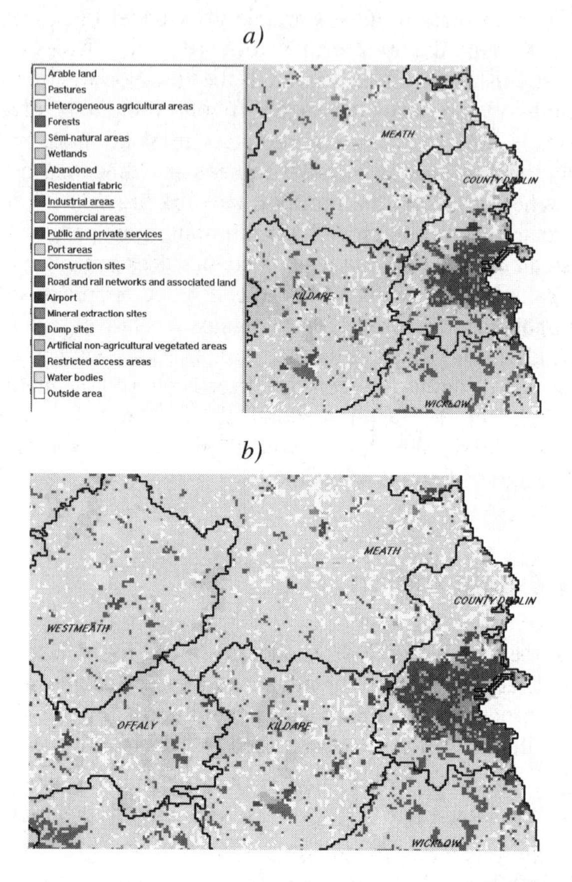

Fig. 4. Dublin area simulation using the variable grid model. Dublin area land use, year 2000. (a) Initial conditions: actual land use, 1990. (b) Simulated land use, 2000.

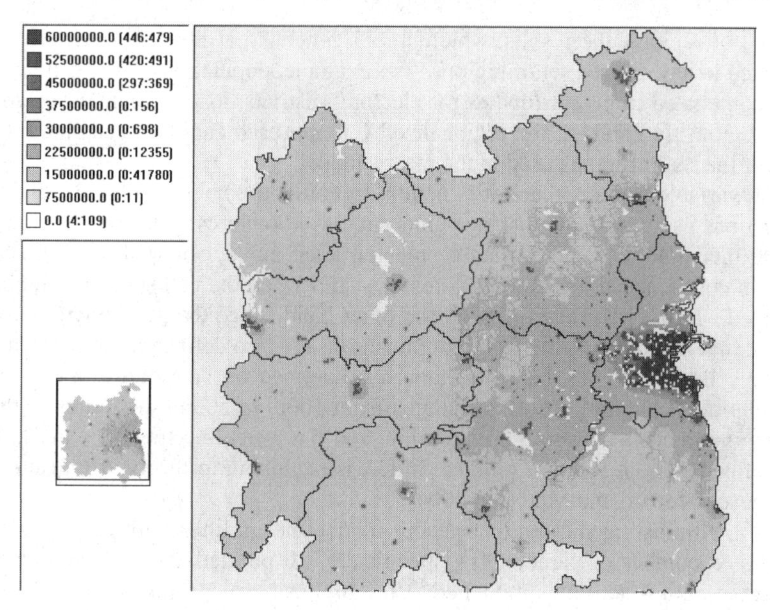

Fig. 5. Cell state transition potential for residential land use, year 2000, generated by the variable grid model.

Nevertheless, even without definitive tests, several advantages of the variable grid approach are already apparent. First, there are no regional boundaries to create artefacts in the form of discontinuities in the potential surface. As an example, in the regionalized CA model of the 9-county Dublin area, much of the growth was allocated to Dublin county, with urbanization spreading along major transportation routes to the west. But at the County Meath border, this growth stopped abruptly, and was diverted to other locations in Dublin county. Meath, having a much lower potential, also had a much lower growth, and furthermore there was nothing in the model to direct that growth to the Dublin edge of the county (Fig. 1a), since cell state transitions depended only on a local neighbourhood of 1.6 km radius. In the present variable grid approach, since there are no regions, locations along major transport routes on the fringe of the Dublin agglomeration have higher potentials than locations away from the routes, and along the routes potentials decrease gradually with increasing distance from the city; there are no discontinuities in the potential surface, so the modelled city can expand naturally across the county border (Fig 1b).

Another advantage is that activity densities (e.g. population density) have approximately continuous distributions, since densities are modelled cell-by-cell. In the regionalized CA model, on the other hand, densities are uniform throughout a county. For example, all residential cells in a particular county are assigned the same population. For some model applications, such as estimating the number of people who will be living within the noise band of a major highway, high resolution density estimates are essential. Similarly, since the model does not actually make use of regions, the results, which are all generated at the cell level, can be aggregated to any desired set of regions. For example, populations at the cell level can be aggregated to give estimates for electoral districts, towns, counties, or any ad-hoc region. In contrast, the regionalized CA approach can only generate estimates for the set of regions used in the macro-model.

In the variable grid approach it is natural to define the neighbourhood weighting functions using absolute distance units such as kilometres, since the functions span the modelled area rather than a predetermined number of grid units in cell space. In consequence, the functions remain valid when the cell space resolution is changed. In the traditional CA, on the other hand, since the neighbourhood is specified in terms of cells, the weighting functions are also defined in terms of distances in cell space. Thus, for a neighbourhood defined with a radius of 8 cells, if the resolution of an application is changed from 100m to 500m, the radius of the neighbourhood changes from 800m to 4km, and the weighting functions must all be re-defined. The resolution problem in CA is beginning to receive some attention (e.g. see Menard and Marceau, 2004).

Finally, eliminating the regional macro model and its links with the CA also eliminates a number of parameters—specifically, 10 per activity. In the Dublin application, with 4 activities modelled at the regional scale, replacing the linked macro model with the variable grid CA reduces the number of parameters by 40.

5. Conclusion

While the variable grid approach to defining the cell neighbourhood was developed for an earlier project (Andersson *et al.*, 2002a,b), no attempt was made there to model activity migration as opposed to land use, nor was the approach tested on real data. Here the variable grid approach has been extended to model spatial dynamics on all scales within a single framework, and simultaneously to model the dynamics of both activities and their associated land uses within that same framework. The unified variable grid approach is much more natural than the hybrid, linked model, regionalized CA framework: it is straightforward, it has high-resolution where resolution is necessary and less elsewhere, and it is essentially lacking in the artifactual discontinuities associated with regional boundaries. It is thus easier to work with, and on the basis of early experiments, it seems likely to give better results.

References

Andersson, C., Rasmussen, S. and White, R. (2002a). Urban Settlement Transitions, *Environment and Planning B* vol. 29 pp. 841-865.

Andersson, C., Lindgren, K., Rasmussen, S., White, R. (2002b). Urban Growth Simulation from 'First Principles', *Physical Review E*, 66, 026604, 1-9.

Engelen, G., White, R. and De Nijs, T. (2003). Environment Explorer: Spatial Support System for the Integrated Assessment of Socio-Economic and Environmental Policies in the Netherlands, *Integrated Assessment* vol. 4, pp. 97-105.

Menard, A. and Marceau, D. (2004). Spatial Scale in Geographic Cellular Automata. Paper presented at the congress of the Canadian Association of Geographers, Moncton, Canada.

Straatman, B., White, R. and Engelen, G. (2004). Toward an Automatic Calibration Procedure for Constrained Cellular Automata, *Computers, Environment, and Urban Systems* vol. 28, pp. 149-170.

White, R. and Engelen, G. (2000). High-Resolution Integrated Modeling of the Spatial Dynamics of Urban and Regional Systems, *Computers, Environment and Urban Systems*, 24, pp. 383-400.

Part four: Cognition and VR

Multi-Agent Models of Spatial Cognition, Learning and Complex Choice Behavior in Urban Environments

Theo Arentze and Harry Timmermans

Abstract. This chapter provides an overview of ongoing research projects in the DDSS research program at TUE related to multi-agents. Projects include (a) the use of multi-agent models and concepts of artificial intelligence to develop models of activity-travel behavior; (b) the use of a multi-agent model to simulate pedestrian movement in urban environments; (c) the use of multi-agent models for simulating the dynamics of land development; (d) the use of multi-agent models to simulate learning and adaptation behavior in non-stationary urban environments and under conditions of uncertainty and information search and (e) the development of computational models linking cognition, choice set formation, activity travel behavior and land use dynamics. The scope, theoretical underpinnings and results of numerical and empirical simulations are presented.

1. Introduction

This paper will provide an overview of some research projects conducted by the authors that relate to the topic of the workshop in the context of the Design and Decision Support Systems in Architecture and Urban Planning (DDSS) research program at the Eindhoven University of Technology, The Netherlands. This program involves the collaboration of the Urban Planning Group, which has a strong emphasis on modeling urban phenomenon and spatial choice behavior, and the Design Systems Group, which has a good reputation in developing computer systems.

One of the main research efforts of the DDSS research program concerns the development of multi-agent systems. One of the potential advantages of such systems concerns their flexibility. It is relatively easy to incorporate appealing behavioral concepts and use these in micro-simulation frameworks. As part of the DDSS research framework, multi-agent systems have been or are currently developed for various types of spatial choice behavior. These projects include:

1. the use of multi-agent models and concepts of artificial intelligence to develop models of activity-travel behavior
2. the use of a multi-agent model to simulate pedestrian movement in urban environments
3. the use of multi-agent models for simulating the dynamics of land development

4. the use of multi-agent models to simulate learning and adaptation behavior in non-stationary urban environments and under conditions of uncertainty and information search.
5. the development of computational models linking cognition, choice set formation, activity travel behavior and land use dynamics

Thus, while some of the early projects were concerned with spatial choice processes of consumers in a static context, more recently the first prototypes of dynamic models and models for agents other than consumers were proposed.

The purpose of this review paper is to present some of these ongoing research projects. The scope, theoretical underpinnings and results of numerical and empirical simulations will be presented. The paper will focus primarily on the projects of the authors. No attempt is made to position the research efforts vis-à-vis related research. Readers are requested to consult the original papers for such positioning and discussion of relevant related work. The focus in this chapter will be on a multi-agent model of activity-travel behavior and how this can be embedded into more general models of spatial cognition and learning, choice set formation and spatial dynamics.

2. A multi-agent model of activity-travel behavior

Recently, research in transportation has shifted away from trip and tour-based models to models of activity-travel behavior. This shift reflected the notion that transport represents induced demand; it is the result of individuals and households organizing their activities in time and space. The utility-maximizing nested logit models still constitute the dominant modeling tradition in this field of research (see e.g. Timmermans, *et al*, 2002), but an increased awareness that multi-agent, computational process models offer a viable alternative can be detected in the literature. The Albatross model (Arentze and Timmermans, 2000, 2004a; Arentze *et al.* 2003), developed for the Dutch Ministry of Transportation, is currently the only available, fully operational computational process model of activity-travel patterns.

The model system especially concerns the scheduling of activities in time and space and can be viewed as a set of software agents to solve the problem. The core of the system is the activity scheduler. The scheduling of activities involves a set of interrelated decisions including the choice of location where to conduct a particular activity, the transport mode involved, the choice of other persons with whom to conduct the activities, the actual scheduling of activities contained in the activity program, and the choice of travel linkages which connect the activities in time and space. Travel decisions represent a sub-decision. Transport mode decisions dictate the action space within which individuals can choose locations to conduct their activities. The organization of trips into chains allows individuals to conduct more activities within a specific time frame. The actual process of scheduling activities is conceptualized as a process in which an individual attempts to realize particular goals, given a variety of constraints that limit the number of feasible activity patterns. Several types of constraints are identified: (i) *situational constraints* impose that a person, transport mode and other schedule resources cannot be at different locations at the same time; (ii) *institutional constraints*, such as opening hours, influence the earliest and latest possible times to implement a particular activity; (iii) *household constraints*, such as bringing children to school, dictate when particular activities need to be performed and others cannot be performed; (iv) *spatial constraints* also have an impact in the sense that either particular activities cannot be performed at particular locations, or individuals have incomplete or incorrect information about the opportunities that particular locations may offer; (v) *time constraints* limit the number of feasible activity patterns in the sense that activities do require some minimum duration and both the total amount of time and the amount of time for discretionary activities is limited, and (vi) *spatial-temporal constraints* are critical in the sense that the specific interaction between an individual's activity program, the individual's cognitive space, the institutional context and the transportation environment may imply that an individual cannot be at a particular location at the right time to conduct a particular activity.

The scheduling model consists of three components: (1) a model of the sequential decision making process; (2) models to compute dynamic constraints on choice options and (3) a set of decision trees representing choice behavior of individuals related to each step in the process model. The first two components are a-priori defined, whereas the third component is derived from observed choice behavior of individuals. The system generates a schedule in terms of an ordered list S_r for each member r of the household simulated. Each element of S_r represents an activity episode described in terms of relevant profile dimensions including activity type, travel party, duration, start time, location, and if traveling to the location is involved, transport mode and travel time. Mandatory activities, such as work or school, are typically fixed in a short time horizon of a day. As the model focuses on daily scheduling, the selection, location, duration and start time of such fixed activities were considered given in the first version of the model, but in the second version these facets were also predicted. The scheduling process model intends to simulate how individuals frame choices and arrange them into a sequence when they schedule their activities. Albatross presently assumes a pre-defined sequence of choice facets based on an assumed priority ranking of activities by type and an assumed priority ranking of activity attributes. An activity is assigned a higher priority the closer it is positioned towards the mandatory end as opposed to the discretionary end on a mandatory-discretionary scale. The activities that are fixed in time and space in the short term constitute a schedule skeleton defining the blocked and open time periods for activities that are more flexible. The process model includes two main stages. In the first stage, the model generates the schedule skeleton. This involves the steps of (1) selecting which fixed activities are included in the schedule for that day, (2) determining the (exact) start time and (exact) duration on a continuous time scale for each fixed activity and (3) selecting the location of each fixed activity. In the second stage, flexible activities are scheduled in open time slots in the skeleton. We will describe the process model for the second stage in somewhat more detail here.

In that stage, the model first decides on the transport mode for the work activity. Mode choice for work is considered the highest-priority decision because this choice determines which person can use the car for a substantial part of the day in cases where there is only one car and more than one driving license available in the household. *Step 2* determines the activity composition of the schedule. For each flexible activity category, a decision whether or not to add an episode of that activity to *S* is made. If an activity is added, travel party and duration of the activity are determined before a next activity category is considered. Duration is determined in a qualitative way as a choice between a long, average and short episode. Each duration class is defined by a time range depending on the activity category under concern. Temporal constraints define the feasibility of both selection and duration decisions in this step. *Step 3* determines the time of day for each flexible activity in order of priority. This is modelled as choosing a time period for the activity based on a given subdvision of the day (e.g., early morning, late morning, around noon, etc.). The time period constrains the start time of the activity. Based on this the model determines a preliminary position in the schedule. *Step 4* determines trip links between activities by choosing for each activity in order of priority whether it is conducted on a Before stop (directly before another out-of-home activity in the schedule), an After stop (directly after another out-of-home activity), an In-between stop or on a single stop trip. The choices made in this stage have several implications for the schedule. First, the activity is repositioned if needed to realize the chosen trip link. After this step, the position of the activity is considered definite. Each activity with a definite schedule position can serve as a basis for a trip link for activities considered next. Thus, trip links can be established between flexible activities as well. Second, in-home activities are inserted where needed to make the schedule consistent with chosen trip links. In this stage, the tours included in the schedule are identifiable as sequences of one or more out-of-home activities that start at home and end at home. *Step 5* involves a choice of a transport mode for each tour assuming that there are no mode changes between trips within tours. Tours including a work activity take on the mode that was chosen for the work activity in the first step. Finally, *step 6* determines the location of each flexible activity in order of priority. For each location choice, the system defines a dynamic location choice-set, dependent on the time-window for the activity, available facilities, opening times of facilities, travel times and minimum activity duration. The decision is modeled as a choice among different pre-defined location categories. The categories are defined as a specific combination of travel-time band and size category of available facilities for the activity. To account for trip chaining, travel time is consistently defined as detour time in comparison to a direct trip between previous and next (known) location. Possibly, there are multiple locations in the choice set that fit a chosen location category in a particular case. In that case, the model selects a location randomly from the candidates.

The model takes possible interactions between persons within households into account by constructing the schedules for persons within the same household simultaneously. This is done by alternating scheduling steps between persons. For each person and each step, the model takes the schedule of the partner as far as developed at the end of the previous step as a condition for the scheduling decisions.

In each step, dynamic constraints determine which choice alternatives are feasible given the current state of the schedule. In many cases there will be a choice left and the decision tree linked to that step is consulted to generate a choice. Socio-economic attributes of the person and household are input to take possible interindividual differences in decision rules into account. State dependence of decisions is taken into account by including outcomes of previous decisions as input to each current decision. In this way, for example, the probability of adding an activity will be strongly reduced when an activity of that category already has been added in a previous step.

The location choice set definition for a given activity is probably the step where the complexity of an activity-based model is most apparent. In Albatross, a location l is considered feasible if the following two conditions are met:

$$\exists g \in G_l, g \in G\{a(\tau)\} \tag{1}$$

$$T_{l_g}^{f\max}(\tau) - T_{l_g}^{s\min}(\tau) \geq v^{\min}(\tau) \tag{2}$$

where, τ is an index of activity episodes in the given schedule, G_l is the set of facilities present at location l, $G\{a(\tau)\}$ is the subset of facilities compatible with activities of type $a(\tau)$, $v^{\min}(\tau)$ is the minimum duration and $T_{l_g}^{s\min}$ and $T_{l_g}^{f\max}$ define the time window for the activity dependent on the current schedule and opening hours of facilities. The latter terms are formally defined as:

$$T_{l_g}^{f\max}(\tau) = \max\{{}_d t_{l_g}^{\min}, T^{f\min}(\tau - 1) + t_l^t(\tau)\} \tag{3}$$

$$T_{l_g}^{s\min}(\tau) = \min\{{}_d t_{l_g}^{\max}, T^{s\max}(\tau + 1) - t_l^t(\tau + 1)\} \tag{4}$$

where, ${}_d t_{l_g}^{\min}$ and ${}_d t_{l_g}^{\max}$ are the known opening and closing times of facilities of type g at location l on day d, $T^{f\min}$ is the earliest end time and $T^{s\max}$ the latest start time of the previous and next activity respectively and t_l^t is travel time to the activity location using the mode chosen in a previous step. Earliest start times and latest end times of activities are calculated by shifting previous activities as far as possible to the right on the time scale and next activities as far as possible to the left within temporal constraints.

Albatross assumes that individuals choices are driven by condition-action rules which have been formed and are continuously adapted through learning based on experience. The general format of a condition-action rule can be described as:

$$\text{if } C_1 \in CS_{1k} \wedge C_2 \in CS_{2k} \wedge \; .... \wedge C_m \in CS_{mk} \text{ then choose alternative } A_k \tag{5}$$

where C_i represents condition variables, CS_{ik} is the state of the i-th condition variable in the k-th rule and A_k is the choice prescribed by the k-th rule. The set notation used implies that a condition state is defined as a subset of the domain of the condition variable. If the condition variable is measured on a nominal scale, the subset may consist of any combination of values from the domain, whereas in the case of an ordinal or metric variable, the condition state specifies a certain sub range of the variable's domain. Each decision step involves a set of decision rules of type (5). To make sure that a rule set generates an unambiguous response for every possible configuration of condition values, the rules must be mutually exclusive and as a set exhaustive. These properties of consistency and completeness are guaranteed by the way the learning mechanism operates in Albatross. Consider an initial situation where the individual has no a-priori knowledge of the domain. Decision-making would be purely random and handled by a single 'rule':

$$\text{if } C_1 \in CD_1 \wedge C_2 \in CD_2 \wedge \; .... \wedge C_m \in CD_m \text{ then choose random} \tag{6}$$

where CD_i represents the domain of condition variable C_i. Since every 'condition state' in this rule equals the entire domain, each variable in effect is irrelevant. Therefore, every possible state in terms of C_i will trigger the rule implying that the model meets the requirements of completeness and consistency. Now assume that through interaction with the environment the individual has learned to discriminate between particular states of a particular condition variable. In the model, this is represented by splitting the domain of that variable and replacing the initial rule by a rule for each partition. For example, if, in rule (6), the domain of C_j is split into states CS_{j1} and CS_{j2}, then rule (6) is replaced by:

$$\text{if } C_1 \in CD_1 \wedge C_2 \in CD_2 \wedge \; .....\wedge C_j \in CS_{j1} \wedge .... \wedge C_m \in CD_m \text{ then choose A1} \tag{7}$$

$$\text{if } C_1 \in CD_1 \wedge C_2 \in CD_2 \wedge \; .....\wedge C_j \in CS_{j2} \wedge .... \wedge C_m \in CD_m \text{ then choose A2} \tag{8}$$

Because the new condition states were achieved by splitting a domain, $CS_{j1} \cup CS_{j2} = CD_j$ and $CS_{j1} \cap CS_{j2} = \varnothing$ and, hence, the changed rule set still meets the requirements of completeness and consistency. Each time the individual discovers new contingencies, the process of splitting is repeated. The result at some given moment in time, may be a highly complex model which still meets the required properties.

Any set of rules that can be obtained by recursively splitting condition states starting with an initial rule of format (6) meets the formal definition of a decision tree. Decision tree induction based on observed choices has received much attention in statistics and AI. Because the number of possible trees increases exponentially with the number of condition variables, work in this area has focused on heuristics to search for the most parsimonious tree that best fits the data. Albatross uses a CHAID-based algorithm to induce a decision tree for a given decision step from activity diary data. The decision trees, thus derived, can be used for predicting scheduling decisions, given that we have a so-called action-assignment rule Consider an action variable that has Q levels and for which CHAID produced a tree with K leaf nodes. In the prediction stage, the tree is used to classify any new case to one of the K leaf nodes based on attributes of the case. An action-assignment rule defines an action (i.e., a scheduling decision) for each classified case. Albatross uses a probabilistic assignment rule. In the basic form of this rule, the probability of selecting the q-th action for a case assigned to the k-th leaf node is defined by:

$$p_{kq} = \frac{f_{kq}}{N_k} \tag{9}$$

where f_{kq} is the number of training cases of category q at leaf node k and N_k is the total number of training cases at that node. This rule is sensitive to residual variance, but fails to take scheduling constraints into account. Scheduling constraints entail that dependent on individual attributes and the state of the current schedule some choice alternatives for the decision at hand may be infeasible. If such constraints are represented in the decision tree, the probabilistic rule would assign zero probability to infeasible categories and the response distribution should not be biased. However, even though it is likely, it is not guaranteed that the induction method discovers constraint rules in data. Therefore, to cover the general case rule (9) was refined as:

$$p_{kq} = \begin{cases} 0 & \text{if } q \text{ is infeasible} \\ \dfrac{f_{kq}}{\sum_{q'} f_{kq'}} & \text{otherwise} \end{cases} \tag{10}$$

where q' is an index of feasible alternatives for the decision at hand. The rule tends to over predict responses that are feasible in the majority of cases (at that leaf node), because the probability of these responses is increased by rule (10) in constrained cases and stays the same as (9) in unconstrained cases.

The Albatross system has been tested in several studies. It turned out that the spatial transferability of the system is good (Arentze *et al.*, 2002), and performance of the model relative to alternative, utility-based models considered in a comparison is favorable (Arentze *et al.* 2001).

3. A multi-agent model to simulate learning and adaptation behavior in non-stationary urban environments under condition of uncertainty and information search

Arentze and Timmermans (2005a, 2005b) suggested a multi-agent modeling framework to simulate the formation of choice sets as a function of knowledge and spatial search processes in the context of activity-travel behavior. At the same time, the approach incorporates learning and adaptation behavior in non-stationary environments. Mental maps are used as a representation of an individuals' knowledge about locations and their attributes in their environment. A Bayesian Belief Network (BBN) is used to model the dynamic knowledge (beliefs) of individuals about locations and their attributes in their environment in response to primary learning processes, that is direct experiences of individuals during the implementation of activity-travel sequences. The approach works as follows.

Assume that the environment is represented as a regular grid of cells. Each cell represents a location where an activity can be conducted and is described by a vector, X_l, of potentially relevant variables, influencing spatial choice behavior. Assume further that a transport network connecting the different locations is represented as a directed graph $G(N, L)$ where N is a set of nodes and L is a set of links. A trip from an origin location l_1 to a destination location l_2 is modeled as a path through the network from the nearest node from l_1 to the nearest node from l_2. Assume that the variables describing locations are discrete. The possible values (or states) of X_{lk} are denoted as x_{lks}, where x_{lks} is a specific value (or 'state') of X_{lk} and l is an index of location. The belief that $X_{lk} = x_{lks}$ is represented by a probability $P(x_{lks})$. These beliefs reflect an individual's uncertainty about the state of the environment. The awareness space or mental map of an individual is modeled in terms of the full set of probability distributions, $\Pi = \{P(Xlk) \mid l = (i, j), i = 1,..., I, j = 1,..., J, k = 1,..., K\}$ where I and J are the number of rows and columns of the grid and K is the number of location attributes.

Full information represents the special case, where the probability is equal to zero for untrue values of the variable and one for the true value.

During implementation of trips and activities, an individual makes observations about certain variables in certain cells. Observations are not necessarily perfect especially not when they are made from a distance. Let X be a particular variable in a certain cell, and let Y denote the outcome of observing the variable in that cell. Let $Y = y_s$ denote the outcome that $X = x_s$ in that cell. Uncertainty about the observation means that the probability of $Y = y_s$ given that $X = x_s$ is not necessarily equal to one and the probability that $Y = y_s$ given that $X \neq x_s$ is not necessarily equal to zero. To account for uncertainty, we propose the well-known Bayesian method as a model of perception updating. Using this method, the updated belief in x_s after making observation y_u is defined as:

$$P(x_s \mid y_u) = \frac{P(y_u \mid x_s)P(x_s)}{\sum_{s'} P(y_u \mid x_{s'})P(x_{s'})} \qquad \forall s \tag{11}$$

where,

$P(x_s)$ is the prior belief in x_s;

$P(x_s \mid y_u)$ is the updated belief after observation y_u;

$P(y_u \mid x_s)$ is the probability of observation $Y = y_u$ given $X = x_s$.

Since the updated belief, $P(x_s \mid y_u)$, is taken as the prior belief in a next observation, learning is incremental.

The conditional probability distributions on the right hand side of Equation (11) represent the individual's perception of the credibility of the observation. This credibility is not constant but will differ depending on how well the particular variable is observable from the point on the route from which the observation is made. The following multinomial logit model is used to predict conditional observation outcome probabilities (COOP) each time an observation made:

$$P(y_u \mid x_s) = \frac{\exp(\theta\beta_{us})}{\sum_{s'=1}^{n} \exp(\theta\beta_{us'})} \qquad \forall u,s \tag{12}$$

where β_{us} are observation-bias parameters and θ is an observation-sensitivity parameter. The sensitivity parameter is inversely proportional to the error scale. The larger the value the smaller the expected error is and vice versa. In the extreme case, where the value is zero, the observation is completely independent of the true value and the posterior belief is identical to the prior belief meaning that no learning has occurred. On the other hand, if the sensitiveness is very large, the conditional probability distributions are very dense at the level of the true value and uncertainty in the updated belief is strongly reduced.

The sensitivity of an observation will vary from case to case depending on many factors. Therefore, the sensitivity parameter is predicted each time an observation is made using the function:

$$\theta = f(S, D, X) \tag{13}$$

where S is the state of the individual, X is the variable being observed and D is the distance between the individual and the object of the observation. The state of the individual may include many things, such as the speed of traveling, the transport mode used, the motivational state, etc. To give an example, sensitivity will be particular big if the variable type is easy to observe, the distance from the cell is small, transport mode is slow and the link is a local road (as opposed to highway).

In the model, mental map Π is updated each time after a trip or activity has been implemented by updating the beliefs for each cell that could have been observed during the event. This involves for each variable and each cell (i) calculating COOPS based on equations (13) and (12), (ii) drawing an observation outcome from distribution $P(Y \mid x_s)$, where $X = x_s$ is the actual cell value, and (iii) updating the beliefs using equation (11).

Making observations is not the only factor in belief updating. In addition, the model takes two additional factors into account. First, belief updating takes place in a network of beliefs representing the individual's causal knowledge. The links in the so-called Bayesian belief network (BBN) represent causal or statistical dependencies between variables. As a consequence of probability propagation through the network, an observation tends to have a wider impact than just the belief to which it is related. Second, the difference between the current belief and the a-priori belief in a certain value of a cell represents specific knowledge about that cell. As a consequence of limited memory retention, we assume that specific knowledge is subject to decay over time and we model this decay by letting posterior probabilities return with a given step size to their corresponding a-priori values. The step size is a parameter determining the speed with which the results of belief updating decay over time.

Location choice and choice set formation are modeled as follows. At every moment in time, the mental map describes the individual's current knowledge about locations. We assume that, at the moment of making a choice, all cells (i.e., locations) compete for the attention of the subject and the cell with the highest perceived value is chosen. This process implies exhaustive search, but does not have the theoretical drawbacks of an equivalent process in conventional (random-)utility-maximizing models. Since the mental map is searched, the strong assumption of full information is prevented. Moreover, the mental map itself can be seen as the choice set, namely one in which awareness of choice alternatives is a continuous rather than a discrete function of the individual's knowledge. The mental map introduces uncertainties the individual should deal with in choosing a location for an activity.

We assume that expected utility, expected information gain and uncertainty are the three main criteria in location choice under uncertainty. We define information gain as the decrease in uncertainty caused by making a trip and conducting an activity at the destination. The subjective value of information gain has three components. First, the information obtained allows the subject to make better-informed decisions in the future. Second, information gain is proportional to the degree of novelty of a choice alternative and, as such, measures the extent to which the choice of the alternative can satisfy curiosity or pleasure of exploring. Third, on the negative side, information gain is inversely proportional to the familiarity of a choice alternative.

Uncertainty, the next criterion, is proportional to the difficulty of predicting the consequences of a choice alternative (in terms of it's utility). As prospect theory emphasizes, the influence of uncertainty on the perceived value of a course of action is complex. In determining a choice, subjects consider a base alternative as a reference point and evaluate alternatives on the expected loss and gain they would yield. Risk aversive persons assign a relatively large weight to expected losses, whereas risk-takers value expected gains higher. Whether personality differences in risk seeking as well as novelty seeking are important factors in choice is unclear as empirical research on this issue shows rather controversial results.

These three criteria are integrated in a utility function of locations. Assuming the origin location, route, transport mode and purpose of the trip as given, the utility of a location choice alternative l can be defined as:

$$U_l = V_l^{trip} + \mu^g E(G_l^{trip}) + E(V_l^{dest}) + \mu^g E(G_l^{dest}) + \mu^- E(V_l^{-,dest}) + \mu^+ E(V_l^{+,dest}) \tag{14}$$

where U_l is the utility of cell l, V_l^{trip} is the utility of the trip, $E(G)$ is an expected information gain, $E(V)$ is an expected utility, $E(V^-)$ is an expected utility loss, $E(V^+)$ is an expected utility gain and μ are weights. The equation assumes that there is no uncertainty related to the transport network so that information gained by traveling arises exclusively from observations of land characteristics. It also means that there is no uncertainty in the assessment of the (dis-)utility of travel. The weights are influenced by the positions of the subject on the familiarity – novelty scale and the risk aversion – risk taking scale. The utility of a trip, V_l^{trip}, is defined in the usual way as a function of distance and possibly other attributes of the route. The expected utility of a destination takes into account the current beliefs in states of the variables within cell l, as given by the mental map. The expected loss and gain are defined based on the notion that, given a reference utility level V^0, the expected utility level can be decomposed as:

$$E(V) = P(V < V^0)E(V - V^0 \mid V < V^0) + \\ P(V > V^0)E(V - V^0 \mid V > V^0) + V^0 \tag{15}$$

where $P(X)$ is the probability of event X, $E(V \mid X)$ is the expected utility given event X and V^0 is the utility of a base alternative. The first and the second product on the right hand side of equation (15) correspond to the expected utility loss and expected utility gain respectively.

The proposed measure of information gain, used in equation (14), is conceptualized as:

$$E(G_{lk}) = H(X_{lk}) - E\{H(X_{lk})\} \tag{16}$$

where $H(X)$ is the entropy of X and $E\{H(X)\}$ is the expected amount of entropy after the observation. The latter is formally defined as:

$$E\{H(X_{lk})\} = \sum_u P(y_u)H(X_{lk} \mid y_u) \tag{17}$$

Note that observation-outcome probabilities in (17) can be found as:

$$P(y_u) = \sum_s P(y_u \mid x_s)P(x_s) \tag{18}$$

and that probabilities of type $P(x_s \mid y_u)$ in (18) can be found by applying Equation (11). Equation (16) provides a measure of expected information gain related to an observation of a single attribute in a particular cell. When making a trip to the destination, however, the subject is able to observe, with distance-varying sensitivity, all attributes in all cells that are observable from the route of the trip. Apart from making observations during the trip, the subject will also make observations during the activity in the destination cell. The total expected information gain of the trip and activity is found as the sum of information gains across all observations made during the event.

4. A computational model linking cognition, choice set formation, activity travel behavior and land use dynamics

Ultimately, many of the above ideas can be put together in a more integral modeling framework. Absolute represents an example of such a system (Arentze and Timmermans, 2004b). It (ultimately) intends to model the location decisions of planners, firms, households and individuals related to various types of facilities: housing, productive facilities and consumptive facilities. On the demand side, individuals schedule their activities and execute their activity schedules on a daily basis. The schedule determines which activities are conducted where, when, for how long and the transport mode used, on a continuous time scale. On the supply side, firms and/or the planning authority decide on where to develop how much of which (consumptive) facilities to support those activities. The present implementation of the system focuses on the dynamics of consumptive facilities and assumes the allocation of housing, industrial and office land used in the study area as fixed and given.

4.1. The system design

To model and micro-simulate the dynamics, a multi-agent system approach is used. In this approach, all demand-side actors (individuals) are represented individually. On the supply side, the system includes an agent for each facility type. It is the objective of a supply agent to develop the largest possible network of facilities under constraints of sufficient viability of each outlet separately, sufficient profits for the chain as a whole and a minimum level of cannibalism between outlets of the same chain. As such, a supply-side agent does not represent a single actor, such as a developer, firm or planning authority. Rather, it represents a group of actors that in interaction determine facility location decisions. The system is driven by the following key principles and notations.

Let there be given a study area, represented as a regular grid of cells that are considered the units of location, a zoning plan for the area determining the allowed facility types in each cell and a population with known home and work locations. The system uses the following classifications and allocation variables:

U is the classification of land-use types used for the zoning plan;

A is the choice set of activity types for individuals;

G is the set of demand types requiring a (consumptive) facility of some sort;

H is the set of (consumptive) facility types;

B is a smaller set of more aggregate demand types, in the following referred to as demand *sectors*.

P_g^a is the probability that activity a involves a demand of type g;

P_e^g is the probability that individuals use heuristic e to select a location for purpose g;

U_b is the subset of land-uses allowing facilities of type $b \in B$ ($U_b \subseteq U$);

G_h is the subset of demand types covered by facility $h \in H$ ($G_h \subseteq G$);

H_b is the subset of facilities belonging to sector b ($H_b \subseteq H$).

As implied by these definitions, we assume that there does not necessarily exist a one-to-one correspondence between scheduled activities, on the one hand, and facility types, on the other. Rather the parameter P_g^a describes for a given activity, a, the probability that the activity involves a demand of type g. Thus, an activity category, a, is seen as a heterogeneous set of more elementary activities that may entail different demands. It is also possible that a specific instance of the activity category does not involve any (consumptive) demand. Examples of the latter are a social activity conducted at the own or someone else's home, work activities and picking up/dropping off persons or goods. Even if such activities would involve a visit to a (consumptive) facility location, they still do not involve a demand in the sense of occupying floor space of the facility and therefore do not classify as a consumptive activity. It means that the sum of probabilities P_g^a across g does not necessarily sum up to one for all activity categories and may even be zero for certain activity categories.

The facility type defines the supply set, G_h, of a facility and, vice versa, the supply set defines a facility type uniquely. Single-purpose and multi-purpose facilities are distinguished. Single-purpose facilities support a single demand, whereas multipurpose facilities cover multiple demands. A facility is said to be of a higher order than another facility if it's supply set includes the supply set of that other facility and has elements not covered by the other facility. A typical example of a higher-order facility is a district shopping center that includes the supply of a neighborhood shopping center and in addition offers supply not included in this facility. Another example is a health care center including elementary facilities such as a physicist, pharmacy, physiotherapist and possibly other medical services that can also be provided by individual facilities. The optional facilities for a specific demand are defined based on sets G_h. Because supply sets G_h may overlap, multiple facility types may be optional for fulfilling a given demand.

To describe the study area at any moment in time, the system uses the following additional set of variables:

U_l is the land use of cell l (defined by the zoning plan)

S_{lh} is the size (square meter floor space) of facility h in cell l;

V_{lg} is the average number of visits per day to cell l for satisfying demand g;

N_{li} is the number of individuals visiting location l for purpose i (residential, working, visitors of certain facility types) per day;

Where U_l is given by the zoning plan, S_{lh} represent the outcomes of suppliers' decisions and V_{lg}, N_{li} represent the aggregate result of individuals' decisions. As the definitions indicate, the present system uses the period of a day as the unit of time. The continuous time scale on which schedules are defined would allow a further (unlimited) disaggregation of time so that an extension in that direction is possible.

4.2. Demand-side agents

Each individual of the population in the study area is represented as an agent in the model. On each simulated day, each agent schedules its activities and executes the schedule. A tailor-made version of Albatross is incorporated as a method in each individual agent to accomplish this. For a given day, the model determines which activities are conducted where, for how long, when, and, if travel is involved, the transport mode used. Hereby, it takes temporal constraints, some socio-economic variables, day of the week and residential location into account. Furthermore, the schedules of different agents may differ because all scheduling decisions in Albatross are stochastic variables. In-home activities are not further differentiated. Thus, an activity generated can be described as:

$$e = (a, t^s, v, t^t, m) \tag{19}$$

where $a \in A$ is the activity type, t^s the start time, v the duration, t^t the travel time and m the transport mode of episode e. The activities conducted by the same person must meet the temporal constraints:

$$\sum\nolimits_{i \in E_{dp}} (v_e + t_e^t) = (24)(60) \quad \forall dp \tag{20}$$

$$t_e^s + v_e + t_e^t = t_{e+1}^s \quad \forall edp \tag{21}$$

where time is expressed in minutes, E_{dp} is the schedule of a person p on day d, defined as an ordered set of activity episodes, and e+1 is the activity succeeding e in E_{dp}. We emphasize that constraint (20) is a logical constraint: since the activity-based model predicts the activities for a day, they should fit into the time frame of a day.

The agents make location choices for out-of-home activities in the sequence in which they occur in the schedule. The schedule defines for each activity the transport mode used for the trip to the activity location and the travel time. Predicted travel times refer to the duration of the trip, but are interpreted here as the time the individual is *willing* to travel. The origin location is given by the location of the previous activity in the sequence. As a consequence of trip chaining, the origin location of an activity does not need to be the home location. As it appears, in a substantial portion of trips the origin location is not the home location.

The combination of origin location, maximum travel time and transport mode determines the locations that are within reach. Before a choice set can be delineated for a given activity a, the demand type is determined by drawing one from probability distribution P_g^a. Then, the choice set is defined as all cells within reach containing supply g. For making a choice, the agent then determines a location selection heuristic. At present, only two simple heuristics are considered as alternatives, namely choosing a cell at random and choosing the nearest cell. We emphasize that in combination with the choice-set delineation and demand-selection rule, the heuristics give rise to more complex behavior than one would expect at first sight. For example, the model could select the nearest highest order location within a maximum travel time as a result of a certain combination of rules.

Selecting a location in this way may fail, however, namely when a facility of the given order is not within reach. If selection fails then the agent tries several ways to overcome the impasse. First, it lowers the demand by accepting also facilities that offer the lower-order service. If this fails, it then relaxes the travel time constraint and searches in a wider area for facilities. In this way, the agent will always find a location for the activity, unless a facility of the demanded type or a lower-order of that type does not exist.

4.3. Supply-side agents

For each demand sector $(b \in B)$ the system implements an agent, which is concerned with developing and maintaining a network of facilities of types H_b. In turn, each of these agents incorporates one or more subagents specialized in a specific facility type involved in that sector $(h \in H_b)$. An agent at the sector level co-ordinates, where needed, the actions of it's subagents, but leaves all tasks involved in developing and maintaining the network to the specialized subagents.

The methods available to each subagent address the problems of assessing the value of a given cell for opening a new outlet and making location decisions. If the (sub)agents would have unlimited knowledge of the activity and location choice of individuals, they would be able to predict exactly the amount of demand a new facility in a specific cell would attract and how much demand would be distracted from existing competing facilities. However, in the model as well as reality, suppliers have only limited knowledge about the behavior of individuals and, therefore, have to rely on estimation methods. In the model it is assumed that, regardless of facility type, all agents perform a catchment area analysis. In this method, a primary and secondary catchment area is defined by drawing concentric circles around the site. Next, the demand attracted from each cell within the circles is estimated taking into account the location of existing competing facilities. Statistics generally available for urban planning are used to set the parameters of the method for each facility type.

In the model of each agent, the catchment-area-analysis parameters include the following:

χ_g^{+} is the maximum cannibalism tolerated for supply type g;

S_g^{-} is the minimum normative floor space size (square meter) for supply g to be viable;

σ_g is the normative ratio between floor space size (square meter) and demand for supply g;

$r_g^{1,i}$, $r_g^{2,i}$ is the radius of the primary and secondary catchment area of supply g related to segment i;

$\pi_g^{1,i}$, $\pi_g^{1,i}$ is the penetration rate of supply g in the primary and secondary catchment area related to segment i;

V_{gj}^{r} is the extra g-demand attracted to the facility if the nearest main road lies within the j-th distance band from the outlet;

V_{gj}^{c} is the extra g-demand attracted to the facility if the city center lies within the j-th distance band from the outlet.

As the parameter definitions indicate, a catchment area (CA) analysis is conducted at the level of elements of the supply set of a facility rather than at the facility level (although the method does take into account efficiency benefits from bundling supply types in singe facilities).

Given these parameters, a CA analysis is conducted for a location (i.e., cell) considered for opening a new facility and results in an estimate of the number of visitors that would be attracted. The penetration rate in the primary CA is larger than in the secondary CA and zero outside the CA, reflecting an assumed decay of demand with increasing distance. The penetration rate defines the proportion of the population in the cell that will be attracted to the location. Hereby, competition is taking into account if the cell is located in the overlapping area between the primary CA and the primary CA of an existing facility offering the same supply. Parameters determine the amount of loss of demand depending on whether the distance to the competing outlet is shorter or longer.

Visitors may originate from different locations, namely from home locations, work locations and other activity locations $(i \in B)$. For each of these origin activities a separate CA analysis is conducted, taking into account that the radius of the CA, the available demand within cells and the penetration rate may differ depending on the origin activity. For example, the radius may be larger and the penetration rate higher for a residential population compared to a work population.

With respect to other activity locations, the settings of radius and penetration reflect the assumed extent to which one facility benefits from the presence of other facilities. For example, a zero penetration regarding a specific other demand sector would mean that no such benefits exist. The size of the population from each origin is estimated using parameters P_i^g, P_g and estimates $N_{l'i}$ (P_i^g is the probability that a g-trip originates from activity location i, P_g is the probability of a g-demand by an individual on a day, $N_{l'i}$ is the number of people present in cell l' for purpose i on a day).

The CA conducted in this way does not only result in an estimate of the number of visitors, but also yields an estimate of the number of visitors a new outlet distracts from an existing competing outlet. The amount of cannibalism is measured as the reduction of penetration rate due to competition in cells in overlapping primary CAs.

The estimates obtained through the CA analysis are estimates indeed. The activity scheduling decisions and in particular the location choices of demand side agents determine how demand is actually distributed across facilities in the study area. We assume that the supply side agents do not have direct access to the rules that drive the choice behavior of the individuals. Rather, the parameter settings used in the CA analysis are best guesses based on long time experience and recordings of actual penetration rates. With respect to available demand within cells we assume that the knowledge of supply agents is more accurate. Specifically, we assume that knowledge regarding P_g terms is accurate, i.e. consistent with actual activity choice probabilities in the population, whereas knowledge regarding P_i^g is tentative, as data of the latter variables is more difficult to obtain. Finally, regarding the N variables we assume that the agents have accurate data about the residential population, the population at work locations and the size of existing facilities in all demand sectors. So, where they can observe N for housing and employment, they derive estimates of facility visitors volumes based on actual facility sizes.

The general objective of each agent is to develop as many facilities as possible within feasibility constraints. The constraints include the following. First, A facility is considered viable only if the normative size is larger than or equal to a minimum size for each supply element. Given an estimate of the number of visitors (per day), the normative size follows from the normative floor-space-to-demand ratio for each element of the supply set. For some types of facilities the minimum size constraint is defined at the facility level rather than the supply-element level. In such cases, a facility may still be feasible even if one or more of the constituent services would not be feasible if it were provided by an elementary facility. Whether such economies of scales are relevant depends on the facility type. Second, the cannibalism incurred by the facility must not exceed a pre-defined maximum.

The sector-level agents (B) are responsible for making location decisions. These agents continuously monitor the study area for opportunities to open new outlets. Hereby, they consider facility types (in their sector) in a certain order of priority, which is a parameter in the model. Giving priority to higher order facilities, generally, would be in line with the objective to develop the largest possible facility network in a sector. Having selected the facility type, an agent considers a cell for possible development only if the land use in the cell allows the type of development and there are no existing facilities competing for the same demand in the cell. Given these cells, the agent selects the cell maximizing V_{lg} within constraints and opens a facility of optimal size at the site. This process is repeated until none of the cells turn out to be feasible. Then, the next facility type in the priority order is considered and the same process is repeated for this facility, and so on.

In addition to a priority order on subagent level, the system also uses a priority order on the higher demand-sector level. In a monitoring stage, the visitor flows are known and each (sector level) agent reconsiders the size and evaluates the viability of existing facilities within their sector. A closure is indicated when the optimal size has dropped below the minimum. The agents do not close more than one outlet at a time, to avoid the risk of closing more outlets than is necessary. Closing an outlet generally improves the market conditions for other outlets competing for the same demand. Therefore, a facility that is not viable in the current time step may become viable in the next time step if a competing facility is closed. Giving higher priority to maintaining larger outlets, an agent ranks currently unviable outlets first on their order and next on their performance, and, then, closes the outlet with the lowest rank. The agents of the different sectors implement such adaptations simultaneously assuming that possible cross effects between sectors can be ignored.

5. Conclusions and discussion

This paper has given a brief overview of some of the work within the DDSS research program on multi-agent systems. The main purpose was to provide a quick glance of the types of models and systems developed or currently under development. The focus has been on underlying concepts, not on empirical results.

The future research agenda of the DDSS program foresees the empirical estimation of the models to the extent this had not been done already, their further elaboration and extension, and their integration in more comprehensive multi-agent systems. Results obtained thus far, however, suggest that multi-agent systems constitute a relatively new and exciting way of modeling complex behavioral processes.

References

Arentze, T.A., H.J.P. Timmermans (2000). Albatross: A learning-Based Transportation Oriented Simulation System. European Institute of Retailing and Services Studies. Eindhoven, The Netherlands.

Arentze, T.A., H.J.P. Timmermans (2004a). A learning-based transportation oriented simulation system, *Transportation Research B*, **38**, 613 - 633.

Arentze, T.A. and Timmermans, H.J.P. (2004b). A micro-simulator of urban land use dynamics integrating a cellular automata model of land development and control and an activity-based model of transport demand. In: *Proceedings 83rd Annual Meeting of the Transportation Research Board*, January 11-15, Washington, D.C. (CD-Rom: 16 pp).

Arentze, T.A. and H.J.P. Timmermans (2005a) Representing mental maps and cognitive learning in micro-simulation models of activity-travel choice dynamics, *Transportation*, forthcoming.

Arentze, T.A. and Timmermans, H.J.P. (2005b), Information gain, novelty seeking and travel: A model of dynamic activity-travel behavior under conditions of uncertainty, *Transportation Research A,* forthcoming

Arentze, T.A. Borgers, A.W.J., Hofman, F., Fujii, S., Joh, C., Kikuchi, A., Kitamura, R., Timmermans, H.J.P. and Waerden, P. (2001). Rule-based versus utility-maximizing models of activity-travel patterns: A comparison of empirical performance, In: D. Hensher (ed.) *Travel Behaviour Research: The Leading Edge*, pp.569-584. Pergamon, Amsterdam.

Arentze, T.A., Hofman, F., van Mourik, H. and Timmermans, H.J.P. (2002). The Spatial Transferability of the Albatross Model System: Empirical Evidence from Two Case Studies, *Proceedings of the Transportation Research Board Conference*, Washington, D.C., January 13-17, 2002.

Arentze, T.A., Hofman, F. and Timmermans, H.J.P. (2003). Re-induction of decision rules using pooled activity-travel data and an extended set of land use and cost-related condition states, *Transportation Research Record*, 1831, pp 230-239.

Timmermans, H.J.P., Arentze, T.A. and Joh, C.-H. (2002). Analyzing space-time behavior: new approaches to old problems, *Progress in Human Geography*, **26**, 175-190.

Cognition and Decision in Multi-agent Modeling of Spatial Entities at Different Geographical Scales

Lena Sanders

Abstract. The modeling of the dynamics of settlement systems can be developed at different geographical scales according to the theoretical framework which is chosen: the micro-level of the households and entrepreneurs, the meso-level of cities and regions, the macro-level of hierarchical and spatial structures. The underlying hypotheses and the links between these three levels are discussed in the case of a multi-agent system (MAS) approach. The question of which are the driving forces of change in a settlement system is raised. Then different ways for building hybrid models combining dynamics referring to different scales are discussed. I refer to the example of SimPop, a MAS model which simulates the emergence and the evolution of a settlement system on a period of 2000 years, in order to illustrate how a function of urban governance that ensures both cognitive and decisional capacities for the evolution of cities can be introduced in a model whose rules are principally built on meso-level regularities.

1. Introduction

The description and understanding of the evolution of a settlement system involves several levels of observation. At a more *global scale* the point of interest is the hierarchical and spatial organization of the settlement system: the degree of centralization, the spatial regularity of the urban pattern, the extension of urban sprawl, the emergence or reinforcement of a polycentric structure for example. The durability or reconfiguration of such macro-level structures have to be questioned. At an *intermediate scale* the questions concern the geographical entities themselves: some sets of places appear to be more attractive at certain periods of time and register then a faster growth than the others, which can lead to a quite long-term trajectory of development for some of them, when others stagnate or even enter an irremediable process of decline. The logic underlying such differentiations has to be identified. At last, there are the *micro-level* entities, households, entrepreneurs, who choose to stay or to leave their place of living, who choose that or that place for a re-installation and the matter is then to show the determining factors of these processes of choice.

The three levels are linked by top down as well as bottom up mechanisms. Indeed a durable convergence in individual preferences (it means a common attraction for some places or kind of environments) will lead to a concentration in the more attractive places and consequently, to the decline of some others. But obviously, if the changes occurring at the different levels are linked, the associated temporalities are completely different. For the individuals change occurs at a precise time t, corresponding to the event of moving. For a geographical entity, a decade is often necessary to identify a real trend of growth or decline emerging from the classical inter-annual fluctuations. Then, these growths and declines at a meso-geographical level may lead in turn (but it is not always the case) to a change in the urban structure at an upper level. A few decades are often necessary to identify a change in the global structure, as the emergence of a polycentric structure in a previously centralized metropolitan region for example. This new structure is the result of convergences in individual decisions (secondary poles more often chosen by new immigrants as places of living than the center, moving from the center to these poles more important than the reverse etc.) and in turn, the characteristics of the global structure, with its new opportunities and inducted effects, will influence individual decisions.

Most often the researcher modeling the dynamics of settlement systems privileges one or the other of these levels according to the theoretical framework which is chosen. If the source of inspiration is the micro-economic approach, the elementary units of the model will be the individuals. In geography, on the other hand, geographical entities as cities, municipalities, labour-markets, are often considered as the elementary objects in a spatial system whose functioning if defined by the interactions (flows of individuals, information..) between places. In this chapter, I discuss the different choices that the modeler is facing in matter of modeling level in the field of agent modeling. The different examples found in the literature lead then to plead for a modeling that integrates hypotheses referring to the different levels at which the driving forces of the dynamics of a settlement system are acting. I will then discuss different ways hybrid models, combining dynamics referring to different scales, can be developed.

2. Agents for modeling the dynamics of a settlement system

The use of multi-agent systems (MAS) appears very "natural" when the aim is to model the actions of individuals in different contexts, and most applications in social science are developed at that level. But MAS is as rich a tool when the agents are used to treat collection of individuals, being assumed that these collectives form entities which make sense. The ontology of the entities has to be reflected upon, through their identification and also the explicitation of the relations between the different levels which are involved. Spatial aggregates as hamlets, towns, cities, which are composed of individuals sharing a same space and contributing collectively to its identity and functioning, are such entities and they can then be usefully formalized with agents. In fact the choice of the adequate geographical level to develop a model and the choice of the best tool have to be better distinguished than they classically are. The question of the level of modeling is conceptual and independent of the chosen tool. In order to understand and to predict the evolution of a settlement system, the question is then to determine if it is more useful to develop a simulation model at the level of the individuals living and acting in the system, or that of the spatial aggregates which organize their activities and interactions. MAS offer new potentials to develop models at both levels, and they are also specially adapted for combining effects of different scales. The kind of hypotheses which are involved in models focusing on the dynamics of settlement systems are examined in next section, first at the level of individuals, then at that of geographical entities, and at last respective advantages are discussed.

2.1. Agents for simulating individuals' decisions in matter of spatial choice

In individual-based models developed in social science the elementary entities are often the individual actors, but depending on the application it could be households, entrepreneurs, urban planners etc. If the point is to simulate the evolution of a settlement system (an urban system, a regional system for example) individuals or households are considered as the elementary units. Decisive events are then arrivals of new inhabitants in the system (birth, immigration), disappearing of others (death, out migration), and particular attention is paid to the process of residential move. Each individual makes choices in matter of place of living, depending on its needs, its preferences and on some constraints, following the principles of the time-geography approach of Hägerstrand. The different kinds of events, changing activity, changing family structure, changing preference, for example, which are at the origin of a move have then to be identified and integrated in the model.

Microsimulation is one family of models which focus on the individuals' decisions and acts. At each time-step of such simulation, each individual is reviewed and a set of rules and probabilities are used to determine if he decides to stay or to leave, and if he leaves, what he chooses as new place of residence... The resultant of these basic choices determines the change which will be observed at the level of the different geographical areas, which in turn determines the eventual change in the form of the settlement system. The two main issues are then to identify and formalize the cause of the decision to move and to focus on the logic of choosing a certain area as destination rather than another. In spatial microsimulation demographic events are often used as the trigger of a mobility : the probability of moving is dependent on the stage in the lifecycle of the individual, and in particular, marriage, divorce, decohabitation from the parents, birth of a child, produce frequently a change of residence. The labour market can also be seen as a driving force in residential change (first job or change of working place). Models of that type are often composed of several linked modules, household formation, housing location, employment location, migration etc. (Holm et al. 2004; Waddell 2002; Moeckel et al. 2002). The objective of these models is to be able to produce and handle information as fine and exhaustive as possible in order to be able to predict the distribution of the population according to different categories, whatever they are, and to build trajectories of different kinds: the evolution of the population at different geographical levels, the evolution of the importance of commuters and of the average distance that they cover, the evolution of different kind of workers of different age groups etc. The ambition is then to simulate an artificial world compound of individuals whose lives are plausible in terms of succession of events, and whose aggregation give usable predictions of the principal trends of change.

Other simulation models using the individuals as elementary units are based on a completely different philosophy. They choose a stylized fact that they try to reproduce using simple mechanisms and a limited number of agents. Different examples using MAS can be quoted, each of them modeling a structural change in the settlement pattern as the consequence of individuals' response to a change in the environment. Trying to understand the reasons why the settlement pattern changed, during some periods between 900 and 1300 in Colorado, from compact villages to dispersed hamlets, Kohler et al. (2000) developed a multi-agent system where each household was represented by an agent, and whose decision to relocate was a function of the characteristic of the environment (access to water resources, changing farming productivity). In a paper focusing on the persistence of the structure of city systems at a macro-level, Batty (2001) uses MAS in order to show how agents' moves between residence and activity (resources or urban activity depending on the model) produce persistent structures at a higher level. The idea was to show the effect of a weak but permanent positive feedback over a long period of time, formalized through a process of learning (the aim of the agents being to find the resources' location) on the final kind of distribution of the population.

In a model simulating the process of peripherisation in cities of Latin America, Barros (2003) uses agents to represent individuals whose locational preferences are all the same, that is to be close to infrastructure, services and job opportunities offered by the central position. Nevertheless the constraints are differentiated through the level of income of the individuals. In this case the driving force for moving is also the propensity to adapt to a change in the environment (proximity of central functions), submitted to some constraints (income). In other applications the decision making of the agents is not only reactive and involves a more cognitive dimension. One of the most classical examples in the framework of settlement systems is the propensity for an agent to avoid or to search for the presence of agents with a specific attribute. Such behaviours were used to simulate the emergence of residential segregation (Shelling, 1971) and are often integrated in location models (Portugali et al. 1997; Torrens, 2001). In a model simulating the location preferences of actors in urban areas, Arentze and Timmermans (2003) introduce the agglomeration principle as a driving force by making the agents' location decisions dependent on the presence of other actors in the concerned site. Aiming to study the formation of land use patterns from micro-level locational rules defined at the level of firms and households, Otter et al. (2001) combine agents with varying behaviours according to their preferences for proximity to working place, to other households, to services, and to an environment of quality. So the examples of such models focusing on spatial patterns emerging from interactions between agents who have a perception of their environment are numerous and convincing. But most of these applications concern artificial worlds limited in terms of dimensions and future developments should deep the question of sensitivity to the size, that is the number of involved individuals in the treated phenomena, most often much more important than that could be simulated with MAS.

2.2. Agents for simulating the dynamics of spatial entities

The other way for modeling the dynamics of a settlement system is to develop the model directly at the *aggregate level of spatial units* which can be considered as having an identity and a certain degree of durability, and which can then be considered as the elementary entities of the model (Pumain, 2000). The hypothesis is that the *interactions* between the spatial entities themselves constitute the driving force in the evolution of the settlement system and that they determine the properties of the system at a macroscopic level. These interactions refer to migratory flows, commercial exchanges, information flows etc, and they operate according to meso-level dynamics which are little sensitive to the diversity of individual decisions. In this approach, the varying individual behaviors, with their convergences and their compensation effects, are seen as producing emergent properties, with regularities, which are not the consequence of any intentionality. So the social identity of a city is not the result of intentions of individuals, it emerges from the long-term accumulation of interactions between different kinds of individuals. And when a certain socio-economic profile has emerged, it tends to maintain itself despite micro-level changes because of the difference of time-scales which is involved, and it acts as a constraint on individual choices. Empirical facts show for example that the social profile of the immigrants to a city is very similar to that of the out migrants and to the cities' profile itself. Most models developed at that geographical level have used simulation tools based on differential equations (Allen 1997; Weidlich and Haag, 1988, for example) or knowledge-based simulation (Page et al., 2001).

The SIMPOP model has been developed at the meso-geographical level, using a MAS approach. A settlement system located in a spatial grid of 20x20 (400 spatial entities) is considered. Each entity is represented by an agent who handles information about local resources, population and activity, who is able to communicate with the other agents, and who brings into play the rules of growth of the spatial entity. Starting from an initial state (time 0, around Roman times) characterized by a quite homogeneous[1] distribution of the population between the inhabited spatial entities, and associated to an economical system based on the agrarian activity only, the model simulates the emergence of a hierarchical system of cities, diversified in their functions (Bura et al., 1996; Sanders et al., 1997) . The aim was to show what are the *basic and minimal rules* which are necessary to simulate the emergence of the fundamental macroscopic properties in terms of hierarchical organization and spatial configuration which are so regularly observed for system of cities over time and space (Bretagnolle et al., 2000; Batty, 2001).

At each time step (every 10 years) the quantitative change (growth of population, sectorial distribution of the active population) and the qualitative one (level

[1] Most models of system of cities start from a known initial situation which is always far-from homogeneity. All studies show how fundamental the initial conditions are in that domain: indeed the hierarchical and spatial organization of a system of cities is very persistent through time. The main focus of SIMPOP being the emergence of an hierarchical structure, it appeared more convincing to start from an initial situation which was not at first hierarchical.

of function) are computed for each entity in relation to the consumption, the production, the process of commercial exchanges. Commercial exchanges between the cities are formalized using the protocol of communication of the MAS: each agent representing a city sends messages to all entities which correspond to its range[2], indicating the kind and the quantity of products which are available for sale. A same entity may then receive several messages of supply and it sends messages back indicating its demand. The cost of the product is function of the distance between the two places. The demands are treated in turn, and new messages are sent indicating the amount still available and so on. This cycle of messages goes on until there exists no more supply which corresponds to a demand and vice-versa. This process doesn't necessarily (and most often not) ends in an equilibrium.

The run of several simulations showed that the model was able to simulate the emergence of a set of towns, some of them progressively becoming cities of higher and higher level. The initially regular pattern transformed in a hierarchical system whose rhythm of evolution as well as final situation in year 2000 are consistent with what can have been observed in many European countries. An example of simulation is given in Fig. 1, which shows a progressive evolution towards a polycentric structure through the mapping of the simulated settlement at six period of times. Other parameter configurations, with for example a greater standard deviation in the growth rate, can lead to more centralized configurations (Fig. 2). The rank-size distribution gives a good representation of the form of the hierarchical organization of the settlement system. For example, the curves on Fig. 3, representing this distribution for ten periods of a simulation between time 0 and year 2000, illustrate the progressive emergence of a hierarchical structure. The slope of the rank-size distribution[3] is a good indicator of the degree of hierarchical organization of the settlement system and the plot of this slope against time shows the characteristics of the evolution of the hierarchical organization of the system. Depending on the set of parameters, the simulated settlement system will register different types of evolution. The settlements could all grow at the same rhythm and their sizes remain then very slightly differentiated (Fig. 4a); at a certain period of time some settlements may initiate a quicker growth than the others and the system enters then a process of hierarchization (Fig. 4c); a period of unequal growth can be followed by a stagnation of the process of hierarchization which stabilizes then at a higher level then the initial one (Fig. 4b). Such maps and plots are useful in order to appraise the nature of the patterns which emerge and the temporality of the process of change, depending on the rules and the parameters which are used

---

[2] In the model, the range depends strictly on the level of the city, and evolves through time ; that means that all cities of functional level 2 for example have a same range.

[3] The function representing the number of inhabitants of a settlement relatively to its size in a bi-logarithmic scale (figure 3) is approximated by a linear expression $ax + b$. The parameter a measures the slope of the relation: if all settlements are of same size, the slope will be of value 0; the higher the value of a, more hierarchical is the associated settlement system. Nowadays European countries register values around 1 (Moriconi-Ebrard, 1993).

for the simulation. Thus it has been possible to identify the fundamental mechanisms which lead to a hierarchical organization.

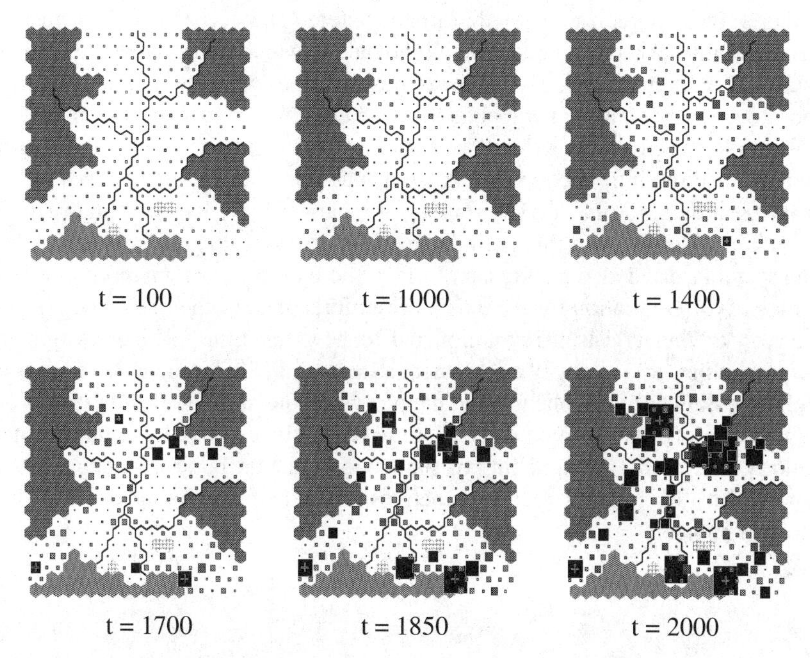

Fig. 1. Maps of the simulated settlement system with the SIMPOP model: emergence of a polycentric urban system

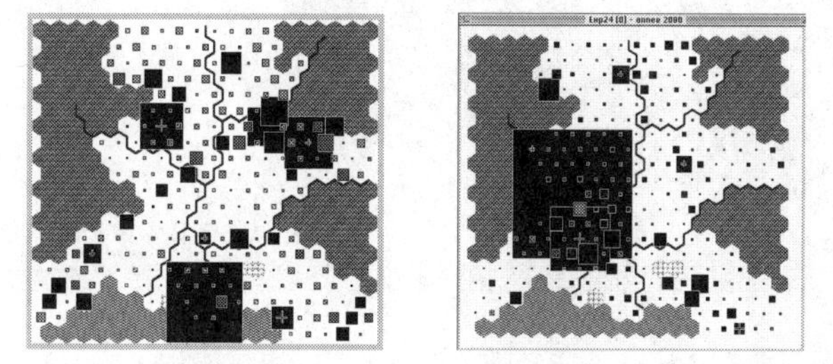

Fig. 2. Two examples of final distribution of the settlement system (at time 2000) for different configurations of parameters

Two main mechanisms showed to be essential, that of competition between cities, and the ability to adopt new innovations. Indeed, if the process of interaction between places is suppressed, towns could effectively emerge each time the ability to create a surplus exists (resource constraint), but the evolution then simply leads to a more and more dense pattern of loosely differentiated towns. At the opposite, when the process of exchange is introduced, a selective process starts through the competition between towns. The best placed get an advantage, grow faster, make the acquisition of new functions which give them the possibility to exchange at larger range and progressively to accentuate their advance. These interactions between places create a positive feedback loop which is at the origin of the spatial and hierarchical organization of the urban system. The second mechanism concerns the ability of a city to adopt new innovations which is formalized by the acquisition of new functions. The constant possibility to get new functions appears to be a necessary condition for obtaining a hierarchization of the urban system. For example, when the model was run without taking into account the industrial revolution and the introduction of industrial functions, the system of cities tended to stabilize around the 16th or 17th century. It means that the hierarchical organization remained stable from this period (type Fig. 4b), and that the total population of the system started also to stagnate. In fact, there didn't exist anymore new possibilities of differentiation for the cities, of reinforcement of their advance. The introduction of the industrial revolution led to new functions, which favored the concerned cities, to a stronger hierarchical structure, and to an increase of the total population. Same kind of phenomena happen when the number of functions is determined a priori, that means, when there exists a higher limit for acquiring new functions. Therefore, the modeling of such system should build in endogeneously the possibility of creation of new functions instead of delimiting a priori their maximum number.

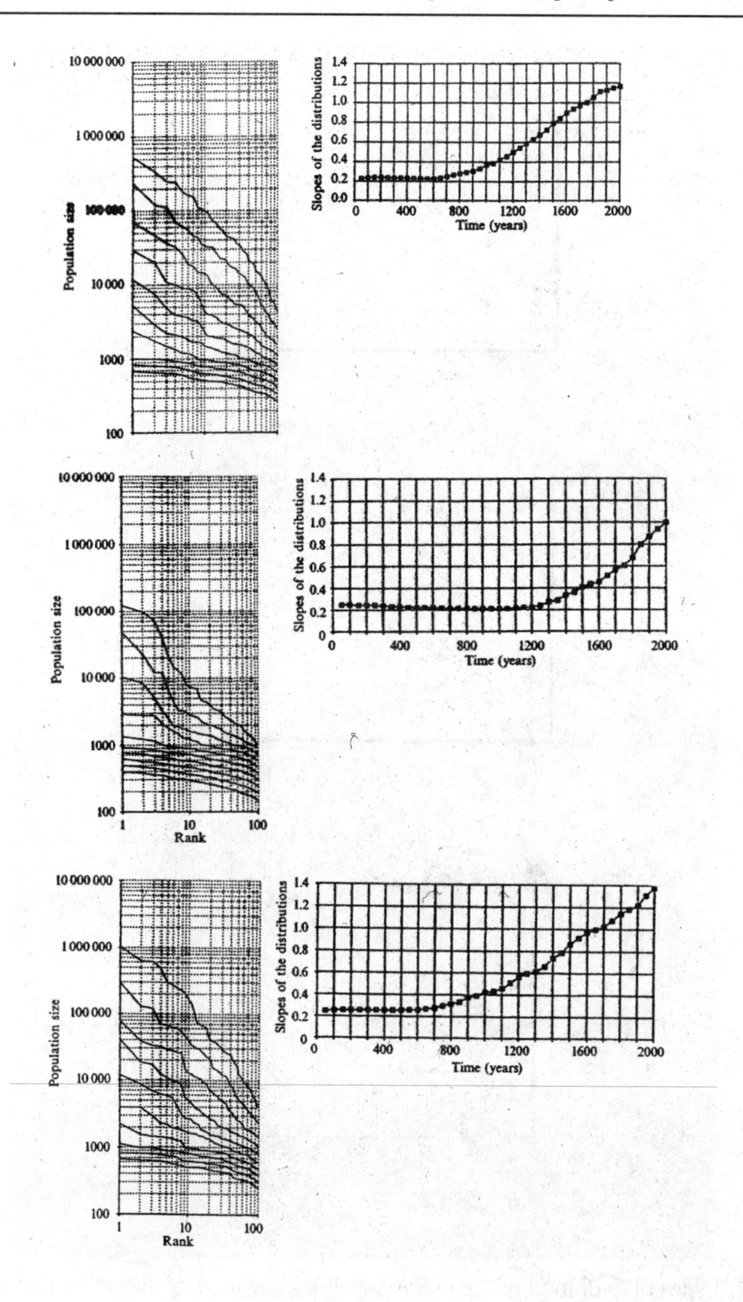

Fig. 3. Rank-size distribution of the simulated settlement system with the SIMPOP model

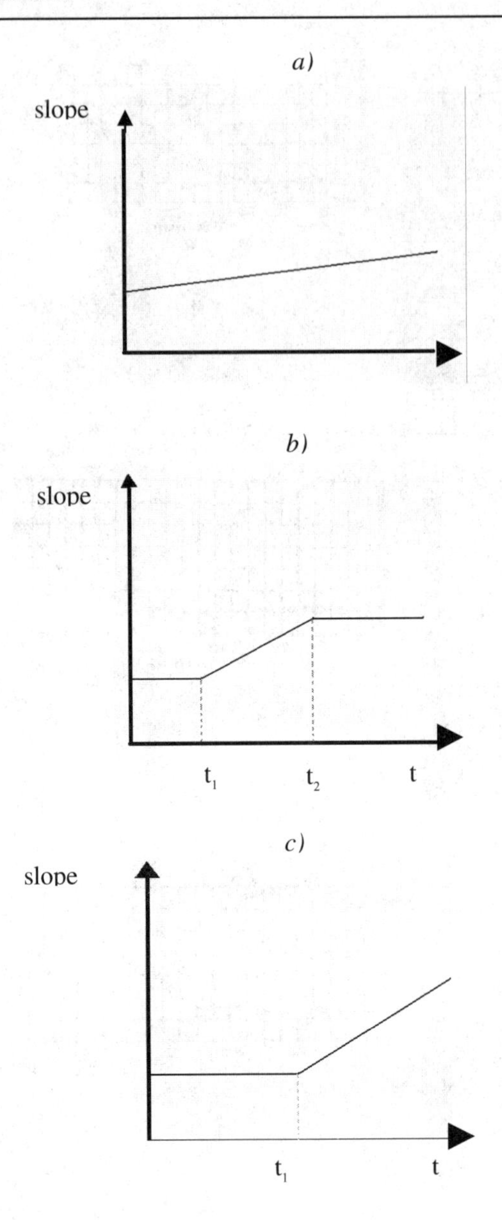

Fig. 4. Types of evolution of the hierarchical organization of the settlement system Representation of the slope of the rank-size distribution againt time

2.3. Where are the driving forces of change in a settlement system?

The dynamics of a settlement system can then be modeled either at the level of individuals, or at that of aggregate geographical entities. Different aspects intervene for the comparison of the advantage of each kind of approach. The first is of conceptual order and concerns the hypotheses which are privileged by the modeler. If one considers that the decisions which are taken at the level of the actors are determining for understanding the changes at the level of the geographical areas, the model would have to integrate the mechanisms of decision of these actors. At the opposite, if one considers that geographical entities have their own logics of evolution, independently of the individual actors' decisions, that it has intrinsic properties that can't be deducted from that of the individuals, then the meso-level model is more appropriate.

A second aspect concerns the reality of the phenomena to simulate. If the objective reasons of move are in a small number and well identified, microsimulation is a good tool. For example, if job location is considered as the main determining factor of an individual's place of residence, microsimulation can model in a coherent way the redistribution of the population in a region through time. Let's imagine that the individual Dupont gets a job at location (x,y), a housing is searched with constraints concerning proximity and housing vacancy in the category corresponding to the profile of the family (in terms of income and family composition). At each step of the simulation the events affecting each individual are modeled with a combination of logical rules and a stochastic dimension (Monte-Carlo method) and the consequences in terms of jobs and housing are stored (a moving implies a housing vacancy for example) and taken into account for defining the potentials offered to the other individuals at following steps. It implies to store and handle individuals, jobs and housing and the relationships between them. The functioning of the model lays on a kind of matching jobs/individuals, and individuals/housing, respecting the coherence of family composition and adequacy between individuals' profile, kinds of job, types of housing (Fransson, 2000).

At an inter-regional level things get more complicated. Vacancy of adequate jobs and housing are not the only determinants to be considered to determine the destination of an individual who has chosen to move. Then the *relative attractiveness* of the different regions has to be taken into account, that is a meso-level logic. Indeed empirical studies support the hypothesis that individuals move more frequently to places that they have practiced in one or other form during their life, and the real spirit of a time-geographical approach would be to integrate this differentiate relation to places in the choice process, but it is difficult conceptually and technically.

More generally, if the housing market is less planified, the logic of vacancy chains is less realistic and the global attractiveness of each place becomes more decisive to handle than the individuals' logic of optimization in front of a little number of alternatives, in order to simulate the settlement systems' dynamics (Sanders, 1999). As soon as the diversity of reasons for a move increases, the advantage of the individual level loose also its weight. This is a kind of a paradox as one of the hypothesis leading to the use of microsimulation is that it is the diversity at the individual level which can lead to structural changes at a higher level. As long as this diversity can be expressed by the combination (even complex) of known features (involving age, matrimonial status, income, education etc.), the individual approach makes it possible to better model change with consistency. But if this diversity goes beyond classical mechanisms of decision, the meso-level dynamics, focusing directly on the emerging structural effects that result from individuals' actions is better adapted.

Another question is to identify which actors are fundamental: the individuals who decide about their place of living, or actors whose decision has an effect on the relative attractiveness of a place? It could be economic actors (firms, entrepreneurs) who decide of the location of the head-office of a new firm, or the opening or closing of a branch, and whose decision depends on and influences the economic and social profile as well as the image of the area. The decisions of such actors may have an effect on the attractiveness and then the dynamics of the different areas. Another level of actors can also be considered, that of the political actors: mayor and other decidors of the municipality. They make decisions (planning, public services) which have an effect on the attractiveness of a commune for economic activities and/or residential purposes, on its accessibility, on its image at a national and international level. So the discussion of the relative weight of individuals' actions versus meso-geographical rules of evolution has to be enlarged. The question is then to determine if the understanding of the decision making of powerful actors is more essential to simulate correctly the dynamics of settlement systems than the identification of meso-level rules. Following example illustrates that it is sometime a matter of interpretation.

Let's consider Nîmes and Montpellier, two cities in Southern France. From the beginning of the century and until the 1960ies the total population of both cities increased slowly, but they both lost regularly places in the urban hierarchy and their relative weight in the urban system decreased (Fig. 5). From the 1960ies there is a complete change of tendency for Montpellier whose curve bifurcates clearly. At the same time the municipality of Montpellier is very dynamic, posting up Montpellier as an "European technopole". One could see there a relation of cause to effect and attribute the difference of growth of both cities to a more offensive politic of Montpellier. But if one considers the economic profile of both cities, it appears that most cities with a profile like Nîmes' (more industrially oriented) have been declining the last decades, and vice-versa for cities with similar profile as Montpellier (more service oriented). This observation leads to put forward the idea that it is the potential of a city relatively to the innovation cycle "en cours" which is decisive.

Our hypothesis is that there exist strong trends of evolution whose dynamics refer to the meso-geographical level of the cities considered as entities. These aggregate logics determine the basic outline of change, they give the bounds of the possible future, given the meso-level properties of the city (size, accessibility, socio-economic profile, cultural characteristics, touristic attractiveness...). Given these bounds, the decisions of "active" actors (political and economic actors) will influence the direction of the trajectory towards one bound or the other. One could say that the meso-level logic gives an “interval of plausibility”, and that the urban actors’ decisions and actions determine where in the “interval” the change will take place.

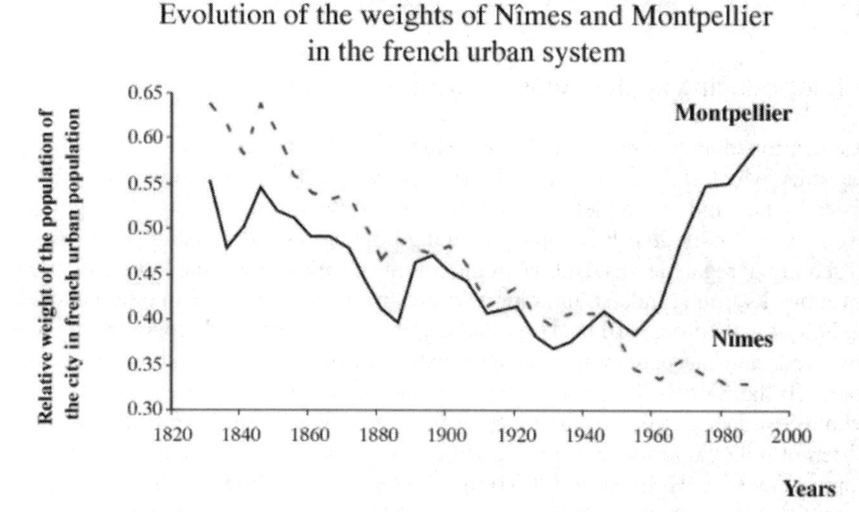

Fig. 5. Evolution of the relative weights of Nîmes and Montpellier in the French urban system

3. Different ways for building a hybrid model combining micro and meso- level approaches

Rather than a dichotomy between meso-level approaches, sometimes perceived as more classical, and micro-level approaches, seen as more innovative, recent methodological developments in the application of artificial intelligence to geography offer the possibility to combine dynamics referring to these different levels through different possibilities. One way consists in a more or less loose coupling of different tools, for example differential equations and cellular automata (Phipps and Langlois 1997) or multi-agent system and cellular automata (Torrens, 2001) for example. Another way is to associate agents representing different kinds of actors in the framework of a multi-agent system. Benenson and Hatna (2003) for example distinguish agents who represent householders or firms, who are located and can relocate, and developers, whose own locations are unimportant, but who take decisions which influence the cities’ infrastructure. In a similar spirit, the model used by Arentze and Timmermans associate agents which are interested in moving to a site (firms, households), with two specific agents who represent respectively the planning authority and a developer.

In the following sections we will focus on two ways of combining meso-level rules and actors' decisions in a same model. In the first meso-level rules are introduced in a microsimulation model where main focus is on the individuals' choice in matter of place of living. In the second, a cognitive dimension is introduced in a meso-level model through the role of an agent representing the urban actors.

3.1. Introducing spatial rules in microsimulation

As mentioned above microsimulation models are well adapted when the chains of causality which link individuals' characteristics and decisions are well identified. Indeed that kind of model give consistent results for simulating demographic events at an individual level and producing population and housing previsions at different aggregate levels. But when migrations have to be modeled, the individual level has its limits. Indeed, migrations result from a quite complex combination of an individual's process of decision, the features of the different geographical areas involved, and the perception and information of the individuals about these features. In the SVERIGE model, where the aim is to simulate the evolution of the whole Swedish population (about 9 millions inhabitants) and to test the effect of different policy options (concerning immigration for example) on this evolution in coming decades, Holm et al. (2004) use a three steps method in order to integrate a multi-level perspective in the simulation of the individuals' migrations. First the decision to move is determined according to life-cycle and socio-demographic variables. Then the choice of the labor market of destination is determined through a classical type of spatial interaction model. This model produces probabilities which are then used in the microsimulation to determine the migrant's choice. Logics of vacancy and social similarity are then used in order to choose the final 100 meters square within the region of destination where the migrating family will relocate. Through these three steps mechanisms of change referring to different geographical levels are combined. The modeling of the decision to move refers to the individual's personal life trajectory when that of the region of destination is related to meso-level regularities. The hypothesis is that larger and more attractive regions exerce a higher attraction on individuals and that there is better chance to migrate to a nearer region than a more distant one. The last step which concerns the intraregional location combine an economical dimension (existence of a vacant dwelling in right category of quality and price) and a cognitive dimension, with a preference for an environment sociologically similar to its own. This framework could even allow to integrate more closely the meso-level rules and the individual preferences by determining the relative attractiveness of places according to the type of the moving family. Presence of a school increases the attractiveness of a place for families with younger children, of medical infrastructure for elder households etc.

A second example is given by a model whose aim was to simulate the population dynamics at the commune level[4] in Southern France according to different scenarios. The logic of vacancy chain wasn't well adapted to the empirical situa-

[4] Commune: elementary administrative entity in France, of 15 km^2 in average

tion, and places of destination were chosen according to the specific attractiveness of the place and its accessibility relatively to the working places of the members of the family. The model then privileged the demand side, that of individual preferences, which is quite driving in that area. But in the first simulations the growth of many suburban communes got overestimated, due to the lack of constraint on the urban development. A factor of diffusion of urban growth was then integrated: when a commune's density goes beyond a certain threshold, the family searching for a housing is located in a contiguous commune to the one corresponding to its demand (Aschan et al., 2000). The hypothesis is that this location represents the second best choice for a household. The high proximity to neighboring commune (four to five kilometers in average) implies most often a similarity in the kind of environment and in the distance to the working place. The quick expansion of urban sprawl in that region is well simulated with this rule. In this model the spatial rules play a role of constraint on the individuals' decisions. That way it was possible to use the advantage offered by the microsimulation approach, with the demographic coherence of the long-term dynamics of the spatial entities, and to take into account the spatial dimension of the process of urban sprawl which emerges as a resultant of individuals' choices but does not correspond to any intentionality from their side. The model is then able to produce consistent spatial trajectories at different geographical levels.

3.2. Introducing cognition in spatial MAS

An inverse way of performing consists in adopting a meso-level model as the main entry, that is that the main rules of evolution are designed at the level of the elementary spatial units, and then in introducing decisions of individual actors at the level where they can be determinant on the systems' dynamics. This direction has been chosen in the development of a new version of the SIMPOP model (described in section 2.2) which will be applied to simulate the dynamics of the European system of cities[5]. The evolution of each city is modeled through a combination of general rules systematically applicable for all the cities of the system and specific rules which depend on the strategic choices of the local urban actors. These strategies are not defined by a single actor as a mayor, but by a more complex and collective entity: some actors who have an intentionality of transforming the image and relative place of the city in Europe (political actors) and others who influence the urban dynamics but whose actions are directed by other forces (economic actors for example).

[5] Two models are in development and the working group associates members of UMR Géographie-cités (A. Bretagnolle, C. Didelon, JM Favaro, H. Mathian, F. Paulus, D. Pumain, L; Sanders, C. Vacchiani) and from LIP6 of university Paris 6 (A. Drogoul, B. Glisse). A long-term model, focusing on the phenomena of emergence, is developed at the European level in the framework of the European program ISCOM (directed by D. Lane, university of Modena). A model centred on contemporary European urban dynamics (EUROSIM), with previsions until 2050, is developed in the framework of the European program TIGRESS (directed by N. Winder, university of Newcastle).

At each time step each town and city is considered and its evolution is determined, in terms of growth or decline of population and wealth, of maintain or change of its functions, and of the sectorial distribution of the active population. Four dimensions enter into account. First an exogenous constraint corresponding to a kind of time axis defines the technical and political context of each period (industrial revolution, speed of the means of communication, composition of the EU for example). The features of that context are taken into account in the model by the value of some parameters, by the list of possible functions to acquire, by barrier effects on some interactions. The three other dimensions are strongly connected : at the base there are mechanistic rules of change, whose effects can be nuanced by a stochastic term, standing for lack of detailed information and all kind of uncertainties, or even modified by the urban actors' strategies.

As in the SIMPOP model, the main driving force of the model is the whole set of commercial interactions that the city has developed during the considered period. These interactions are formalized through the exchanges between agents at different geographical levels. First there are the exchanges which take place with the towns and cities located in the considered cities' neighborhoods (at different geographical levels), and whose importance and nature depend on the cities' economic profile relatively to the other towns and cities of the neighborhood. The spatial organization of the settlement system is then essential. It determines the accessibility of the different places and the density of their neighborhoods, which define together the potential of interactions of the different towns and cities. Besides these exchanges which function the same way as in SIMPOP, there are exchanges related to specialized functions of higher level (finance, high technology, tourism of high level etc.) which take place in the framework of networks of different size and range. The ability of a city to sell successfully its production through these different exchanges increase its wealth and attractiveness, which induces in turn a relative growth of its population. An important difference with the classical version of the model is that the acquisition of new functions and the changes in the sectorial distribution of the active population are not only the consequences of some systematic rules associated to given thresholds. An element of strategy, which reflects the policy chosen by the urban actors, is formalized through the cognitive dimension. It means that a given strategy will be chosen according to, on the one hand, the own situation of the city, its economic balance during previous periods, the situation of the cities with which it interacts and, on the other hand, some local behavioral features such as imitation, avoidance, anticipation, risk taking, ... The agent representing the considered city can for example privilege an investment in a specific sector, and create that way an increase in the associated productivity. The change of function is then determined either from a hierarchical principle (when a certain threshold of population is passed, a function of higher level is adopted), either as the result of a strategic choice of the urban actors. It could be for example the decision to adopt, with a certain delay, a new function when cities of same size, or similar profile, or same region, have adopted it previously. It could be, at the opposite, to avoid a specialization if a few neighboring cities are equipped.

This model introduces a feedback between the interactions and the specialization in innovative activities. It means that two cities with same size and resources at a certain period may have different kinds of evolution, due to different strategies. Such strategies have of course a cost, and imply that the city has cumulated a certain level of wealth to be able to act. On the other hand they are expected to increase the cities' future wealth. Simulations will allow to test in what measure differences of strategies will affect the evolution of the cities and the system of cities relatively to what would happen through the only use of mechanistic rules. They will help to emphasize which strategies may have or not have an effect in the long term according to the characteristics of the local context. The combination of agents representing different levels of collective entities offers then promising ways for showing the respective effects of macro-economic phenomena, regional context and local decisions related to cognitive features, on the dynamics and spatial patterns of urban systems.

References

Allen, P., (1997). *Cities and regions as self-organizing systems; models of complexity*, Gordon and Breach Science Publishers, Amsterdam.

Arentze, T., Timmermans, H., (2003). Modeling agglomeration forces in urban dynamics: a multi-agent system approach, *Proceedings of the 8th International Conference on Computers in Urban Planning and Urban Management*, Sendai, Japan (www.ddss.arch/tue/nl/people/pages/theo/publications/AGGLOMERATION.pdf)

Aschan, C, Mathian, H., Sanders, L., Mäkilä, K., (2000). A spatial microsimulation of population dynamics in Southern France : a model integrating individual decisions and spatial constraints, in (Ballot and Weisbuch, eds.), *Applications of Simulation to Social Sciences*, Hermes, Paris, p109-125.

Barros, J., (2003). Simulating Urban Dynamics in Latin American Cities, Proceedings of the 7th International Conference on GeoComputation, Southampton (www.geocomputation.org/2003/)

Batty, M., (2001). Polynucleated Urban Landscapes, *Urban Studies*, vol. 38, 4, p635-655.

Benenson, I., Hatna, E., 2003, Human choice behavior makes city dynamics robust and, thus, predictable, Proceedings of the 7th International Conference on GeoComputation, University of Southampton.

Bretagnolle, A., Mathian, H., Pumain, D., Rozenblat, C., (2000). Long-term dynamics of European towns and cities : towards a spatial model of urban growth, *Cybergeo*, 131 (www.cybergeo.presse.fr), 17p.

Bura, S., Guérin-Pace, F., Mathian, H., Pumain, D., Sanders, L., 1996, Multi-agents system and the dynamics of a settlement system, *Geographical Analysis*, 28, 2, 161-178

Fransson, U., (2000). Interrelationship between household and housing market: a microsimulation model of household formation among the young, *Cybergeo*, 135 (www.cybergeo.presse.fr).

Holm, E., Holme, K., Mäkilä, K., Mattson-Kauppi, Mörtvik, (2004). The microsimulation model SVERIGE; content, validation and applications, SMC, Kiruna, Sweden (www.sms.kiruna.se)

Kohler, T., Kresl, J., Van West, C., Carr, E., Wilshusen H., (2000). Be there then: a modeling approach to settlement determinants and spatial efficiency among late ancestral Pueblo populations of the Mesa Verde Region, U.S. Southwest, in (Kohler and Gumerman, eds.), *Dynamics in human primate societies; agent-based modeling of social and spatial processes*, Santa Fe Institute Studies in the Sciences of Complexity, Oxford, p145-178.

Moeckel, R., Schürmann, C., Wegener, M., (2002). Microsimulation of urban land use, Proceedings of the 42nd Congress of the European Regional Science Association (ERSA), Dortmund.

Moriconi-Ebrard, F., (1993). L'Urbanisation du Monde depuis 1950, Paris, *Anthropos*, 372p.

Otter, H., van der Veen, A., de Vriend, H., (2001). ABLOoM: Location behaviour,spatial patterns, and agent-based modeling, Journal of Artificial Societies and Social Simulation, vol.4, n°4, http://www.soc.surrey.ac.uk/JASSS/4/4/2.html

Page, M., Parisel, C., Pumain, D. and Sanders, L. (2001). Knowledge-based simulation of settlement systems. Computers, *Environment and Urban Systems*, 25, 2, 167-193

Phipps, M., Langlois A., (1997). Spatial dynamics, cellular automata, and parallel processing computers, *Environment and Planning B*, 24, 193-204

Portugali, J., Benenson, I., Omer, I., (1997). Spatial cognitive dissonance and sociospatial emergence in a self-organizing city, *Environment and Planning B*, 24, 263-285

Pumain, D., (2000). Settlement systems in the evolution, *Geografiska Annaler*, 82B, 2, 73-87

Sanders L. Pumain D. Mathian H. Guérin-Pace F. Bura S. (1997). SIMPOP: a multiagent system for the study of urbanism. Environment and Planning B, 24, 287-305.

Sanders, L., (1999). Modeling within a self-organizing or a microsimulation framework: opposition or complementarity, Cybergeo n°90 (www.cybergeo.presse.fr).

Shelling, T., (1971). Dynamic model of segregation, *Journal of Mathematical Sociology*, p143-186.

Torrens, P.M., (2001). Can geocomputation save urban simulation? Throw some agents into the mixture, simmer and wait..., *CASA working paper*, 32, http://www.casa.ucl.ac.uk/paper31.pdf

Waddell, P., (2002). UrbanSim: Modeling urban development for land use, transportation and environmental planning, *Journal of the American Planning Association*, 68, 3, 297-314

Weidlich, W., Haag, G. (eds.) (1988). Interregional migration, *Dynamic theory and comparative analysis*, Berlin, Springer Verlag, 387p.

Cognitive Modeling of Urban Complexity

Sylvie Occelli and Giovanni A. Rabino

Abstract. New model potentials exist for coping with the complexities of today's cities. These are related to the cognitive mediation role that modeling allows one to establish between the abstraction process (internal loop) and the external environment to which a model application belongs (external loop). The focus is turned to the two main aspects involved in that role, i.e. the modeling task and the technological interface. As far as the first is concerned, there are claims that model building in geography involves three main components: a syntactic component (how are the mechanisms underlying the functioning of the system accounted for?), a representational (semantic) component (what kind of urban descriptions are conveyed by the model?) and a purposive investigation project component (what is the modeling activity intended for?). As they increasingly rely on computing technology, models as cognitive mediators are not just simple, autonomous entities, but active complex objects. A model can therefore be understood as an ALC (Action, Learning, Communication) agent, capable of performing a certain course of Action, and permitting a certain Learning ability, which, because of its cognitive mediating role, Communicates with other kinds of agents (other models). This notion is then related to the various aspects of model-building in geography as originally introduced in the early seventies. These aspects are re-interpreted in light of the above characteristics. We conclude the paper with some remarks about the implications which may be derived as far as the harnessing of complexity in urban systems is concerned.

1. Introduction

The many facets of the relationships between modeling and urban complexity have been a major theme of interest in the authors' recent works (Occelli, 2001a, 2001b, 2003, Occelli and Rabino, 2000a, 2000b, Rabino, 1998). The following points were central to that discussion:

- Complexity issues are disclosing new potentials for modeling.
- These potentials are related to a general structural-cognitive shift that affects the overall modeling activity.
- In this shift a new role for modeling is created. Modeling functions as a cognitive mediator, between an *internal loop*, related to the conventional steps underlying a process of abstraction, and an *external loop* representing the general context of a modeling activity. As the observable is no longer the only link between the two loops, modeling as a cognitive mediator, allows us to establish and investigate the various relationships likely to be established between them.
- The form of cognitive mediation enabled by modeling is not unique but could be instantiated in several ways.

In this paper, we'll deepen the arguments given above by trying to emphasize their implications as far as spatial analysis and policy-oriented knowledge are concerned. The discussion will be organized in two main parts. In the first part we focus on the role of cognitive mediation played by models. We discuss its main features and the main components of model building which are entailed, i.e. the syntactic, representational and knowledge project components. Insofar as models increasingly rely on computing technology all these components are also affected. This makes models unique artefacts with their own hardware and software identity. It is in this respect that a model can be understood as an ALC agent, i.e. an entity endowed with its own representational attitude towards the world and teleonomic drive, capable of performing a certain course of Action, and providing a certain Learning ability, which, because of its cognitive mediating role, Communicates with other kinds of agents (other models), and therefore may influence them and/or modify itself in the process.

To better illustrate these arguments, in the second part of the paper, we will refer to the classical aspects of model building in geography as originally introduced in the early seventies and discuss how they can be re-interpreted in light of the above characteristics.

To conclude the paper, we identify a few points which, according to this view of modeling as a cognitive mediating activity, can assist us in harnessing the complexity in urban systems.

2. Modeling as an ALC agent

Since the second half of the 1990s, a number of epistemological, operational and socio-cultural changes are taking place in the whole field of geography and urban modeling (Batty, 1994; Clark, Perez-Treio and Allen, 1995; Occelli, 2001a; Wegener, 1994). As a result, a broader view on the very concept of modeling is called for. We have given the term *structural-cognitive shift* to the move toward this broader view (Occelli, 2001a, b, Occelli and Rabino, 2000a, 2000b). Its essence is a shift from viewing a model as a simplified representation of urban phenomena and the ways they are produced to a view in which modeling is an activity for testing, exploring, creating and communicating knowledge about certain urban phenomena. Here models are a means of representing how our knowledge hypotheses work and their outcome. It can be seen that the shift is from the classical (realistic and axiomatic) approach to modeling to the constructivist approach, in which modeling is used to uncover interpretive keys for problem definitions (Pidd, 1996).

The structural-cognitive shift entails implications for the modeling process (internal loop) that concerns the underlying process of abstraction, as well as the modeling domain (external loop) representing the general context in which the modeling occurs (Fig.1).

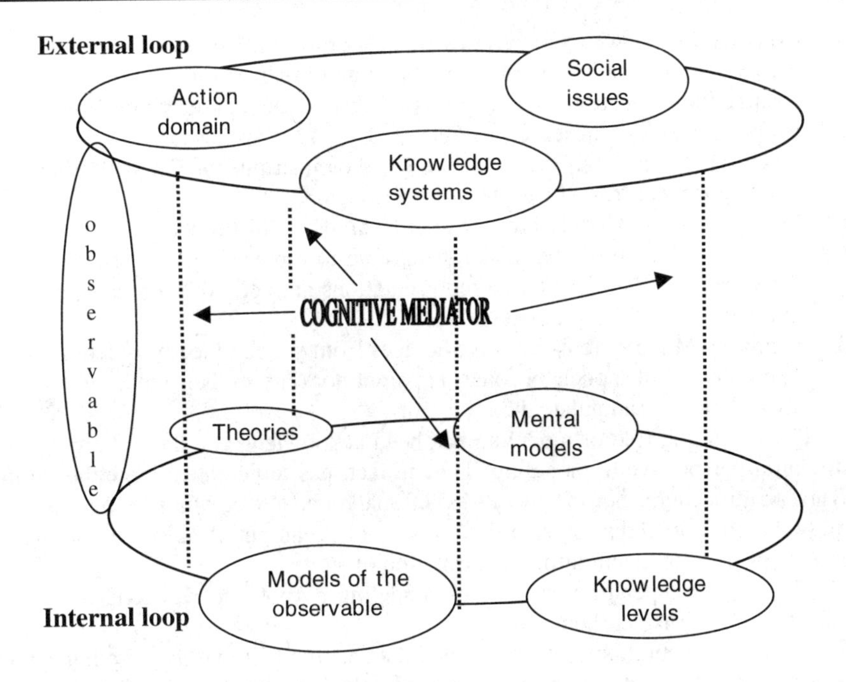

Fig. 1. Modeling as a Cognitive Mediator

As a result of the above shift, the model acts as a cognitive mediator between the two loops. In fact, it (a) does not limit itself to the internal loop; (b) it provides the means to reinforce, to articulate better and clarify the relationships between the two loops; and (c) it becomes a kind of interface between the two loops, playing the role of a *cognitive mediator* between them. To appreciate the potential of this role one has to consider the very notion of a model in geography and the knowledge likely to be obtained from the modeling activity.

The notion of a model is largely disputable. Here we can recall Dupuy's definition (2000, p. 29) in which "a model is an idealization (usually formalized in mathematical terms), that synthesizes a system of relations among elements ..." Building on this definition, a model can therefore be: (a) an abstract form instantiated by various phenomena; (b) an equivalence class establishing equivalence relationships among domains of real phenomena; (c) an abstract form enjoying a transcendent position.

The definitions above ultimately share a general idea of the mediating role of modeling. Such an endowment is derived from the intrinsic autonomy of models from both our ability to conceptualize and our perception of the urban phenomena. In particular, as maintained by Morrison and Morgan (1999), models can be used as instruments of exploration in both theories and in the world. A model's autonomy, in particular, should be related to the following basic elements.

a. Construction. If we examine how models are built, we realize that they involve neither just theory nor data, but both of them. Modeling therefore is something in-between our capacity of abstraction and the perception of the phenomena we observe.
b. Function. Models can function like tools or instruments. As such, they are independent from the things they operate on.
c. Representation. Models are not simple kinds of instruments. They are tools of investigation that make it possible to represent some aspects of the world and theories. These representations are able to teach us something about the things they represent.
d. Learning. Models are explorative devices from which to learn something. The potential of models becomes apparent not only in their building, but also when we manipulate them.

The mediating role of modeling has been acknowledged only partially by the structural perspective to modeling. This, in fact, has turned attention either to the syntactic procedure, i.e. the model as an analytical device whose function is to study the city, or to the relevant descriptions of urban phenomena, i.e. the model as a simplified but meaningful representation of a city.

The cognitive perspective towards modeling purports to reconcile these two aspects and to consider them jointly.

The claimed autonomy of models and the ensuing fact that they are human artefacts, *natural* objects having their own complexities, has been mostly unnoticed in spatial analysis. Apart from the general exhortation of the complexity approach to enrich conventional analysis insofar as "achieving comprehensive "stereoscopic" views of most complex systems usually demands that they be described in more than three dimensions…" (Cowan and Pines, 1994, p. 710), the kind of knowledge to be expected or gained from this 'kind of complex artefact' has remained largely unquestioned.

If, however, we share the view that the *explanatory search for understanding* is an intrinsic feature of human beings, who, 'explain themselves and their circumstances while operating as observers' (Maturana, 2000), then two major perspectives permit to grasp the range of possibilities enabled by the role of cognitive mediation (Occelli, 2002a, 2003).

a. A first one, which we will call *action-oriented*, involves considering a multiplicity of views of the urban phenomena in a coordinated way. Such an approach involves what Zeleny (1996) has indicated as "relating descriptions of objects into coherent complexes. The relationships among objects are not simply out there to be captured, but are being continually constructed and reconstructed and re-established by the knower" (p. 212). The model, therefore, is an active entity to be operated on in order to build these coherent complexes. In this endeavour, the analytical/procedural task of modeling (as conventionally understood in the structural approach to modeling) extends to the explorative/ cognitive one.
b. A second dimension, which we will call *meaning-oriented*, reflects a need to reconcile those contrasting views conventionally held when relating human experience and the external worlds, i.e. subjective and objective

approach, tacit and explicit knowledge, local and global knowledge, hard and soft sciences, common sense and scientific knowledge. Meaning is obtained through a process of learning in which knowledge of cities is permanently fuelled and opened to confrontation and critical revision. Several representations of urban phenomena can be derived, updated and progressively enriched. Modeling as *a cognitive mediator* means that in order to achieve a shared representation of the observable, shared form-meaning pairs (i.e. a language) are also required. Language is an essential mechanism to both distinguish between relevant objects in an environment and to "coordinate our action in a social domain" (Zeleny, 1996, p. 212).

The discussion carried out so far allows us to emphasize how, in a geographical domain, models as cognitive mediators are confronted with three main components (Occelli, 2002b):

a. A syntactic component concerning the methodological aspects of modeling. Modeling therefore entails a method of analysis (i.e. a coherent set of steps of enquiry) which, through an abstraction process (the encoding and decoding process), yields an explanation of the mechanisms underlying the observed urban phenomena. As discussed elsewhere (Occelli and Rabino, 2000b), this component also presupposes making the underlying epistemological background explicit, i.e. the 'degree of explanation' we are deemed to achieve in our enquiry and the level of insights we (as analysts or urban system experts) should be satisfied with;
b. A representational (semantic) component, related to the sense associated with the representations of the urban phenomena provided by the model. The kind of system structure we refer to in our view of the city, and the meaningfulness of the relative urban descriptions conveyed by the model are crucial aspects of this component. Apart from the different notions of representation underpinning its use in many domains of the social sciences (Lassègue and Visetti, 2002), by representational component, here we understand the range of prototypical images which are produced by our perceptions and the categorization of cities (i.e. the kind of Inter-Representation Network linking internal and external maps, discussed in Portugali, 2000);
c. A knowledge project component, associated with the purports of the investigation project underlying a certain model application (i.e. aims of the model application, resources required for the model implementation, expected results, etc.). Besides reflecting on the kind of awareness of the relationships existing between the syntactic and semantic components, this component plays a crucial role in both instantiating the model application and steering the accompanying learning process.

Not only models possess the attributes of technology (Morrison and Morgan, 1999), thus revealing their strength as they are used, but they increasingly rely on computing technology. Models as cognitive mediators, therefore, are active complex artefacts having their own hardware and software identity. The tremendous progress in the latter significantly affect the components mentioned above.

This is particularly evident in the recent development of simulation and its unprecedented diffusion in the social sciences (Ballot and Weisbuch eds. 2000;

Conte, Hegselmann and Terna, 1997; Couclelis, 1998; Epstein, 1999; Gilbert and Troitzsch, 1999; Marney and Tarbert, 2000; Parisi, 2001).

To suggest a label for this novel definition of modeling, a model can be eventually understood as an agent, which is able to instantiate the evolving relationships among the three components mentioned above. In particular, it may be defined as an ALC agent, an entity endowed with its own representational attitude towards the world and teleonomic drive, capable of:

a. Performing a certain course of Action, thus enabling the realization of a certain project of investigation of spatial phenomena. This most directly involves the relationships between the syntactic and representational components of modeling;
b. Enabling users with a certain Learning ability, thus generating stimuli in revising both the external and internal loops of the modeling activity likely to trigger new series of investigations;
c. Communicate with other kinds of agents (other models), thereby influencing them and/or modifying itself in the process.

3. The ALC agent in action

To sharpen our discussion in this section, we examine the classical aspects of model building as originally introduced for the design of urban models in the early seventies (Wilson, 1974) and we re-interpret them in light of the above arguments.

These aspects which substantially refer to the internal loop of the modeling activity (Fig.1), constitute the basic ingredients necessary to the operational implementation of an urban model. They can be summarized in the following check list:

1. Purpose: What are the aims of the model application in relation to the problem issues for the study area?
2. Conceptualisation: How are the observed urban phenomena, i.e. the problem issues, accounted for in entities and represented by the model constitutive objects (i.e. variables, parameters, relationships)?
3. System Control: Which model variables are relevant for taking the decision-makers and/or planner's control into account?
4. Variable Definition: Which level of spatial and categorical articulation should be retained in specifying the variable?
5. Treatment of Time: How are time and its correlates introduced and managed?
6. Theories: Which consistent collections of hypotheses describing urban phenomena (i.e. the views of cities as developed in literature) are pertinent for dealing with the problem issues addressed by the model?
7. Methodologies: Which technical approaches (i.e. statistical and spatial methods)are most convenient in developing the model?
8. Data Gathering and Information Processing: Which kind of spatial and socioeconomic data are available or need to be retrieved?
9. Verification: How good is the model's adherence (i.e. calibration of parameters, validation of the model outputs) to the study area?

To some extent, the above aspects can be related to a process of moulding which starting from an initial draft of the model, progressively gives form, contents and shape to the model artefact. Of course, the listed aspects are not strictly sequential, and have multiple links and feed-backs (i.e. conceptualization is obviously linked to both theories and data).

A fundamental requirement is that the treatment of each aspect should be consistent within the overall model construction. This also implies that the choices made in dealing with a certain aspect influence and/or constrain those made for the others. They affect, therefore, the whole course of the modeling process and can ultimately condition the success of the modeling enterprise.

Whereas this consistency requirement still holds for an ALC model, all the aspects mentioned are considerably enriched insofar as they are transformed from simple items of a check list into more active autonomous objects. Besides having its own functions, each of these objects is therefore endowed with syntactic, representational and knowledge project components. These contribute to the overall model activity, but can also produce research by-products and foster additional modeling activities.

Our review of these extensions will not be exhaustive and is limited to highlight the most salient features, see Tab.1. To support our arguments, some comments will be made regarding two model experiences, the PF.US and SimAC models, recently carried out at IRES, as they may be instructive in illustrating the different objects of an ALC model (Occelli and Landini, 2002):

- The PF.US model stems from a revision of classical operational urban models. These are aimed at providing a comprehensive view of the functioning of a city by means of the interdependences of its constitutive sub-systems (economic activities, services, population and transport). The social account and matrix and spatial interaction approach underpin the methodological basis of the model. The model assumes that a set of functional and spatial interdependences (i.e. expressed by means of activity multipliers) link the various urban sub-systems. Through them, changes in a sub-system are propagated to all the others and determine an adjustment in each other sub-system. A two- level system description, at the regional and sub-regional level is given. Each level is supposed to have a perception of the state of the other level. Such a perception, in turn, affects the level's own state. Therefore, in the model, a complete (although simplified) description of urban structure is given (i.e. as stated by a set of mathematical equations), which sets the rules of interaction of the agents at both levels (the various economic sectors, population types and local systems). The main purpose of the model development was to simulate socio-economic scenarios, in order to assist us in reasoning about the likely future of Piedmont regional systems.
- The model for the Simulation of ACcessibility (SimAC Model) deals with the exploration of a novel view of accessibility, i.e. accessibility is not simply an entity derived from transportation demand, but a resource associated with the many interaction fields existing in an urban system. As in the previous model, both a micro and macro level of the system exist, although in this case the

agents at both levels are conceived as reactive agents capable of updating their behaviour as a result of changes in their perceptions of accessibility. In the model there is no quest to give a complete description of urban structure. Agents, at both individual and collective levels, are autonomous entities and can modify their behaviour (i.e. choose a different travel path in their journey-to-work) according to their varying representations of accessibility. There is no explicit system control. The agents are reactive entities whose behaviour depends on limited cognitive ability (i.e. their representations of accessibility are local and updated over a certain time span). Besides the cognitive underpinning of the model exercise, simulation is an essential condition for the unfolding of spatial behaviour of the individuals over time.

	Modeling as a Plain Mediator Agent	***Modeling as an ALC Agent***
Purpose: What the model is intended for	Problem solving: WHAT to do	Problem-making: How to think about WHAT do to, i.e. what to do IF…
Conceptualization: Framing the problem and identification of variables	Entity organization by taxonomical approaches (one-to-one mapping)	Entity organization as a feasibile set of alternatives (mapping one-to-many, geno-pheno mapping)
Control Variables	Identification of the mechanisms controlling the responses of the system	Search for the range of possible actions to steer the system's behaviour
Level of Aggregation of Variables	Definition of the most encompassing/synthetic spatio-temporal data sets	Paying attention to the representation levels of phenomena, windows of observation
Treatment of Time	Internal clock marking the progression steps of dynamics	Irreversibility of time, influence of the initial conditions
Theories	Search for the truthful approximation of the real world	Critical thinking and interpretive analysis
Methodological Approaches	Exploitation of the existing tools and search for the best ones (optimizing attitude)	Exploring new tools and using the available ones (satisfying attitude)
Data	Measurements of phenomena and quest for comprehensive information (the myth of data)	Significant information and perception (virtual data and quality of information)
Validation	Parameter calibration and consistency with real phenomena	Moving up the level of validation: emulation and simulation

Table 1. Modeling Aspects in an ALC Agent

The *aim* of a model is the aspect which most clearly reflects the structural-cognitive shift in modeling as we mentioned earlier. Both the action-meaning perspectives are involved in being highly sensitive to the current trends of epistemological, socioeconomic and cultural changes. As increasing attention is turned to the need to identify the 'right questions' to be addressed by decision-making (i.e. the planner, urban stakeholders, general public), in an ALC model, scopes are less interested in problem-solving and become more focused on problem-making.

Therefore, an ALC model is more widely recognized as an activity that assists us in both obtaining deeper insights and creating a 'new kind of knowledge' about a problem. The type of modifications occurring in modeling scopes is highlighted in Tab.2. It shows the view of a city in which dynamics and unexpected changes are advocated as major urban characteristics, and also the traditional distinction between the predictive and prescriptive perspectives in modeling purposes that become blurred and meaningless.

<table>
<tr><td></td><td colspan="2">Analyst's Points of View About the System</td></tr>
<tr><td>General Attitude Towards the System</td><td>The system is stable and weakly reactive</td><td>The system is dynamic and can exhibit unexpected changes</td></tr>
<tr><td>Predictive (Positive)</td><td>To help predict the impact of alternative strategies for developmental policies</td><td rowspan="2">To explore the set of actions likely to form viable strategies for improving the survival of the system (i.e. for devising socially shared sustainable policies)</td></tr>
<tr><td>Prescriptive (Normative)</td><td>To investigate the best performing actions for optimal policy strategy</td></tr>
</table>

Table 2. Broadening the Modeling Scopes in an ALC Model

In this context, the investigation of the set of actions likely to form viable strategies, i.e. the IF term of the IF THEN condition of conventional impact analysis, becomes a general requirement in applying an ALC model.

Of course, a relatively wide range of possibilities exist. As shown in the IRES experiences, for example, a model application can be aimed to provide new ways of thinking about practical questions, i.e. to make a framework available to give coherence to a scenario's analysis, as in the PF.US model. In other cases, it can focus on new means for theoretical reasoning, i.e. to implement a novel artefact allowing us to deepen our understanding of agents' spatial behaviour, as in the SIMAC model.

This general broadening of model scopes clearly affects most of the other objects. For the *conceptualization* object, in particular, attention tends to shift:

a. From the undertaking of an efficient filtering of the problems at hand in order to obtain a fully defined set of model entities, i.e. to identify a one-to-one mapping of the problem at hand with the model entities and their representations in quantitative variables;
b. To a questioning of the generation process of the model entities themselves, i.e. of how these entities are likely to reveal a multiplicity of forms that are even quite heterogeneous.

Even at a preliminary stage of model sketching, in fact, it is recognized that model entities may show different forms, whose viability can be unlike and differentially effect the outcomes of the system evolution. In an ALC model, therefore, greater attention is paid to the mechanisms and processes of entity definition, paralleling, to some extent, the same kind of quest undertaken by biologists studying the geno-pheno mapping of living systems.

The IRES studies are illustrative in this respect. In the PF.US model, in particular, the extension of the conceptualization object relates to the variable organization. A two-level description of the regional areas has been built, which although providing distinct profiles, it allows the latter to influence each other. In the SIMAC model it concerns the introduction, though in an elementary form, of novel types of variables, i.e. the cognitive ability of individual agents and speech acts.

As control is no longer the major preoccupation of planning, and planning as well as policy making is undergoing deep transformations, the identification of the 'correct and most appropriate' control variables in a model is losing importance. In the ALC model, emphasis is shifted from a purely engineering type of approach to control, i.e. an approach in which one sees to regulating the system by means of exogenous and well-behaved mechanisms, to an approach in which steering and guiding become general rules. Ultimately, control becomes something which should result from adaptation and co-evolution of an intelligent agent's behaviour.

In this context, the object pertaining to *the definition of the aggregation levels of variables* is confronted with a set of different questions from those commonly raised. The search for the most detailed variable articulation which would guarantee the best socioeconomic and spatial representation of a problem, has to recognize that phenomena possess their own representation levels, i.e. they have several levels of intelligibility and therefore have different windows of observation according to which particular features become observable and measurable.

The enormous progress in computer and information technology has both moved up and extended these levels of intelligibility. In this respect, GIS can be considered one of the domains of geographical analysis most clearly connected with the ALC model's object, which has considerably improved many aspects of its representational components (Egenhofer and Golledge, eds.1998).

Most of the previous observations are pertinent also for the object dealing with *the treatment of time*. One major observation is that in the ALC model, time becomes more substantial. Besides highlighting the relevance for urban evolution of features of the temporal dimension, i.e. irreversibility and uniqueness of a time trajectory, there is less emphasis on the syntactical aspects of their treatment, as addressed in the comparative static versus dynamics debate, and more on their intrinsic aspects, i.e. time-budgets, time resources, spatio-temporal accessibility (Merz and Ehling eds.,1999).

Again, the experience of the IRES models is useful in illustrating this argument. In the PF.US model the substantial aspect of time is outside the model itself, i.e. it is accounted for in the anticipatory view underlying the socioeconomic scenarios used for the model simulations. The syntactic dimension is related to the algorithmic procedure allowing for the adjustments of the model variables to the impact of the socioeconomic scenario. In the SIMAC model, time is an intrinsic component of both model architecture and functioning. The unfolding of time organizes the daily routines of urban activities and gives perspective to the actions undertaken by agents. A computer module dealing with the time management of events is also included in the simulation platform.

As far as *theories* are concerned, our earlier arguments about the action and meaning perspectives involved in the cognitive mediation are cogent for revising the role of this object in the ALC model. In particular, one major aspect emphasized on the epistemological ground is the need to reconcile contrasting approaches conventionally held in relating human experience and the external world, i.e. the subjectivist's approach where what matters is the individual subject and its experience and the objectivist's approach where a world reality is postulated to be observer-independent. In this context, theories lose their unique role of "depository" of scientific truths, called for in legitimizing the formation of individuals' mental models. They acquire a more pragmatic role as means for both suggesting interpretive views of cities, motivating comparison and critical discussion and providing coherent collection of hypotheses for validating models.

One major point worth underlining is that theories can greatly benefit from the action of an ALC model. If the formation of new concepts is a major drive in the modern role of theories, then, an ALC model makes a powerful device available for exploiting the various mechanisms underpinning it, see Table 3.

	By Discovery	*By Instruction*
Development of New Concepts	Formation of new concepts Generalisations Generation of hypotheses	Introduction of new terms Discussion of experimental results Comparison of alternative hypotheses
Substitution of Existing Concepts	Use of algorithms to select sets of consistent hypotheses	Arguments aimed to investigate explanatory coherence

Table 3. Mechanisms for the Generation of New Concepts (Adapted from Thagard, 1992)

The changes in the roles of theories also have consequences for the object concerning the *methodological approaches*. Whereas in classical modeling, methodological development was mainly driven by a general quest for best performing methods, either from a mathematical and/or statistical point of view, or from a set of optimality criteria of model output, in an ALC model the quest is for an approach that is able to cope with the problems raised in a given situation. The question, then, becomes one of assessing the possible alternatives, in relation to the existing cultural, technological and informational context. While adopting a satisfactory attitude, the possibility to integrate existing explanatory styles and explore novel ones is an essential additional alternative. The tension between exploitation and exploration thus characterizes this object of the ALC model. This kind of tension is reflected, for example, in the recent computation - dynamics debate in cognitive science and, namely, in the plea for their rapprochement in order to have a full explanatory account of cognition (Mitchell, 1998).

The range of possibilities likely to be derived from acknowledging the exploitation/ exploration tension can be grasped also from the IRES applications. Whereas in the PF.US model a quest for the integration of existing approaches predominates, in the SIMAC model an effort is made to explore novel methodological capabilities.

The model object concerning *data* has evident connections with those related to both *the identification of variables* and *the definition of their aggregation levels.* In this respect, the generation and representation level of phenomena have been pointed out as major aspects involved in the definition of model variables. They hold the main responsibility for the fundamental data hungriness of models. More often that not, in fact, data requirement has been a crucial issue, often hampering model developments and applications. For the ALC model, data availability is no longer a crux. Data are primarily meaningful measurements of phenomena, as derived from the multiple ways humans perceive the world surrounding them. The long-standing debate on social, urban and environmental indicators has contributed to elucidate this point. Today, virtual data represent an additional possibility.

The last model object refers to the *verification* of the model. Conventionally this requires a discussion on very technical aspects of calibration, i.e. parameter estimation to obtain the best fits between model output and data and testing, i.e. deciding whether these fits are good. Statistical techniques and indicators of "goodness-of-fit" are typically involved. For an ALC model, the relevant questions of verification have a different nature. They are no longer limited to comply with the methodological procedures of standard analysis, i.e. as developed in the modeling internal loop, but need to consider the *model purpose* object, i.e. all the steps in the modeling external loop. Model verification therefore requires a broader framework in which:

1. The model's course of Action is scrutinized in relation to the ethical aspects of social acceptability and relevance and consequences for the recipients;
2. What the model has taught us in terms of thinking about the world, i.e. the Learning from the model is assessed in relation to the existing domains of scientific knowledge;
3. The ability of the model to Communicate with other kinds of models and more generally with the social context is examined, i.e. to expose citizens and stakeholders to its representations of urban phenomena.

These are also crucial points in the recent debate about simulation, insofar as simulation is not only associated with the operational realization of a modeling activity, but it is co-determined by it. On the one hand, as the role of cognitive mediation reinforces the modeling potentials, the functions of simulation are also extended from technical/algorithmic to explorative. On the other hand, as the progress in the technological and information background makes simulation and the computer artefact more powerful, the role of a modeling activity is progressively effected, thus improving our knowledge gains and action capabilities to deal with urban complexities (Occelli, 2003).

4. Concluding remarks: what an ALC model can teach us in dealing with city complexities

In this paper an effort was made to illustrate aspects of the new strengths existing in today's modeling. We argued that these are related to the cognitive mediation role for modeling makes it possible to establish between the abstraction process (internal loop), and the external environment to which a model application belongs (external loop).

In particular, we emphasized how this role entails three main components of model building. Firstly, a syntactic component, aimed to deal with the mechanism underlying the functioning of cities. Secondly, a representational (semantic) component, related to the sense associated with the representations of the urban phenomena provided by the mode. Thirdly, a knowledge project component, associated with the purports of the investigation project underlying a certain model application.

We also pointed out that as they increasingly rely on computing technology, models as cognitive mediators are not just simple autonomous entities, but active, complex and unique artefacts. Because of this intrinsic property a claim was made that a model can be understood as an Action, Learning and Communicating agent, whose ultimate role is to strengthen our capability to act.

To sharpen this contention, we then recalled the classical steps of model building in geography and examined, how their functions are modified in an ALC model.

Insofar as it entailed how we organize the ways to apprehend urban phenomena, i.e. how we know the city, our discussion was fundamentally epistemological. On the one hand, the claim about the cognitive mediation role of modeling may be considered as a kind of hub allowing us to enter several paradigms of enquiry of city complexities (i.e. dynamic analysis, multi-agent approaches, statistical techniques, hermeneutic discourse, etc.).

On the other hand, it may be understood as an exhortation to look for sounder, more innovative and satisfactory means to grasp the complexities of the city. In this respect, our discussion indicates that these are not only technological but, as implied by an ALC Model, require a more general language, involving the underlying ways to look at and apprehend urban complexities (Casti, and Karlqvist eds.

1986; Pattee, 1986; Van Gigch, 2002). The case of the diffusion of GIS, without acquiring such language, may be instructive in this respect (Batty, 2002).

Maybe, one major lesson which has been learnt so far, is that underlying these complexities, no matter how they are disclosed or approached, there are surprises and more trivially something we did not expect. This calls for constant attention towards innovative actions to address the problems at hand. As shown in discussing the different objects of the ALC model these are made of a number of operational and practical ingredients.

References

Ballot, G. and Weisbuch, G. (eds.) (2000). *Applications of Simulations to Social Sciences,* Hermes, Paris.

Batty, M. (1994). A Chronicle of Scientific Planning. The Anglo-American Modeling Experience, *Journal of the American Planning Association*, 60, 7-16.

Batty, M. (2002). Editorial. A decade of GIS: what next?, *Environment and Planning , 29, 2*,157-158.

Casti, J.L., Karlqvist A.(eds.) (1986). *Complexity, Language, and Life: Mathematical Approaches*, Springer, Berlin.

Clark, N., Perez-Treio, F. and Allen, P. (1995). *Evolutionary Dynamics and Sustainable Development. A system Approach,* Edward Elgar, Aldershot.

Conte, R., Hegselmann, R. and Terna, P. (eds.) (1997). *Simulating Social Phenomena. Lecture Notes in Mathematical Systems 456,* Springer, Berlin.

Couclelis, H. (1998). Geocomputation in Context. In: P. Longley , S.M. Brooks., R. McDonnell R. and B. MacMillan eds. *Geocomputation: A Primer.* New York:: Wiley, 17-29.

Cowan, G.A., Pines, D. (1994). From Metaphors to Reality ?, in Cowan G.A., Pines D. and Meltzer D. eds., *Complexity. Metaphors, Models and Reality*, Proceedings Volume in the Santa Fe Institute Studies in the Science of Complexity, Addison Wesley, Reading MA.

Dupuy, J.-P. (2000). *The Mechanization of the Mind,* Princeton University Press, Princeton.

Egenhofer, M.J., Golledge, R.G. (eds.) (1998). *Spatial and Temporal Reasoning in Geographic Information Systems*, Oxford University Press, New York.

Epstein, J. (1999). Agent-Based Computational Models and Generative Social Science, *Complexity, 4, 5,* 41-60.

Gilbert, N. and Troitzsch, K.G. (1999). *Simulation for the Social Scientist,* Open University Press Philadelphia.

Lassègue J. Visetti Y.M. (2002). Introduction :What is left of representation?, *Intellectica, 2, 35,* 7-26.

Marney, J.P., Tarbert, H.F.E. (2000). Why do simulations? Towards a working epistemology for practitioners of the dark art, *Journal of Artificial Societies and Social Simulation, 3(4):* www.soc.surrey.ac.uk/JASSS/3/4/4.html

Mitchell, M. (1998). A Complex-System Perspective on the "Computation vs. Dynamics" Debate in cognitive Science, WP 98-02/017 Santa Fè Institute, Santa Fè.

Merz, J., Ehling, M. eds. (1999). *Time Use – Research, Data and Policy*, Nomos, Baden Baden.

Maturana, H. (2000). The Nature of the Laws of Nature, *Systems Research and Behavioural Science, 17*, 459-468.

Morrison, M., Morgan, M.S. (1999). Models as mediating instruments, in Morrison M. and Morgan M.S. eds., *Models as Mediators. Perspectives on Natural and Social Science*, Cambridge University Press, Cambridge, 10-37.

Occelli, S. (2001a). La cognition dans la modélisation: une analyse préliminaire, in Paugam-MoisyH., Nyckess V., Caron-Pargue J. eds. *La cognition entre individu et société*, ARCo'2001, Hermes, Paris, 83-94.

Occelli, S. (2001b). Why modeling: the Cognitive Drive, Paper presented at the INPUT Meeting, Isole Tremiti, 26-29/6/2001.

Occelli, S. (2002a). Facing urban complexity: towards cognitive modeling. Part 1: modeling as a cognitive mediator, Paper presented at the XII European Colloquium on Theoretical and Quantitative Geography, St.Valery-en-Caux, 7-11 September, 2001. www.cybergeo.fr.

Occelli, S. (2002b). L'uso dei modelli nelle Scienze regionali: da strumenti analitici a mediatori cognitive, Relazione presentata al Seminario AISRE, Scienze Regionali ed Interdisciplinarietà, 31 Maggio, Torino.

Occelli, S. (2003). Simulation for Urban Modeling, *Actes des Journées de Rochebrune sur la Simulation dans les Systèmes Complexes*, Rochebrune, 27 Janvier, 1 Février, ESNT 2003, S 001, Paris, 175-199.

Occelli, S., Landini S. (2002). Le attività di modellizzazione all'Ires: una rassegna e prime considerazioni., WP 160, IRES, Torino.

Occelli, S., Rabino, G.A. (2000a). Rationality and Creativity in Urban Modeling, *Urbanistica, 113,* 2000, p.25-27.

Occelli, S., Rabino G.A. (2000b). *Modeling for the sustainable city. A contribution of thinking to action*, Paper presented at the World Congress 'Humankind and The City. Towards A Human and Sustainable Development', Naples, 6-8 September 2000.

Parisi, D. (2001). *Simulazioni,* Il Mulino, Bologna.

Pidd, M. (1996). *Tools for Thinking. Modeling in Management Science*, Wiley, New York.

Portugali, J. (2000). *Self-Organization and the City*, Springer, Heidelberg.

Rabino, G.A. (1998). Selected issues in urban planning, in Bertuglia C.S., Bianchi G., Mela A. eds., *The City and its Sciences*, Physica Verlag, Heidelberg, 577-598.

Wegener, M. (1994). Operational Urban Models. *Journal of the American Planning Association, 60*, 17-29.

Thagard, P. (1992). Conceptual Revolutions, Princeton University Press, Princeton.

Van Gigch, J.P. (2002). *Comparing the Epistemologies of Scientific Disciplines in Two Distinct Domains: Modern Physics versus Social Sciences*, Systems Research and Behavioral Science, *19*, 551-562.

Wilson, A.G. (1974). *Urban and regional Models in Geography and Planning*, Wiley, London.

Zeleny, M. (1996). Knowledge as a coordination of action, *Human System Management, 155*, 211-213.

Navigation in Electronic Environments

Stephen C. Hirtle

Abstract. The ability to locate information in a complex information space requires specialized tools to support searching and browsing behavior. Inherent in browsing is the ability to navigate through informational items, while retaining a sense of orientation. A tripartite theory of navigation is presented based on cognitive studies of navigation in physical spaces, which divides navigation into three levels: planning, procedural and motor. The last two levels become critical for virtual reality, while the first two levels are critical for the traversal of more abstract information spaces. The analysis leads to various insights for information designers, which are demonstrated in two different environments. First, for hypertext navigation, it is argued that the inclusion of structural components, such as neighborhoods and landmarks, can improve the navigability of electronic spaces for browsing and non-directed search. Second, for spatial information kiosks, the use of text, images and maps, are shown to improve the accessibility of the information. Together, these two examples highlight the benefits of grounding information design in theories of wayfinding and spatial information processing.

1. Introduction

A basic principle of all human behavior is the ability to seek and find locations of particular interest or meeting particular needs (Golledge, 1999). From our basic needs of substance and shelter to more complex needs of data and information, humans must store a vast array of spatial knowledge for quick access and manipulation. Computational advances have dramatically altered the ability to provide information about spatial locations on a real-time basis. In fact, the World Wide Web and other non-mobile information sources are providing the first, and often the only, source of information to travelers.

Today we move about not only in physical space, but also within worlds of information. The ability to locate information in a complex information space requires specialized tools to support searching and browsing behavior. Browsing and searching are two fundamental processes in locating information (Korfhage, 1997). Browsing is an undirected search, in which either there is no clearly specified goal ("I wonder if there is anything interesting here") or the user is interested in determining the scope of coverage ("What kinds of information can I find here?"). In contrast, when searching is used if there is a clear goal. Search engines, such as Google, are more useful in the latter than in the former, although the search engine may help the user find the best neighborhood to begin browsing or may provide a tool to browse at the meta-level of data.

It is interesting to note that in physical space there are two similar processes related to wayfinding. That is, in directed navigation there is clearly specified goal and the task of the user is to move as quickly, efficiently, or aesthetically to that goal (Jul and Furnas, 1997). In other cases, such as wandering, one has no clear destination in mind, but rather is interested in understanding the nature of a space. Inherent in browsing is the ability to retain a sense of orientation or position, regardless of whether or not one is discussing an informational space or an environmental space.

One approach to the problem of electronic navigation, which we have found to be beneficial, is to compare navigation in the physical world with navigation in electronic worlds, with a focus on the underlying cognitive structures and the implicit metaphors that are adopted by the navigators of the space. Kim and Hirtle (1995) have argued that some of the difficulty in traversing electronic spaces is due to the lack of what can be termed cognitive structures, such as identifiable neighborhoods and notable landmarks, which provide a user with a sense of place. Appropriate formal analyses can also lead to the development of intelligent views of a space, such as modified fisheye views and other "you-are-here" pointers for electronic worlds. In the next section, we review principles from the cognitive studies of navigation and then suggest methods for incorporating these principles into the design of navigational tools for electronic spaces.

2. A View from the real world

Research on cognitive mapping has examined the ability of individuals to acquire and use spatial information. The acquisition of spatial knowledge has been shown to be based on the use of organizing principles, such as the use of hierarchies, reference points, rotational and alignment heuristics and other related principles. These organizing principles, in turn, result in what Barbara Tversky (1993) has coined a "cognitive collage" of multimedia, partial information. Inherent within this collage is the ability to extract slices of information sources, such as visual cues, route information or linguistic labels. The collage necessarily operates at multiple levels, allowing one, for example, to discuss and plan a route, using highway systems or one's back alley with equal ease. In our own lab, we have shown that the need to structure space is so strong that subjects will impose overlapping clusters on an otherwise homogeneous distribution, which results in biased judgment of distance and orientation (McNamara, Hardy, & Hirtle, 1989).

In addition, while real space is organized hierarchically into multiple levels, these levels are not well represented by a tree structure. Christopher Alexander was the strongest advocate of this position. In a classic paper, entitled "The city is not a tree," he argued that a tree is too constrained to represent organic cities, which have evolved naturally to contain many cross-linkages and overlapping nodes. Furthermore, planned cities, which are designed to be fit a strict hierarchy are either unlivable or adapt to a more fluid structure, such as a semi-lattice. If one uses real-space a metaphor for electronic spaces, Alexander's position suggests that there would be great benefit for an interface which breaks down any strict hierarchical organization. Thus, most common interfaces, while supporting a hierarchical structure for storage of information, includes many non-hierarchical alternatives for navigation, such as links in hypertext systems, transporting in virtual reality systems, or short-cuts in windowing systems. Each of these modifications can not only improve the usability of a system, but, in contrast to intuitions, can increase the correspondence between the structure of real-worlds and the structure of electronic worlds.

Navigation in the physical world then acts upon this complex representation. The process is not independent of the representation, but rather explicably tied together. One of the more interesting models how to relate the navigation processes, the external environment and cognitive representation of that environment is given by the self-organized inter-representation networks (Portugali, 1996).

3. Application I: hypertext navigation

As one moves from real world spaces to electronic spaces, such as hypertext or the World Wide Web (WWW), how does navigation evolve and adapt? First, a framework for when navigation is used must be established. Just as one often uses surrogates to aid in real world travel, electronic environments provide a number of surrogates to minimize the need for navigational expertise. One might arrive at an airport in an unfamiliar city and simply take a taxi directly to a pre-identified hotel. Likewise, there are occasions when a query is well formulated and specific enough to result in a single target page. However, more often a successful search query on the WWW would be directed to a collection of pages, images and links. A search engine, which fails to recognize that interconnectedness of pages, might result in a long list of pages from the same site, which would be better represented as a single hit. Likewise the traveler, who is looking for a good place to eat dinner in this unfamiliar city, might be directed to neighborhood known for culinary excellence, without a single target location in mind.

Furthermore, one often needs to navigate within sites and among nearby connected sites to locate the specific information that is needed. That is, the result of a search query is to move to the appropriate neighborhood, where navigational strategies will need to be adopted to move efficiently through the neighborhood. In turn, navigational overload may lead to the problem of getting lost (see Kim & Hirtle, 1995, for a review). This "lost-in-hyperspace" phenomenon occurs for several reasons. First, real space has real constraints, whereas hyperspace does not. Nodes might join in a strict linear order, a tree, a network, a cycle or any number of other topologies. Some topologies are indicative of a book, others of a museum, and others of an unorganized wilderness. You-are-here maps are either absent or uninformative when present.

The WWW provides a particularly interesting electronic environment, given the immense size, inherent complexity, and dynamic nature of the Web. The ability to find information in the WWW is dependent on a variety of inter-related factors, including the navigability of the space, the transparency of the information, and the expertise of the user. Tools must support both browsing and searching activities and these should be complementary. Fixed classification schemes were developed for storage, rather than browsing, and should be not be viewed as the solution to the complex problem of finding useful information.

3.1. Graph theoretic approach

Navigability of the graph has been formalized by Furnas (1997) among others. Furnas defines an Efficiently View Traversable (EVT) graph to a graph in which the number of outgoing links is small compared with the size of the graph and distance between pairs of nodes is small compared to the size of the structure. By example, he shows that modification of linear graph into a tree or into a fish-eye view will result in a increase in the traversability of the graph. Here a fish-eye view is taken to mean links from a site to other major headings within nearby neighborhoods are available from a site. Watts and Strogatz (1998) have also shown that adding only a few shortcuts dramatically decreases the distances in a network, but that providing additional shortcuts will not improve the efficiency of traversal further.

Furnas (1997) argues that traversability is not enough to make it navigable as the navigability depends on the ability to follow a cues. In particular, a space is view navigable (VN) if every node has good residue at every other node and the amount of out-link information is small. Residue is reflected in the semantic content of link labels. Thus, a dictionary has good residue, since in moving from one entry to another, it is unambiguous whether to scan forwards or backwards in the entries. Together, these two ideas, EVT and VN are needed for what Furnas (1997) calls Effective View Navigability (EVN). That is, a space must have an efficient structure and appropriate labels to lead a user to find the information that is needed.

3.2. Cognitive map approach

A second approach to the problem of electronic navigation is to turn to the heuristics that people use to navigate physical space. There are different types of spatial knowledge, such as route and survey knowledge. In simple spaces, individuals begin to acquire survey knowledge upon the first exposure to the space, whereas in complex spaces, such as hospitals, survey knowledge is rarely acquired even after years of experience.

Furthermore, aspects of the representation can be generalized to the characteristics of the physical space. For example, architects and urban planners have learned that undifferentiated spaces are harder to learn than rich environments. Even an idea as simple as using different colors on different levels of a parking garage will increase the likelihood of recalling where your car was parked upon return. Thus, aids in helping the user structure space and differentiate neighborhoods should lead to fewer errors and greater satisfaction with hypertext systems (Kim & Hirtle, 1995).

One might also consider the metaphor that users adopt in hyperspace (Gray, 1990; Kim & Hirtle, 1995). Here the focus is on the relationship between the virtual space and the users' understanding of the virtual space. A critical observation is that the virtual space need not have a physical correlate to be easily traversed, and the inclusion of a physical correlate does not guarantee avoiding disorientation. For example, understanding the mapping of a video game that assigns the top row to the bottom row, and the left edge to the right edge is easily understood and visualized, even if it is physically impossible in real-space. Likewise, as people may find themselves lost in a museum of interconnected rooms, the corresponding hyperworld would be equally disorienting (Foss, 1989), even though such a space obviously exists in the real world. Instead, disorientation is often the result of either adopting the incorrect metaphor or the lack of an appropriate metaphor.

On-line aids, such as history trees, maps, and fish-eye views, can assist the user both in developing an appropriate metaphor and locating one's self in the virtual space. Pointers with some degree of redundancy will tend to more useful. However, the exact methods, which prove to be of the most use in a given situation, will depend on the structure of the virtual space and the preferences of the user. Rarely do most information systems build on both of these factors.

In our own lab, we have most recently begun to explore these hypotheses in hypertext navigation, by examining the role of imposing structural cues in the virtual space. Such studies highlight the benefits and problems in generalizing about navigational behaviors between real space and virtual space. To test whether the ability to navigate in space in dependent upon cognitive structures, we have begun to examine in depth two structural characteristics of the environment: landmarks and regions. Since the writings of Lynch (1960), landmarks and regions have been identified as critical components for organizing space. The problem of how to transfer these concepts to electronic worlds is the focus of the two studies.

3.2.1. Landmarks

The ability to navigate in an environment is dependent upon one's ability to form a spatial representation of that environment, and landmarks play a key role in the creation of such a cognitive map. A landmark is an object or location external to the observer, which serves to define the location of other objects (or locations). Heth et al. (1997) describe two ways landmarks are fundamental to navigation. First, landmarks are the memorable cues, which are selected along a path, particularly in learning and recalling turning points along the path. Second, landmarks enable one to encode spatial relations between objects and paths, enabling the development of a cognitive map of a region. This distinction can also be described as landmark-goal relationships, where landmarks are cues along a path to a goal, and landmark-landmark relationships, which provide a global understanding of the environment (PastergueRuiz, Beugnon, & Lachaud, 1995). Sorrows and Hirtle (1999) argued that landmarks are important for navigation in both real and electronic environments.

Navigation can be considered in both open terrain and networked environments, and these environments may be either physical or electronic spaces. The term 'networked environment' refers to an area where movement is restricted to particular paths, such as cars driving on developed roads or a person following links in a hypertext environment. Open terrain environments are not restricted to movement along predefined paths, for example orienteering, open terrain robot navigation, or visualization interfaces for document spaces. In each of these environments, the purpose of navigation could be any of a variety of tasks or goals, such as directed at arriving a known goal, searching for a possible but uncertain goal, or meandering/browsing in the environment. This leads to the question of what tasks and in what environments landmarks are either beneficial or necessary and what types of landmarks work best in different environments.

In the World Wide Web, Mukherjea and Hara (1997) define a landmark as a node which is important to the user because it helps to provide an understanding of both the organization and the content of that part of the information space. Glenn and Chignell (1992) describe landmarks as part of a symbol system which is both visual and cognitive, and in which the visual and cognitive functions are intricately tied. Although these and other definitions of landmarks in the WWW seem compatible, a key problem exists in how to determine specifically what nodes are landmark nodes. Algorithms have been proposed which use the connectivity of a node, the frequency of use of a node, and the depth of the node in the local WWW directory structure. Sorrows and Hirtle (1999) and Sorrows (2004) have extended the typologies of landmarks to include three distinct categories: visual, structural and semantic. The categories are shown to apply to both real and virtual environments.

3.2.2. Neighborhoods

In many ways, neighborhoods form the dual of landmarks. Whereas landmarks represent a important beacons and/or decision points, regions suggest common constraints, such as, navigation tools, home pages, and indices in the case of electronic worlds.

Hirtle, Sorrows and Cai (1998) contrasted navigation through a hypertext space, with and without implicit neighborhoods defined, to show that the inclusion of neighborhoods increased the navigability of the space. In this study, neighborhoods were induced by coloring the background of a set of pages to be consistent with the content and structure of the pages in an academic department. For example, faculty pages might be blue and course information might be yellow. Search times were compared with sets of pages where the background was either monotone or colored randomly. Consistent with a theory of spatial information, the spaces where neighborhoods were indicated by color were easier to search.

4. Application II: spatial information kiosks

Informational kiosks for the display of geographical information have proliferated in past years. The ability to display images, maps, and text in dynamic displays has lead to the ability to provide complex directions with new clarity and precision (e.g., Tufte, 1997) or new interactive tools for exploring space in desktop settings (e.g., http: //www.news.harvard.edu/tour/qtvr_tour/index.htm). In contrast to paper-based cartographic maps, computer-based maps have the flexibility to present three dimensional images from a user centered perspective and provide zoom, pan, and indexing capabilities (e.g., Masui, Minakuchi, Borden and Kashiwagi, 1995). Furthermore, electronic displays can include visual components, which Barbara

Tversky (1993) has argued are included in our 'cognitive collage,' of an environment. By combining the foundations of cognitive mapping with the principles of good design, we have been able to build an effective, off-line, navigation system.

The initial Library Locator (LibLoc) system is a web-based browser that was designed to locate satellite libraries on the University of Pittsburgh campus (Hirtle and Sorrows, 1998). There are 17 small libraries on the campus and while most students would be familiar with a few of library locations, many of library locations are not well known. Furthermore, many satellite libraries are located in isolated locations deep inside an academic building. To further assist students in finding libraries, new versions of the LibLoc system have been constructed. In each case, the system consists of four frames as shown in Figure 1. The upper-left quadrant is the spatial frame that contains a map of the campus, a 2D floor plan, or a 3D model of the building. The upper-right quadrant shows a key image or sequenced image, such as the front door of the target building or the inside of the library. The lower-right quadrant gives verbal instructions as to the location of the library and the lower-left window provides instructions for the use of the system. All information is presented in a structured, hierarchical manner, to allow users to explore within buildings, as well as around the campus, with equal ease.

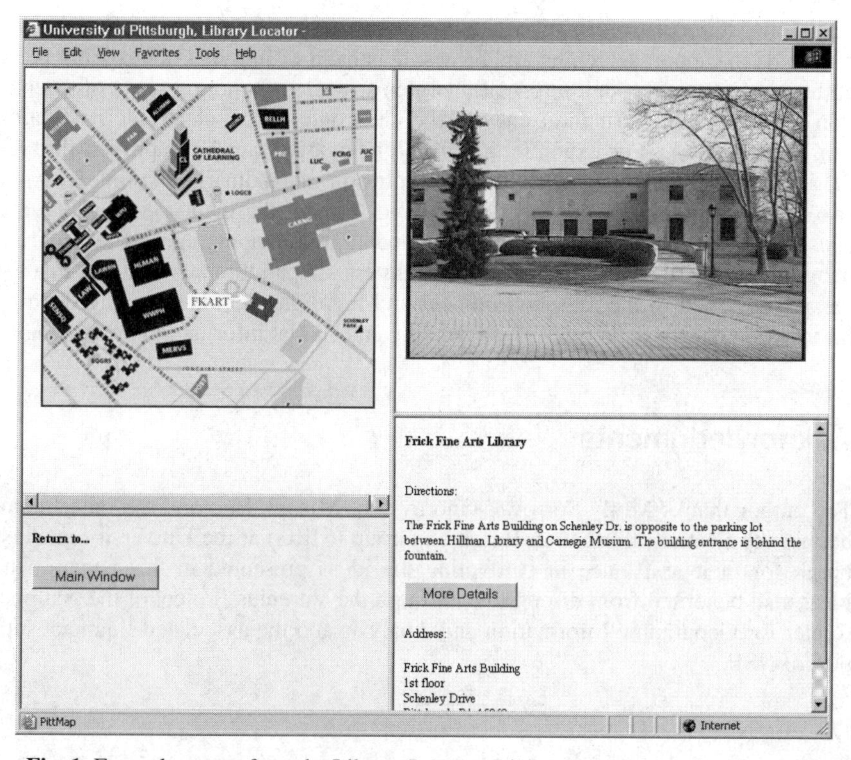

Fig. 1. Example screen from the Library Locator (LibLoc) showing maps, image and text in a yoked system.

LibLoc can be viewed as a research platform for exploring issues related to the construction of a cognitive collage.

For this reason, several versions of the LibLoc prototype have been constructed. For example, users may be presented with automatic slide shows of through the building or users may initiate image changes by tracing a path with the mouse. Thus, the LibLoc versions vary in terms of the richness of the representation and the level of cognitive load for the user. The automatic slide show present a rich set of slides of the environment, but requires greater a cognitive load to assimilate all the information. Another version of the system minimizes the cognitive load, but at the expense of the continuity of the visual environment. Together, the various versions argue that a navigation system for learning about a spatial layout can be improved by the use of redundant, multi-media information. Furthermore, the inclusion of a theoretical framework based on Tversky's (1993) cognitive collage will constrain and direct some of the possible design considerations for such a system.

5. Summary

The ability to locate information in a complex information space requires specialized tools to support searching and browsing behavior. Inherent in browsing is the ability to navigate through informational items, while retaining a sense of orientation. By providing structural cues to the electronic space, electronic navigation can be facilitated. For example, electronic navigation can be improved through the induction of neighborhoods and the inclusion of landmarks within electronic spaces. Information kiosks can be improved through the inclusion of visual images, maps, and directions, to provide redundant, overlapping cues of an environment. Together, these studies point to a richness of spatial representations that are inexplicably tied to the real environment and highlight the benefits of grounding information design in theories of wayfinding and spatial information processing.

Acknowledgments

The author thanks Molly Sorrows, Guoray Cai, Misook Heo and the other members of the Spatial Information Research Group (SIRG) at the University of Pittsburgh for their assistance in cultivating the ideas presented in this paper. The ideas also benefited from discussions through the Varenius Project of the National Center for Geographic Information and Analysis and the associated financial support of NSF.

References

Foss, C.L. (1989). Detecting users lost: Empirical studies on browsing hypertext, Technical Report No. 972, INRIA, Sophia-Antipolis.

Furnas, G.W. (1997). Effective view navigation, In Human Factors in Computing Systems CHI '97 Conference Proceedings, ACM.

Glenn, B.T. & Chignell, M. H. (1992). Hypermedia: Design for browsing. In H. R. Hartson & D. Hix (Eds.) *Advances in Human-Computer Interaction,* Vol. 3. Norwood, NJ: Ablex Publishing Corp.

Golledge, R.G. (1999). Wayfinding behavior : Cognitive mapping and other spatial processes. Baltimore: Johns Hopkins Press.

Gray, S. (1990). Using protocol analyses and drawings to study mental model construction during hypertext navigation, *International Journal of Human-Computer Interaction,* 2, 359-377.

Heth, C.D., Cornell, E. H. and Alberts, D. M. (1997). Differential use of landmarks by 8- and 12- year-old children during route reversal navigation, *Journal of Environmental Psychology,* 17, 199-213.

Hirtle, S.C. & Sorrows, M. E. (1998). Designing a multimodal navigational tool for locating buildings on a college campus, *Journal of Environmental Psychology,* 18, 265-276.

Hirtle, S.C., Sorrows, M. E., & Cai, G. (1998). Clusters on the World Wide Web: Creating neighborhoods of make-believe. In the *Proceedings of Hypertext'98.* Pittsburgh, PA.

Jul, S., Furnas, G.W. (1997). Navigation in electronic worlds, *SIGCHI Bulletin,* 29 (4), 44-49.

Kim, H. & Hirtle, S.C. (1995). Spatial metaphors and disorientation in hypertext browsing, *Behaviour & Information Technology*, 14, 239- 250.

Korfhage, R. (1997). Information Storage and Retrieval. John Wiley.

Lynch, K. (1960). *The Image of the City.* Cambridge, MA: MIT Press.

Masui, T., Minakuchi, M., Borden, G. R. and Kashiwagi, K. (1995). Multiple-view approach for smooth information retrieval, *ACM Symposium on User Interface Software and Technology,* Pittsburgh, 199-206.

McNamara, T.P., Hardy, J.K., & Hirtle, S.C. (1989). Subjective hierarchies in spatial memory, *Journal of Experimental Psychology: Learning, Memory and Cognition*, 15, 211-227.

Mukherjea, S. & Hara, Y. (1997). Focus+context views of world-wide web nodes. *Hypertext '97: The Eighth ACM Conference on Hypertext* (Southampton, UK). New York, NY: ACM Press.

PastergueRuiz, I., Beugnon, G., & Lachaud, J.P. (1995). Can the ant Cataglyphis cursor (Hymenoptera: Formicidae) encode global landmark-landmark relationships in addition to isolated landmark-goal relationships?, *Journal of Insect Behavior*, 8(1), 115-132.

Portugali, J. (1996). Inter-representation and cognitive maps. In J. Portugali (Ed.), *The Construction of Cognitive Maps*, Dordrecht: Kluwer, pp. 11-43.

Sorrows, M. E. (2004). *Recall of landmarks in information space,* Unpublished doctoral dissertation, University of Pittsburgh.

Sorrows, M. E. & Hirtle, S. C. (1999). The nature of landmarks for real and electronic spaces. Freksa, C., & Mark, D. M. (eds.) *Spatial information theory: Cognitive and computational foundations of geographic information science.* (Lecture Notes in Computer Science 1661). Berlin: Springer.

Tufte, E. R. (1997). *Visual Explanation*, Graphics Press, Cheshire, CT.

Tversky, B. (1993). Cognitive maps, cognitive collages, and spatial mental models. In (A. U. Frank and I. Campari Eds.), *Spatial information theory: Theoretical basis for GIS*, Springer-Verlag, Heidelberg-Berlin.

Watts, D. J. and Strogatz S. H. (1998). Collective dynamics of 'small-world' networks. *Nature*, 393, 440-442.

Enhancing the Legibility of Virtual Cities by Means of Residents' Urban Image: a Wayfinding Support System

Itzhak Omer, Ran Goldblatt, Karin Talmor, Asaf Roz

Abstract. In this paper we present an operative Wayfinding Support System (WSS) for a virtual city using the virtual model of Tel Aviv for targeted and exploration wayfinding tasks. The WSS was developed under the assumption that a design of a virtual city should allow a transfer of spatial knowledge from a real city to its virtual representation. Accordingly, the information for this system was obtained from an empirical study on Tel Aviv residents' urban image by using their city sketch maps. The WSS uses the topological structure between the urban elements in these sketch maps to decide which elements would be highlighted to the virtual city user, according to the observed urban environment and to the user's real time log navigation parameters (scale and perspective).

1. Introduction

Virtual Environment (VE) is a real-time simulated environment that enables the end user to virtually walk through or fly over a specific terrain. Generally speaking, however, although VEs differ in scale, realness and levels of immersion, their users are more than likely to experience difficulties locating their current and desired destinations and have problems maintaining knowledge of locations and orientation (Darken, 1995; Stuart and Lochalen, 1998).

These problems can be related to several characteristics of VEs. First, the real time nature of VE, which is characterized by high speed of locomotion, different viewing perspectives and varying geographical scales, making it an unfamiliar and non-intuitive experience. Second, many of the VEs, and mainly desk-top VEs, require movement in the virtual space using standard input devices (mouse, keyboard etc.), which can affect accuracy of how far and in which direction the user has moved. Third, lack of "presence" and restricted fields of view may also contribute to poor performance in VE navigation (Witmer et al., 1996; Waller et al, 1998). The degree of these problems depends on the VE type, the wayfinding task in question and the user's ability, but also on the design of the VE. Proper VE design and the availability of navigational aids can allow for effective and efficient wayfinding task performance.

In this paper we concentrate on the design of a virtual model of a real city to support wayfinding tasks. Due to the large scale and high density of a virtual city, the wayfinding difficulties one can experience are concerned mainly with direction, orientation, location and distance between places and the relative positions of places. This kind of knowledge, which is known as configurational or survey knowledge (Golledge, 1992), allows people to navigate in a real geographical environment (Montello, 1998; Kitchin, 1996) as well as in a virtual environment (Janzen et. al., 2001).

Configurational knowledge is taken into consideration in VE studies by adding navigational tools, constructing guided navigation systems and designing the simulated environment e.g. local and global landmarks, mostly according to theoretical principles or the cartographic intuition of the designer, and virtual cities are no exception (Batty et al., 1998). However, the uniqueness of creating virtual models of real cities lies in the possibility of knowing how their residents perceive them i.e. which elements are more imageable and how they interrelate, and to use this knowledge to design a more legible simulated environment. This approach of using empirical information, we believe, could help the transfer of spatial knowledge from a real city to its virtual representation in different wayfinding tasks. We refer here to the term wayfinding as the actual application of the knowledge the users have in a process of navigation i.e. the process of determining and traveling along a path through an environment (Darken & Silbert, 1993). Wayfinding comprises both targeted and exploration tasks. *The first refers to a* movement with respect to a specified target *whose location may or may not be known, and the second refers to a passive or active movement* in an effort to learn about what objects are involved and the spatial relationships between objects (Kitchin and Freundschuh, 2000).

The aim of this paper is to present an operative Wayfinding Support System (WSS) for a virtual city using the virtual model of Tel Aviv for targeted and exploration wayfinding tasks. The information for this system was obtained from an empirical study on Tel Aviv residents' urban image by using their city sketch maps. We used the Q-Analysis method to identify the topological structure between the elements in these sketch maps. The WSS uses this topological structure to decide which elements would be presented to the virtual city user, according to the geographical context and to the user's real time log navigation parameters: perspective and scale.

In the next section we describe the applied methods aiming to improve the legibility of virtual environments for supporting navigation. The last section describes the identification of structural relations in Tel Aviv residents' urban image, the principles of WSS and its implementation for the virtual model of Tel Aviv.

2. Principles and tools for supporting VE wayfinding tasks

Intensive efforts are being made to improve navigation and element recognition in VEs. These efforts can be broken down into design principles of the simulated environment, navigational aids, manipulating display parameters and guided navigation systems.

The principles suggested in the literature for improving the legibility of the simulated environment can be categorized on the basis of Lynch's framework. Accordingly, landmarks were found critical for navigation suggesting that any VE should include several memorable landmarks on all scales in which navigation takes place (Charitos and Rutherford, 1996; Vinson, 1999; Jansen-Osmann, 2002). Combining paths, "places" (a distinct, recognizable location), nodes and edges in a VE were also found to be essential for effective navigation (Darken & Silbert, 1996). In addition, the importance of the relation between the elements for navigation enhancement is emphasized in several VE studies: paths should have a clear structure and start/end points, or nodes, such as paths junctions, can provide a structure for placing landmarks (Vinson, 1999). This spatial structure aims to help the user to mentally organize the VE representation (Darken & Silbert, 1996).

Navigation tools such as map and compass are mostly necessary to display the user's position and orientation. A compass contributes to spatial orientation by adjusting the simulated geographical environment to a frame of reference, while a map assists in spatial orientation and in collecting spatial information from the surrounding environment (Chen and Stanney, 1999). In addition, a simulated physical cue such as a virtual sun, viewed from any vantage point, improves performance on a search task, mainly due to its relative immobility and its visibility (Darken and Silbert, 1993).

Another dimension of VE design concerns the manipulation of display parameters: field of view (FOV), graphics eyepoint elevation angle (EPEA) and viewing perspective heights (VPH). The use of a larger FOV allows the user to integrate larger amounts of spatial information more quickly for search tasks (Tan et al, 2001; Nash et al, 2000), and is necessary for a complete and accurate sense of space. However, larger FOV might produce distortions in the evaluation of spatial relations. Therefore, as Kalawski (1993) states, it is important to determine what FOV is required according to the task at hand. EPEA defines the perspective on, and the distance from a given object. When a large positive EPEA is used, producing a top-down view, it makes it difficult to judge the elevation differences but, on the other hand, small positive EPEA causes compression of the altitude dimension. Likewise, the relation between EPEA and FOV can affect judgment of elevation and azimuth (Barfield et al, 1995). The VPH is often determined by the area the user wants to see, but the most effective viewing height would be one enabling both landmark recognition and a large-scale view (Witmer et al., 2002).

Based on the above principles for simulated environment design, and taking into account the implications of display parameters, varied guided navigation systems were constructed. The advantage of a guided navigation lies in the reduction of the cognitive load from the user, enabling a keener focus on the tasks rather than spending efforts on navigation. Some of them are simply a guide for defined locations. An example of such a system was constructed by Haik et al (2002). The guided navigation consists of a simple map and navigation arrows positioned in the environment in potential problem areas. Clicking on an arrow takes the viewpoint to the related area of interest.

More sophisticated and interactive guided navigation systems were constructed by utilizing the user's real time log navigation and by fitting one or more of the displaying parameters accordingly. Bourdakis (1998) introduces a theoretical model for urban environment, based on the recognition that at any given position there must be a minimum amount of information i.e. sensory cues. The basic idea behind the model is to deal with the many problems that arise when employing a "flying" based navigation mode in large geographical virtual worlds. The model uses a series of gravity-like nodes that pull the viewer towards the ground when approaching an area of high density in sensory cues. Similarly, in a low-density area, a global negative gravity pulls the viewer towards the sky until there are enough sensory cues. The cues presented are determined according to the reference location and the user's real time log movement. Another system suggested by Tan et al (2001) defines viewpoint motion by position, orientation and speed. The technique offeres links speed control to VPH and EPEA, allowing the users a seamless transition between and navigation within local environment views and global overviews. A logarithmic function was implemented for speed control so that the speed was scaled with respect to the user's distance to a target in the environment - the faster the user moves forward, the higher the VPH i.e. the zoomed-out position.

The design principles, navigational aids, display parameters and guided navigation systems presented above can be used for improving VE navigation, depending on the VE type, quality of the simulated environment, wayfinding tasks and the user's ability. In the next section we describe the aim of the WSS we suggest according to the type of VE and the potential users in the system.

3. WSS for the virtual model of Tel Aviv: the rationale and methodology

The specific aim of the WSS constructed for this model is to support wayfinding in the virtual model of a real city where the city is familiar to the users. We developed the system for the virtual model of Tel Aviv city, an area of about 50 squared km. The virtual model is a 3D desk-top VE built from actual terrain data by Skyline® software. This software integrates DTM and GIS features to generate a geographically detailed virtual environment. The model is constructed from DTM in a resolution of 50 m grid, digital orthophoto at a resolution of 0.25m pixel and GIS layers of building height and street network.

The trigger for the present study is the difficulty to navigate within this model. The WSS constructed for this model was based on information of structural knowledge acquired from Tel Aviv residents' urban image. The process of collecting information and identifying structural relations, the principles of WSS and its implementation are presented below.

3.1. What is the appropriate design for enhancing wayfinding in a virtual model of a real city?

The specific difficulties of navigation in a large-scale virtual model of a real city lie, on one hand in the many great possibilities available to users due to the real time movement in open terrain: they can move at high speed in all directions with no constraints, and view the terrain in different perspectives and scales. On the other hand, users face unfamiliar restrictions that characterize all desk-top VEs as described above. These characteristics can lead to disorientation in wayfinding tasks.

The incorporation of navigation aids, such as compass and map, and highlighting important elements like local and global landmarks, paths and nodes clearly helps improve orientation in the virtual model of a real city. However, the main question is how to design the simulated environment, that is, which elements are necessary for navigation on different scales and in different observed areas of the city?

One method for deciding which objects are important for enhancing legibility, and thus will be displayed in the simulated environment, is to gather the data from individual sketch maps into one aggregate map. In order to do so, 32 Tel Aviv residents were asked "to draw a map of Tel Aviv and to sign/draw the dominant elements in it (no more than 15 elements)". Then an aggregate map was extracted representing the urban image or the aggregative external cognitive representation (see Figure 1), as suggested by Lynch (1960) and as used recently by Al-Kodmany (2001).

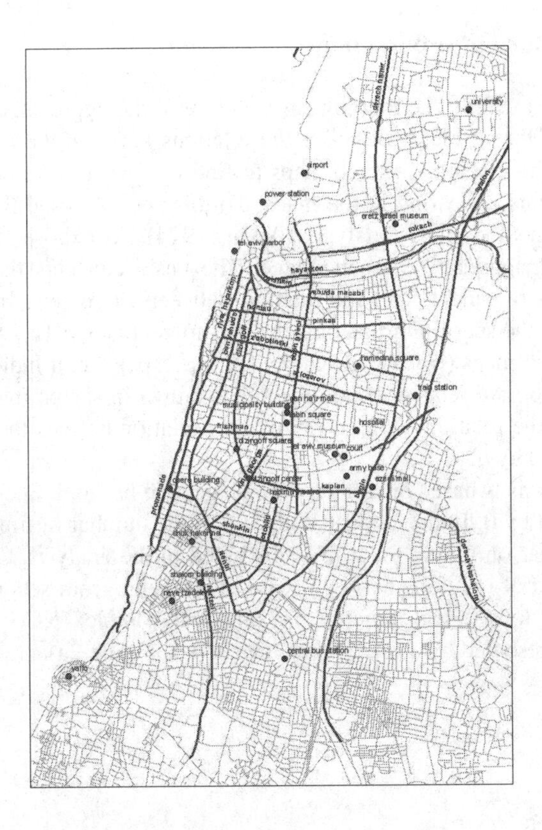

Fig. 1. The urban image of Tel-Aviv.

Though its efficiency for improving the legibility of simulated environment has been proved, this method is not appropriate when it comes to estimating the structure and quality of the connections between these elements. Lynch was aware of this problem and noted that "there was a lack of information on element interrelations, patterns, sequences and wholes. Better methods must be evolved to approach these vital aspects" (Lynch, 1960, p.155). Moreover, there is also a need to reveal how this knowledge is processed and used in wayfinding practice (Kitchin, 2002, p.15).

We assume that the structure and quality of the connections between these elements could help the users to form a spatial representation of a given observed environment. Namely, to help them organize the relations between urban elements in a local and global context. Such spatial representation is necessary for navigation - it enables the user to place his location with reference to other locations to help him know exactly where he is and how to arrive to other locations. It is worth noting that this is a working assumption at this stage, because as yet we don't have a clear idea on the navigation strategies of the user when he or she transfers spatial knowledge from a real city to its virtual model. If this assumption is acceptable however, we need a method by which we can verify that.

3.2. The structure of urban image

Our approach is to study the topological structure in the aggregative urban image to find out how the residents organize the relations between the urban elements. Thus, we used the residents' sketch maps to find the structure of connections between the elements appearing in the maps. To this end, we used the multidimensional scaling method of Q-Analysis (Atkin, 1974), to expose the topological structure in that aggregative urban image. The basic concepts that underlie Q-analysis are sets of objects and the relation between these sets. In terms of our context, let C be the set of m urban objects (i=1..m) so that $C=\{c_1, c_2, \ldots, c_m\}$, and P a set of n sketch maps (j=1..n) so that $P=\{p_1,p_2,\ldots, p_n\}$. Let μ indicate that a pair of elements (c_i, p_j) are related. If an object c_i is drawn in sketch map p_j, then c_i is related to p_j by the relation μ: $(c_i, p_j) \in \mu$. This relation defines the dimension of an object denoted by q.

Since Q-analysis is based on a binary language, one has to define the slicing parameter, denoted by θ. In the current study $\theta = 3$; the number of times an object is required to appear on a map for it to be included in the analysis. On the basis of presence/absence of relations between pairs of elements from sets C and P (incidence matrix) a topological structure or simplical complex $KC(P; \mu)$ was constructed and represented by a shared face matrix (see Figure 2a).

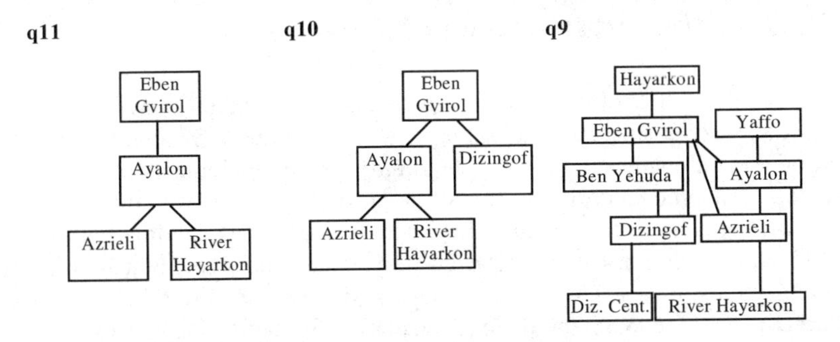

Fig. 2a. The shared face and topological representation of parts of the Tel Aviv urban image objects at dimensions 9-11

With the shared face matrix we can see the direct relation between objects in each dimension (q-near) e.g. 3-near denotes that the objects appear together in q+1 (4) sketch maps. q-connectivity means the objects could be connected transitively by different q+1 sketch maps i.e. q-connectivity indicates that the objects appear in q+1 sketch maps directly or indirectly. In this way, the topological structure KC(P; μ) enables us to adjust the simulated environment to the user's knowledge by identifying the set of objects connected directly and indirectly to each object and its place in the topological structure, as illustrated in Figure 2b.

AM	DS	HS	RS	DC	HR	Aya	EG	Diz	BY	Hay	
5	0	1	3	3	3	7	9	5	5	11	Hayarkon (Hay)
5	3	2	2	6	5	6	9	9	10		Ben Yehuda (BY)
8	5	3	5	9	7	8	10	15			Dizingoff (Diz)
9	5	3	6	7	7	11	19				Eben Gvirol (EG)
12	4	6	4	8	11	18					Ayalon (Aya)
9	5	5	4	7	14						Hayarkon Rive (HR)
8	5	7	8	16							Dezingoff Center (DC)
4	3	3	11								Rabin Square (RS)
5	3	10									Hamedina Square (HS)
5	10										Dizingoff Square (DS)
17											Azrieli Mall (AM)

Fig. 2b. The shared face and topological representation of parts of the Tel Aviv urban image objects at dimensions 9-11

4. WSS for the virtual model of Tel Aviv:

To improve the ability of the user to construct coherent spatial representation during targeted movement and exploratory movement, the WSS construction was based on the following guideline: to highlight the urban elements in a given observed environment according to the topological structure to which the reference objects belong. The system enables three wayfinding modes: landmark search, address locator (both constituting targeted movement) and free flight (constituting exploratory movement). The following section describes and illustrates the operation principles of the WSS and its implementation for navigational tasks.

4.1. The operation principles of the WSS

The operating height range of the WSS is between 250-7,000 m. The minimum viewing perspective height (VPH) of 250 m is defined as a local scale according to the terrain resolution. At this height (scale) all the objects that are included in the topological structure should be displayed once the reference object is identified. Above 7,000 m a constant global spatial structure is presented, independent of perspective or a reference geographic object.

Since increasing VPH above local scale would likely result in high density of objects, a generalization method is needed to produce optimal visual complexity. For that purpose, four intermediate hierarchic levels Li (i=1…4) are defined according to the relation between VPH and the topological structure KC(P; μ): L1(q_{2-4}, 300-600 m); L_2 (q_{5-6}, 600-1,500 m); L3 (q_{7-8}, 1,500-3,000 m) L4 (q_{9-10}, 3,000-7,000 m). Accordingly, let L_{local} (q_1, 200-300 m) indicate the local-scale spatial structure (minimum VPH) and L_{global} ($q_{>11}$, >7,000 m) indicate the global spatial structure. Based on this hierarchy, a generalization process determines which objects will be presented according to the following steps:

1. The reference object identification process varies according to the wayfinding mode at hand; in the landmark search, the chosen landmark constitutes the reference object. In the address locator and free flight modes, the "nearest element" method is implemented. In this case a desired location is reached, the system finds the urban image element nearest to it, and identifies the urban image element as the reference object.
2. When movement stops, the reference object is identified and the VPH is set, the objects that belong to the relevant hierarchical level are displayed. Let Lc indicate the current displayed hierarchical level.
3. For connecting the reference object to the spatial structure of Lc, additional objects are needed to create hierarchical spatial structure. These objects are chosen from L_{c-1} objects that are q-near with the reference object (where q is the highest dimension in which the reference object has a direct connection with at least k objects of L_{c-1} level). The parameter k denotes the minimum amount of objects from the adjacent hierarchical level that are needed to be presented. For the application we used in the current study, we decided that k=3. In this way, a topological connectivity was established between a given

reference object and a more global structure in a certain scale/VPH. That is, a vertical connectivity between the dimensions of hierarchic levels L_{c-1}–Lc and between the higher dimensions' objects of L_{c+1}–L_{global}.

Aiming to achieve continuity and to prevent visual distortions, we decided that the display parameters: geometric field of view (FOV) and the eyepoint elevation angle (EPEA) would be constant at 53° and 30° respectively (in free flight mode the EPEA varies). This relation between FOV and EPEA was chosen considering the resolution of the simulated urban environment and the elevation differences, and it also corresponded with results of empiric experiments (Barfield et al, 1995). Hence, the VPH and distance from the referenced geographical object are interactively changed according to the geographical context.

4.2. The implementation of WSS for navigational tasks: illustration

The interface of WSS includes several components (see Figure 3). The main screen displays the 3D simulated environment including a compass (can not be seen in the Figure). In the lower left side of the screen, a 2D map representing Tel Aviv residents' urban image is located. This map is synchronized with the observed environment, and the user's position is indicated on the map at all times. In the upper left side of the screen the Wayfinding tool is located.

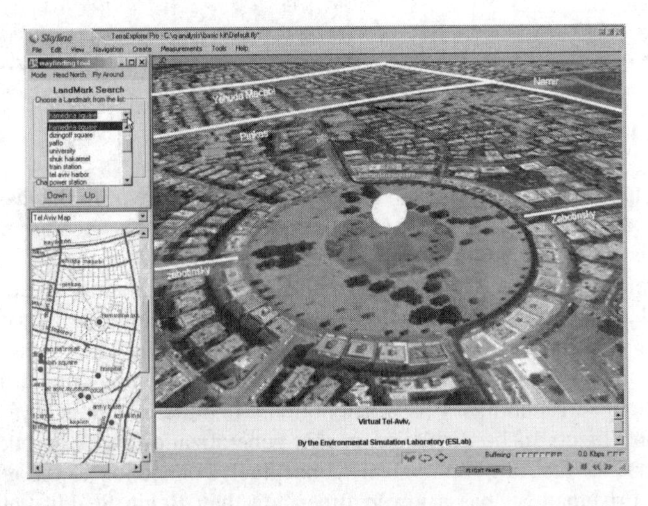

Fig. 3. WSS interface including the main screen displaying the 3D simulated environment, a 2D map representing Tel Aviv residents' urban image and the wayfinding tool.

This tool allows the user three Wayfinding modes: landmark search, address locator and free flight (see Figure 4): The 'landmark search' mode enables the user to choose an urban image element (landmark) from a predefined list. The user can easily move up or down the scales using the 'change scale' buttons (Figure 4a). The 'address locator' mode enables the user to reach any address in the city using an address locator engine. The 'change scale' buttons are applicable in this mode as well (Figure 4b). In the 'free flight' mode the user can fly in the simulated environment manually by using the mouse or the keyboard. At any given time, the user can stop his motion and ask for relevant information by clicking the 'info' button (Figure 4c). In addition, two orientation aids are available at all times. The 'Head north' option changes the viewing perspective to a northerly direction and the 'Fly around' option enables viewing the reference object in a circle pattern "automatically".

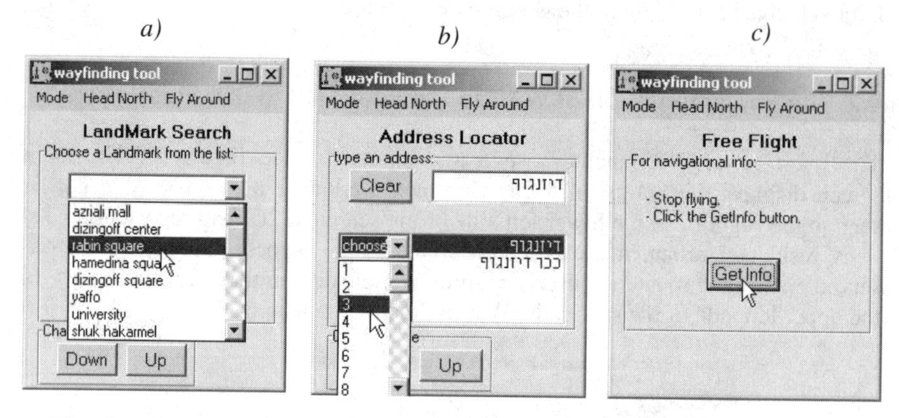

Fig. 4. Three Wayfinding modes of WSS: (a) Landmark search; (b) Address locator; (c) Free flight

The implementation of the WSS is illustrated in Figure 5. In this case a specific address was requested, indicated by the white arrow. The nearest urban image element is the court building and is therefore identified as the reference object. The movement between scales results in changes in the quantity and quality of the displayed elements; more high-dimension objects appear (turn on) while low-dimension objects disappear (turn off) in order to maintain a hierarchical topological connection between the court building (the reference object) and the observed geographical surroundings. It is worth noticing that there is no need for the higher dimension objects to be displayed in this generalization process, but only those objects that have strong topological (conceptual) connection to the reference object, e.g. Frishman St. has a lower dimension than Begin St. but only Frishman appears in L_2 scale, because it has heigher q-connectivity than Begin St.

Fig. 5. Changing the simulated environment according to scale changes and to the topological and geographical connection between the reference object (the court building) and the observed area

5. Discussion

The uniqueness of the operative approach to support wayfinding in a virtual ci
proposed in this paper is its contextual character: using empiric spatial knowle
for design of the simulated environment and the ability to change this environm
according to the observed geographical area and the real time display paramet

The imageable city objects displayed by the WSS could help the virtual city user to maintain knowledge on position and orientation due to the hierarchal structure and the topological (conceptual) connectivity. A network of recognizable urban objects that provide the relative distance and information among them is the basis for a representation which can permit an effective navigation. Though the transfer of spatial knowledge between real and virtual environments is not yet clear (Peruch et al., 2000), the displayed spatial structure has the potential to support "piloting" (position-based) and "path-integration" strategies that are usually involved in human navigation (Loomis et al. 1999). For example, when employing a "flying-based" navigation mode, the topological structure of the streets allows the user's orientation, relying on global and local spatial relations. This structure enables the user to move along a certain street or to follow the relative order of streets and, by that, to update distances and relative bearings of objects.

However, it is important to remember that the choice of a certain navigational strategy not only depends on the cognitive representation of the environment or on the quality of environmental knowledge but also on the design method of the simulated environment e.g. the displayed or highlighted urban elements and their spatial structure. Hence, further study is needed to evaluate the efficiency of the conceptual and operative approach we presented in this paper for supporting wayfinding, and to clarify its influence on navigational strategies.

References

Al-Kodmany, K. (2001). Supporting imageability on the World Wide Web: Lynch`s five elements of the city in community planning, *Environment and Planning B: Planning and Design,* vol. 28, pp. 805-832.

Atkin, R.H. (1974). *Mathematical Structure in Human Affairs.* London: Heinemann.

Barfield, W., Hendrix, C. and Bjorneseth, O. (1995). Spatial performance with perspective displays as a function of computer graphics eyepoint elevation and geometric field of view, *Applied Ergonomics*, Vol 26, No. 5.

ıtty, M., Dodge, M., Doyle, S. and Smith, A. (1998). Modeling virtual environments. In (Longley, P.A., Brooks, S.M., McDennell, R and Macmillan, B., eds.) *Geocomputa- ion: A Primer.* Chichester: John Wiley, 139-61.

kis, V. (1998). Navigation in Large VR Urban Models, J.-C. Heudin (Ed.): Virtual lds 98, LNAI 1434, pp. 345-356.

). and Rutherford, P. (1996). Guidelines for the design of virtual environments. Virtual Reality Special Interest Group Conference, Leicester, Proceedings. ality Special Interest Group and Contributors, UK, 93–111.

ney, K.M., (1999). A theoretical model of wayfinding in virtual environ- sed strategies for navigational aiding. *Presence: Teleoperators and Vir- nts*, 8 (6), 671–685.

J. (1993). A toolset for navigation in virtual environments, *Proceed- mposium on User Interface Software and Technology, Atlanta. GA.* ww.movesinstitute.org/darken/publications/toolset.pdf

ding in large-scale virtual environments, available on: hi/chi95/Electronic/documnts/doctoral/rd_bdy.htm

Darken, R. and Silbert, J. (1996). Navigating large virtual spaces, *International Journal of Human-Computer Interaction*, 8 (1).

Golledge, R. G. (1992). Place Recognition and Wayfinding: Making Sence of Space. *Geoforum,* 23(2): 199-214.

Haik, E., Barker, T., Sapsford, J. and Trainis, S. (2002). Investigation into effective navigation in desktop virtual interfaces. Available on: 'http://delivery.acm.org/10.1145/510000/504513/p59-haik.pdf

Jansen-Osmann, P. (2002). Using desktop Virtual Environments to investigate the role of landmarks, *Computers in Human Behavior,* **18,** 301-311.

Janzen, G., Schade, M., Katz, S. and Herrmann, T. (2001). Strategies for Detour Finding in a Virtual Maze: The Role of the Visual Perspective, *Journal of Environmental Psychology,* Volume 21, Issue 2, Pages 149-163.

Kalawski, R. S. (1993). The Science of Virtual Reality and Virtual Environments. Addison-Wesley, Reading, MA.

Kitchin, R.M. (1996). Increasing the integrity of cognitive mapping research: appraising conceptual schemata of environment- behavior interaction. *Progress in Human Geography,* 20, pp. 56-84.

Kitchin, R. and Freundschub, S. (2000). In Kitchin, R. and Freundschub, S. (eds.) *Cognitive Mapping*. Routledge Taylor & Francis Group, pp.1-9.

Kitchin, R.M. (2002). Collecting and analysing cognitive mapping data, In (Kitchin, R. and Freundschuh, S., eds.) *Cognitive Mapping: Past, present and future,* Routledge, 9-24.

Loomis, M. (1999). Dead reckoning (path integration), land-marks, and representation of space in a comparative perspective. In (R. G. Golledge, ed.), *Wayfinding Behavior Baltimore*, Johns Hopkins University Press, pp.197-228.

Loomis, J.M., Klatzky, R.L., Golledge, R.D. and Philbeck, J.W. (1999). Human Navigation by path integration. In (Golledge, R.G., ed). *Wayfinding Behavior.* The Johns Hopkins University Press, 125-152.

Lynch, K. (1960). *The Image of the City.* CambridgeL MIT Press.

Montello, D. (1998). A new framework for understanding axquisitin of spatial knowledge in large-scale environments. In (M. J. Egenhofer , C.G. Volledge, eds.), *Spatial and temporal reasonig in geographic information systems*, New York: Oxford Universtity Press, 143, 154.

Nash, E.B., Edwards, G.W., Thompson, J.A. and Barfield W. (2000). A Review of Presence and Performance in Virtual Environments, *International Journal of Human–Computer Interaction*, 12(1).

Peruch, P., Gaunet, F., Thinus-Blanc, C. and Loomis, J. (2000). Understanding and Learning Virtual Spaces. In (R. Kitchin, and S. Freundschub, eds.) *Cognitive Mapping.* Routledge Taylor & Francis Group, 108-125.

Stuart, C.G and Lochlan, E.M. (1998). Contribution of propioception to navigation in virtual environments, *Human Factors*, Vol. 40 (3).

Tan, D.S., Robertson, G.G. and Czerwinski, M. (2001). Exploring 3D Navigation: combining speed-coupled flying with orbiting. Available on: http://www-2.cs.cmu.edu/~desney/publications/CHI2001-final-color.pdf

Vinson, N.G (1999). Design Guidlines for Landmarks to Support Navigation in Virtual Environments, published in *Proceedings of CHI `99, Pittsburgh, PA., May, 1999*

Waller, D., Hunt, E., and Knapp, D. (1998). The Transfer of spatial knowledge in virtual environment training. *Presence: Teleoperators and Virtual Environments,* 7(2).

Witmer Bob G., Balley John H. and Knerr Bruce W. (1996). Virtual Space and real world places: transfer of route knowledge, *Int. J. Human-Computer Studies,* **45,** 416-428
Witmer, B.G, Sadovsky, W.J. and Finkelstein, N.M (2002). VE-Based Training strategies for acquiring Survey knowledge, *Presence*, vol. 11 no. 1.

Small World Modeling for Complex Geographic Environments

Bin Jiang

Abstract. This paper aims to provide some insights into geographic environments based on our studies using various small world models. We model a geographic environment as a network of interacting objects - not only spaces, places and locations, but also vehicles and pedestrians acting on it. We demonstrate how geographic environments might be represented as a form of networks and be illustrated as small worlds. Furthermore we try to shed light on the implications of small world properties from various application perspectives.

1. Introduction

Conventional geometric-based networks have various limitations when modeling geographic environments. Essentially they are geometric representations, and thus interrelationships between locations or objects are not well defined and represented. Therefore it is difficult for the model to deal with issues such as dynamics, growth and evolution. Now researchers tend to view geographic environments as a complex network, and treat them as such with a range of features such as nonlinearity, interdependence and emergence. For instance, a topological based network representation is suggested as an alternative model for the study of the evolution of street networks (Jiang and Claramunt 2004). In modeling a built environment and terrain surface, the concept of a visibility graph is introduced to show the interrelationship of individual locations in terms of visual accessibility (Turner et al. 2001, Jiang and Claramunt 2002). These efforts represent a new wave of studies of geographic environments using a topological-based network view following the main stream of study on complex networks.

We model geographic environments as a network of interacting objects – not only spaces, places and locations, but also vehicles and pedestrians acting on it. This view goes beyond the conventional geometric-based network, in the sense of dealing with neighboring, adjacency and relationships, and it provides an alternative representation of geographic environments with respect to geographic modeling. For instance, agent-based and cellular automata modeling can be based on the topological based representation (O'Sullivan 2001). This topological model has been considered in a variety of disciplines in the study of complex networks involving the Internet, cells, scientific collaboration, and social networks, to mention a few examples (see Strogatz 2001 for an overview). The nature of geographic environments, as a complexity system, can be illustrated via the underlying networks. We can examine whether or not there is a hidden order behind a geographic environment, on the one hand, and study how global properties are emerged via the interaction of individuals modeled as the vertices of the underlying network, on the other. It helps understand that complexity systems are more than sum of individual components, and this extends the reductionism towards interrelations of the individual components. This has been a major development stream in the current study of complexity theory (Holland 1995).

Recent advances in the study of complex networks have been tremendous. The availability of various datasets of real-world networks and powerful computers has made possible a series of empirical studies since the seminal work by Watts and Strogatz (1998). The concept of small world networks has featured in many scientific journals and conferences, and has been becoming an increasingly interesting topic being investigated as an emerging science (Barabasi 2002, Watts 2003).

The small world network is a network with a kind of hidden order between regular and random networks in analogy with the small world phenomenon observed in social systems (Milgram 1967). The small world phenomenon states that in a large social system, e.g. a population of a country, the distance between any two randomly chosen persons, e.g. yourself and the president of your country, is just about six persons away, so called "six degrees of separation". Whether the number really is six remains a matter for debate (Kleinfeld 2002), but most real world networks are indeed small enough, as evident in many real systems mentioned above. A real world system with which the underlying network demonstrates the small world phenomenon is likely to be an efficient and stable system in terms of the diffusion of a variety of phenomena ranging from information to terrorist networks and AIDS epidemics. Although originally observed as a property of social systems, the small world phenomenon has been examined in a variety of natural systems with which basic units are interconnected as a complicated network.

This paper aims to investigate how small world models can be used for modeling and understanding geographic environments from a structural point of view. We start with an introduction to small world networks, followed by Watts and Strogatz's small world model and several other models that expand on the small world concept in various ways. We then examine how small world properties are demonstrated with geographic environments and further discuss the implications of the small world properties from various application perspectives, in particular linking to the issue of search efficiency with a geographic environment, and of designing a geographic environment with high search efficiency. This chapter concludes with a summary and outlook for future work.

2. Small world networks and models

2.1. Small world networks

To explain what a small world network is, let us consider a visibility graph in terms of how each point location is visible to every other within, for example, a downtown area. We impose in a grid fashion a set of point locations between buildings and examine how these point locations, represented as nodes, are visible to each other. If there is no obstacle between two point locations (nodes), then there would be a link between the two nodes. In this way, we would see that those visible locations from a given location are likely to be visible as well. In other words, the visibility structure is highly clustered. This is the first characteristic of small world network, and it is measured by the *clustering coefficient*. If visible locations from a given location are not visible to each other at all, the cluster coefficient equals 0. On the other hand, if visible locations from a given location are all visible to each other, then the clustering coefficient equals 1. The ratio of actual links over all possible links among a set of visible locations (for n visible locations, all possible links is $n(n-1)/2$) from a given location is defined as clustering coefficient. The average of all locations' clustering coefficient is that of an entire network.

There is another important property that characterizes a small world network. In most cases, two locations are not directly visible, but they may be visually accessible via another location. In this case, these two locations are visually separated by a distance of two. In some cases, two locations may be separated by a distance of three or four etc. The distance is actually the notion of *shortest path length*. The average of the shortest path length between all pairs of locations for a visibility graph is called the *characteristic path length*, which shows the separation between any two randomly chosen locations.

In summary a small world network is the network with a high clustering coefficient and a low characteristic path length. It resembles a regular graph in the sense of high clustering and resembles a random graph in the sense of short path length. The two properties determine an efficient structure for a small world network in terms of information propagation. In the following subsections, we will briefly introduce various models that illustrate a number of features of the real world networks.

2.2. Rewiring simulation (W-S model)

A regular graph has a high clustering coefficient and a long characteristic path length. On the other hand, a random graph has a low clustering coefficient and a short characteristic path length. These properties of both regular and random graphs prompt exploration of small world behaviour. Watts and Strogatz (1998) designed a simulation through rewiring a few links of a regular graph, i.e. to introduce a few random links to replace the original neighbouring links (Figure 1). They define a parameter to control the level of randomness. When the parameter equals to 0, it represents purely ordered graph, not random at all; when the parameter equals to 1, it represents purely random graph. Imagining a regular graph represented as in Figure 1, the simulation starts rewiring a few links, which means that fix one end of the links and randomly rewire to another node in the ring. Through carefully controlling the parameter, a range of graphs with different levels of randomness is created. Between the two extreme graphs, it is found that there is a range where networks maintain both high clustering coefficient as the regular graph and a low path length as the random graph. The networks that combine the two properties are the small world networks.

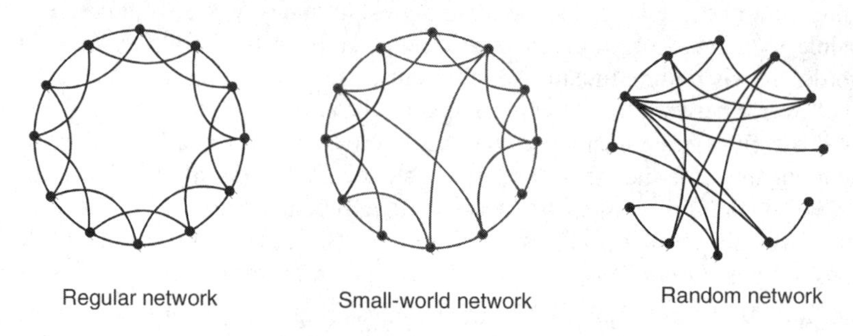

Fig. 1. Regular, small world and random networks

2.3. Efficient behaviour of small world networks (M-L model)

Marchiori and Latora (2000) investigated various small world networks and suggested a generalized concept of connectivity length toward a better understanding of small world behaviour. The concept is defined by the harmonic mean (rather than the arithmetic mean) of the shortest distances among all pairs of nodes within a graph. The model also relaxed several constraints set by the initial small world model by Watts and Strogatz (1998) and extended it to weighted and unconnected graphs. What is important for the model is that it brings an efficiency view into the small world networks, i.e. a small world network is an efficient network with both local and global efficiency in terms of information propagation.

2.4. Scale free property of small world networks (B-A model)

In studying the small world property of World Wide Web, Barabasi and Albert (1999) noted that the degree distribution of individual web pages is extremely uneven. Some pages are extremely well connected, while most pages have nearly the same low level of connection. This distribution is called scale-free, as it is different from normal distribution, which has "a characteristic *scale* in its node connectivity, embodied by the average node and fixed by the peak of the degree distribution" (Barabasi 2002, p. 70). The scale-free property can be used to explain the growth mechanism of real-world networks, i.e. the rich get richer and preferential attachment in Barabasi's term. Most dynamically evolved networks investigated by various researchers demonstrate the scale-free property. Note that a network with scale-free property is a small world, but not vice versa.

2.5. Directed search (W-D-N model)

The W-S model proves existence of short chains of intermediaries with many real world networks. However, how to find a short chain by individuals constitutes another challenge. This is actually the concept of directed search rather than broadcast search in analogue with Breadth First Search (BFS), which is impossible in practice (Watts 2003). With W-S model, each node is chosen at a uniform randomness for rewiring, while Kleinberg (2000) realized that in reality people use different senses of distance such as geographic distance, professionals, and race to decide to whom they want to make an acquaintance. Based on this observation and insight, Kleinberg constructed a two-dimensional cellular space in which each cell is connected by four immediate neighbours plus a random link. He found through the study that some networks that meet some particular condition are searchable (see Kleinberg 2000 for details). Watts et al. (2002) took a step further and integrated all what they called social dimensions (or identities) into search strategies. Eventually Watts and his colleagues concluded that most social networks are searchable because that various identities are involved in determining a next target of a short chain by individuals. It sets a significant difference from Kleinberg's finding where the condition is hard to meet for a network to be searchable.

3. Geographic environments as small worlds

3.1. Small world properties of geographic environments

Small world properties can be considered as a perfect combination of regularity and randomness, and they seem available in geographic contexts. On the one hand, spatial processes are far from a random process, as investigated for instance by spatial autocorrelation (Cliff and Ord 1973); on the other, common sense appears to suggest that geographic environments and spatial processes are not regular or ordered at all. The first law of geography elegantly states that "everything is related to everything else, but near things are more related than distant things" (Tobler 1970). The law appears to suggest that with geographic environments everything has many links to neighboring things (regularity), but a few distant links between the distant things (randomness). The regularity can also be said to be a sort of high clustering. Taking a transport network for example, the block-by-block street network constitutes a sort of regular network with a high degree of clustering.

Regarding randomness, various transport systems, such as bus and underground networks, imposed on the top of a street network provide some randomness or shortcuts. This is the very small world mixed with the nature of regularity and randomness used in Kleinberg's model. Sui (2004) made a similar remark in a forum on Tobler's first law of geography with AAG annuals. Batty (2001) has also commented how cities might be treated as small worlds, in particular, how new technologies have shrunk the cities with new transport means such as underground and high speed trains for global cities. It should be noted that near things should not be understood only in the sense of Euclidean geometry. For instance, two distant locations could be near things if they are visually accessible, or two GIS people are near people as they are in the same field, no matter how far they are in physical space. It is worth noting that the small-world problem was studied in the '70s and '80s in a geographic context. One of the earliest works is about the small-world problem in a spatial context (Stoneham 1977), although some issues like integration and segregation can now be better measured by clustering coefficient, a basic measure in the W-S model.

Geographic environments can be represented as topological networks in which vertices represent geographic entities and edges represent possible links or relationships between entities. For example, from a topological perspective, we represent a street network as a topological network based on a "named street"-oriented view (Jiang and Claramunt 2004), i.e. all the named streets are represented as nodes, and street intersections as links of a graph (Figure 2). Compared to the conventional geometric view, this representation provides a complementary view to street networks for modeling purposes, as it is defined at a higher level of abstraction. It can be used to study morphological structure and evolution of urban street networks. Apart from street network, a built environment and a terrain surface can be represented as a visibility graph as briefly introduced in section 2.1.

With the topological representations or street topologies, we have found that small world properties appear with urban street networks. Thus streets are highly clustered on the one hand, and are separated by a short chain of intermediate streets on the other. In a similar way, we examined visibility graphs with a built environment and a terrain surface (Figure 3). All the studies show that these geographic environments are indeed small worlds (Jiang 2005, Jiang 2004). In the context of the visibility graph, it means that two randomly chosen locations within an area are visually separated by a short chain of intermediate locations, no matter how many point locations one imposes on the spaces.

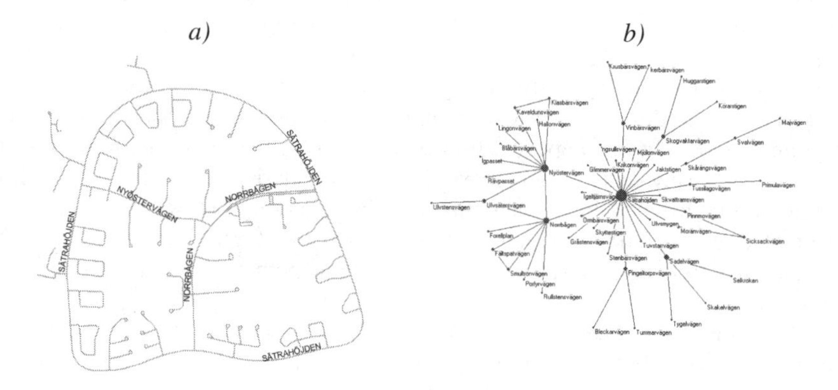

Fig. 2. A small street network (a) and its connectivity graph (b)

(Note: every node in (b) is labelled by the corresponding street name, and the size of nodes shows the degree of connectivity of individual streets)

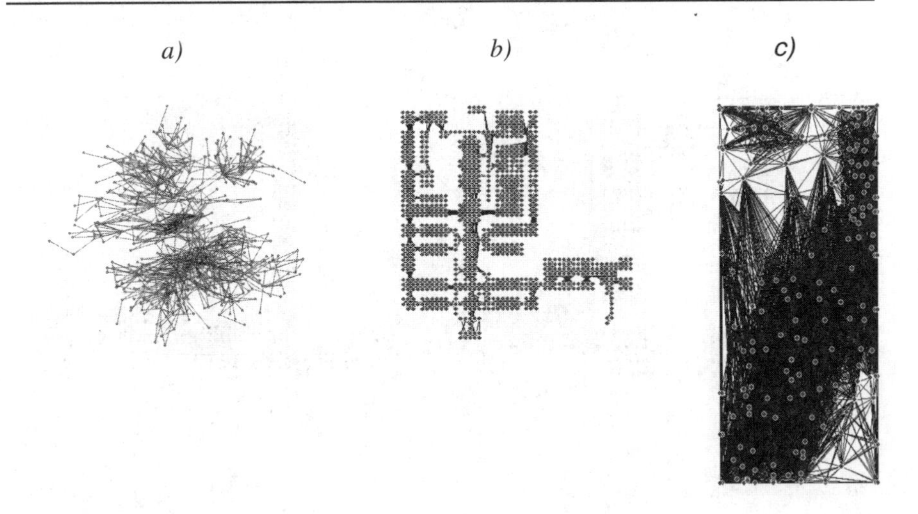

Fig. 3. Street-based topology of Gävle city (a), visibility graphs with the Tate Gallery (now called Tate Britain) (b) and a terrain surface (c)

3.2. Efficiency distribution for geographic environments

The exhibition of small world properties supports the idea that urban street net works are efficient at both local and global levels in terms of traffic flow. In a similar fashion, the Tate Britain and the terrain surface are efficient for search and navigation. It means that locations are visually reachable in an efficient manner with a built environment due to a high clustering coefficient at a local level and a low path length at a global level. This provides theoretic evidence whether or not a geographic environment in general is efficient. However, the efficiency view is with respect to an entire system and it can be extended to individual levels.

It is interesting to assess how these measures differentiate among individual geographic objects, locations within a geographic environment. Thus spatial effect for the small world properties of a geographic environment can be examined. For instance, using the M-L model, the efficiency for each street and each point location can be computed to illustrate its distribution. As an example, Figure 4 shows the distribution of both local and global efficiency for the Tate Britain. We can note that those locations with large dots have a higher efficiency.

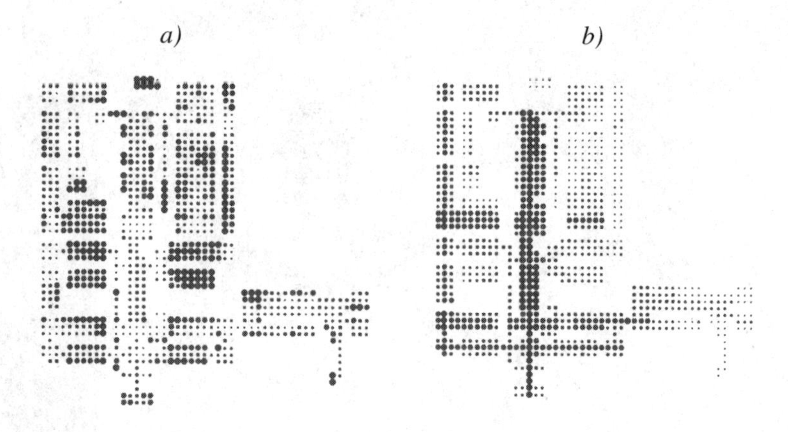

Fig. 4. Distribution of local (a) and global (b) efficiency for the individual locations with the Tate Britain

The distribution of efficiency provides new insights into geographic environments from both analysis and design perspectives. From an analytical perspective, we can examine the questions as follows: which parts are more efficient than others? When would a street network be evolved into most efficient? From a design perspective, it would be interesting to study how to create an environment with both local and global efficiencies, which are often desired. Additionally we can introduce weights into the graph representations. For example, in the case of street network, we did not represent multiple intersections between a pair of streets, which could be defined as a weight for a link, i.e. the more intersections, the stronger ties between a pair of streets. In the similar way, visibility link can be weighted in terms of distance, i.e. the shorter the distance between two point locations, the stronger the visibility link.

3.3. Re-examination of scale-free property

True scale-free property is only applicable for infinitely large networks, but in practice, most networks have a cutoff with their power law distribution (Watts 2003). This leads us to re-examine the finding we made in our recent work (Jiang and Claramunt 2004). Taking the case of Gävle street network for example, its log-log plot is indeed not strictly linear as shown in Figure 4a. However, a linear tendency seems exist if you make an appropriate cutoff, in particular when compared to the log-log plot for visibility degree (Figure 4b) that shows a stronger exponential distribution. We could characterize street networks as a broad-scale network, i.e. connectivity distribution has a power law regime followed by a cutoff, according to Amaral et al. (2000). The distribution difference shown in Figure 4a and 4b suggests the fact that scale free property is indeed a signature of dynamic networks, since street networks are evolved dynamically while visibility graphs are static in essence. It is important to note that it is these well-connected streets within a network, or hubs as they are often called, that keep a network small. For example, some streets are connected up to 29 other streets in the Gävle network and some locations are visible up to 135 other locations among a total of 960 locations in the Tate Gallery. These streets and locations constitute the hubs for the respective networks.

a)

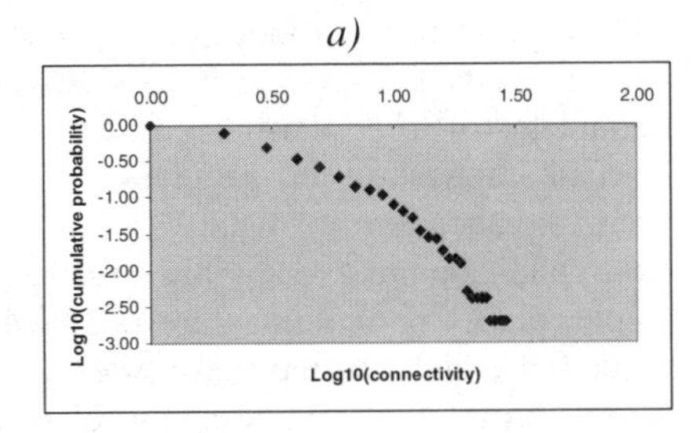

b)

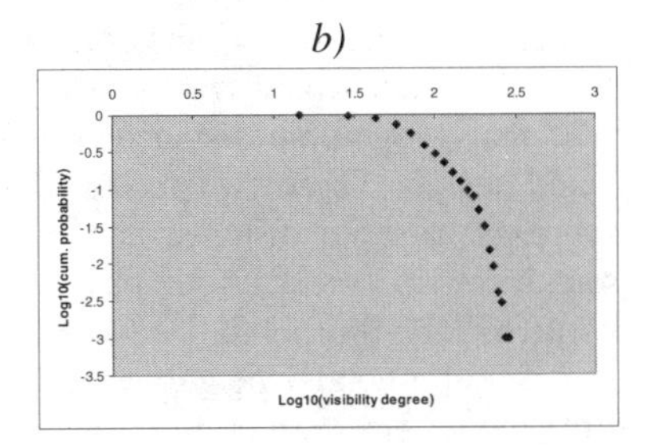

Fig. 5. Log-log plot of street connectivity versus cumulative probability (a) and visibility degree versus cumulative probability (b)

Based on the remark that the scale-free property is the patent signature of self-organization in complex systems (Barabasi 2002), we tend to suggest that street networks or street topologies are self-organized because of the scale-free property with street topologies - another signature that cities are self-organizing systems as elaborated by Portugali (2000). In this connection, the evolution of street network is much similar to that of the Internet and Web. Two important laws governed the development of the Internet and Web (Barabasi 2002) seem to be applicable to the street topologies as well, that is, growth - earlier streets have more time to be linked by other streets, and preferential attachment - preferences are given to the already well connected streets. The two laws together generate the scale-free property, or a power law of the street topologies.

3.4. Directed search in geographic environments

Search and navigation in geographic environments have been a research issue for a long time in the fields of behavioral geography and environmental psychology. How to design an artificial environment that is easy to navigate has also been a basic requirement in architectural and urban design. The small world models in general and the W-D-N model in particular provide some insight into various issues involving search and navigation in a geographic context. The visual separation within a geographic environment is short. However, short visual separation does not imply that an individual can easily find a short path that leads to a specific target. This is the issue of *directed search*. Directed search is a decentralized search strategy, in contrast to broadcast search in analogue with BFS algorithm.

Put differently, individuals use only local information about the network to determine the next link and eventually a short chain is found with the collective effort of individuals. Although it is defined in the context of social networks, we believe that people conduct directed searches from time to time in a geographic context, in particular when one comes to a new place with frequent search within an airport, an unfamiliar city and even a building complex.

The W-D-N model assumes that individuals are hierarchically organized within a social network, and it relies on the height of the hierarchy to reach a target. On the other hand, Kleinberg's model assumes a constant degree of the nodes, and finds that geographic distance is an important factor to be considered in the directed search. These two models have special implications to geographic environments, as both distance and hierarchical organization are natures of a geographic environment. Both physical worlds and cognitive maps are organized hierarchically (Portugali 1996). For example, places or buildings are organized into neighborhoods, neighborhoods into districts, and districts into cities. The same hierarchical structure is available in human internal representation of geographic environments, so called cognitive maps.

A geographic environment for human navigation can be constructed as a search network in which nodes are individual locations and edges are links between the locations. The links could be in different ways like a visual link (visibility), transport link (a bus line), or a road link. Therefore visibility is one of the factors that impact on search and navigation. Several other factors such as available transportation means, maps, and guided information given by someone are also involved in search decision. Within the network, landmarks, defined as having key characteristics to be easily recognizable and memorable in an environment, act as hubs within a search network. Sorrows and Hirtle (1999) suggested three types of landmarks that involve visual, cognitive and structural landmarks. The final structural landmarks can be considered as those point locations with the highest visibility degree. Although cognitive and visual landmarks are not so visually accessible compared to structural landmarks, they are often treated as hubs and are easily reminded because of its distinction in meaning. We can say that because of various landmarks of an environment it becomes searchable.

4. Discussion and conclusion

A network of interconnected and interacting objects provides an alternative representation and modeling approach to geographic environments. The representation is complementary to the conventional geometric representation in the sense of representing relationship of interconnected things. Therefore from a modeling perspective, it facilitates detection of the hidden structure of small world, i.e. most geographic environments are neither ordered nor random, but somewhere in between. Geographic environments seem to show high efficiency at both local and global levels. It provides new theoretical evidence that most geographic environments are efficient in search and navigation, as studied in behavior geography and environmental physiology.

This chapter took an exploratory approach to small world modeling for a better understanding of geographic environments. The ideas discussed here need some further research and experiments. The demonstration of small world properties with various geographic environments is a first step towards the understanding of geographic environments. Our finding of small world properties in the city context appears to support Alexander's famous contention that "a city is not a tree" (Alexander 1965). Indeed, the street topologies and visibility graphs with the Tate Britain tend to suggest the kind of semilattice structure rather than a tree. Some ongoing studies with small world modeling in a geographic context have taken a step forward. For instance, the idea of efficiency has been applied to the study of information and knowledge diffusion in the context of interactive learning (Morone and Taylor 2004). The small world structure and representations have been implemented in various agent-based modeling software platforms (Dibble and Feldman 2004).

A challenging issue from a small world perspective is how to design a geographic environment that is easy for search and navigation, or alternatively how to make a geographic environment closer to a small world. As a small world network shows both local and global efficiency, this should be an ideal design target. A simple principle appears to be: neither regular nor random but somewhere in between. With this principle, we still need to measure the degree of in-between in a quantitative manner. For instance, with the above-mentioned case studies, typical path length is a bit longer than that of equivalent random graphs. In the case of street topology, the average path length for the three levels of detail is 5.3, while the average path length for the equivalent random graphs is 4.1. It implies that such geographic environments could be better designed towards a more random one, i.e. to be a more efficient system at a global level. This can help in the course of design adjust a design into a right direction: to be more random or more regular. Another alternative approach could be using a randomness measure to show the level of randomness and regularity. These ideas need further experimental studies.

We have not spent much time in this chapter on how mobile vehicles and people are networked and interacting with each other and to geographic environments, therefore it is worthwhile to add some speculations on the issue. Nowadays, the world and people are more wired than ever before because of pervasive use of cell phones, tracking and positioning devices. This is particularly true in the city context. In the cities, more and more vehicles are equipped with GPS receivers and tracking systems, so it is convenient for online logistics and fleet management. On the other hand, pedestrians or people in general are interconnected with cell phones and mobile devices. This is the very concept of SwarmCity (Mitchell 1999) refers to. In many aspects, geographic environments are similar to the Internet space and many other real world networks. This similarity opens up possibilities and continues to give new inspirations to geographic modeling from the network point of view.

Acknowledgements

The author would like to thank the participants of the ESLab's international workshop who provided valuable inputs and comments to an earlier version of chapter when it was presented there. Thanks are also due to Daniel Z. Sui and Juval Portugali who suggested some valuable references.

References

Alexander, C. (1965), A city is not a tree, *Architectural Forum*, 122 (1+2), pp. 58 – 62.

Amaral, L.A.N., Scala, A., Barthelemy, M. and Stanley, H.E. (2000), Classes of behaviour of small-world networks, *Proceedings of the National Academy of Sciences*, 97, 11149 – 11152.

Barabasi, A.L. (2002), *Linked: The New Science of Networks*, Perseus Publishing: Cambridge, Massachusetts.

Barabasi, A.L. and Albert, R. (1999). Emergence of scaling in random networks, *Science*, 286, pp. 509 – 512.

Batty, M. (2001), Editorial: cities as small worlds, *Environment and Planning B: Planning and Design*, Pion Ltd, 28, pp. 637 – 638.

Cliff, A.D. and Ord, J. K. (1973), *Spatial autocorrelation*, London: Pion.

Dibble, C. and Feldman, P.G. (2004), The GeoGraph 3D computational laboratory: network and terrain landscape for RePast, *Journal of artificial societies and social simulation*, 7(1), available at http://jasss.soc.surrey.ac.uk/7/1/7.html (accessed 2004-04-05).

Holland, J.H. (1995) *Hidden Order: How Adaptation Builds Complexity*, Addison-Wesley Publishing Company, Reading, MA.

Jiang, B. and Claramunt, C. (2004), Topological analysis of urban street networks, *Environment and Planning B: Planning and Design*, Pion Ltd, 31, pp. 151 – 162.

Jiang, B. and Claramunt, C. (2002). Integration of space syntax into GIS: new perspectives for urban morphology, *Transactions in GIS*, Blackwell Publishers Ltd, 6(3), pp. 295 – 309.

Jiang, B. (2005), A structural perspective on visibility patterns with a topographic surface, *Transactions in GIS*, Blackwell Publishers, 9(4), pp. 475 – 488.

Jiang, B. (2004), Small world properties of visibility graph with a built environment, *University of Gävle Working Paper*, 15 pages.

Kleinberg, J. (2000), Navigation in a small world, *Nature*, 406, pp. 845.

Kleinfeld, J.S. (2002), The small world problem, *Society*, 39(2), pp. 61 – 66.

Marchiori, M. and Latora, V. (2000), Harmony in the small-world, *Physica A*, 285, 539 – 546.

Milgram, S. (1967), The small world Problem, *Psychology Today*, Vol. 2, pp. 60 – 67.

Mitchell, W.J. (1999), *e-topia: Urban Life, Jim--But Not As We Know It*, Cambridge: The MIT Press.

Morone, P. and Taylor, R. (2004), Small world dynamics and the process of knowledge diffusion: the case of the metropolitan area of greater Santiago De Chile, *Journal of artificial societies and social simulation*, 2(2), available at http://jasss.soc.surrey.ac.uk/JASSS.html

O'Sullivan, D. (2001), Graph-cellular automata: a generalised discrete urban and regional model. *Environment and Planning B: Planning & Design* **28**(5), pp. 687 – 705.

Portugali, J. (1996), *The construction of cognitive maps*, Kluwer Academic Publishers, Dordrecht.

Portugali, J. (2000), *Self-organization and the city*, Springer, Berlin.

Sorrows, M. and Hirtle, S. (1999), The nature of landmarks for real and electronic spaces, in C. Freksa and D. Mark, editors, *COSIT 99: Spatial information theory: cognition and computational foundations of geographic information science*, Springer, pp. 37 – 50.

Stoneham, A.K.M. (1977), The small-world problem in a spatial context, *Environment and Planning A*, **9**, pp. 185 – 195.

Strogatz S. (2001), Exploring complex networks, *Nature*, 410, pp. 268 – 276.

Sui, D.Z., (2004), Tobler's first law of geography: A big idea for a small world? *Annals of the Association of American Geographers* 94(2): pp. 269 – 277.

Tobler, W.R. (1970), A computer movie simulating urban growth in the Detroit region, *Economic geography*, 46. pp. 234 – 240.

Turner, A., Doxa, M., O'Sullivan, D. and Penn, A. (2001), From isovists to visibility graphs: a methodology for the analysis of architectural space, *Environment and Planning B: Planning and Design*, Pion Ltd., pp. 103 – 121.

Watts, D.J. and Strogatz, S.H. (1998), Collective dynamics of 'small-world' networks, *Nature*, 393, 440 – 442.

Watts, D.J., Dodds, P.S. and Newman, M.E.J. (2002), Identity and search in social networks, *Science*, 296, pp. 1302 – 1305.

Watts, D.J. (2003), *Six degrees – the science of a connected age*, W. W. Norton & Company: New York, London.

Part five: Planning

Planning and Designing with People

Michael Kwartler

Abstract. This paper details the use of visual simulations by the Environmental Simulation Center, Ltd. (ESC) and its collaborators to involve citizens in the neighborhood/city/regional visioning and planning process. It does so by examining three visioning projects undertaken in the last three years. The case studies demonstrate how to obtain citizen input regarding their values and group identity through their participation in designing the place in which they would like to live. Fully integrating 3D/Geographic Information System-based simulations and visualizations into the visioning process makes it possible for citizens to better understand their choices at both a policy and experiential level and arrive at consensus for the future of their communities.

1. Overview

The use of digital simulation and visualization is explored through the experience of three planning projects:

- *Vision 2030: Shaping our Region's Future Together* for the five-county Baltimore (Maryland) Regional Transportation Board (January 2003)
- *Southwest Santa Fe City/County Master Planning Initiative* for the City and County of Santa Fe, New Mexico (April 2002)
- Near Northside Economic Revitalization Planning Process for the City of Houston, Texas (September 2001)

The impetus for these plans derives from and is a negative reaction to current development practice, generally automobile based-planning manifesting itself in traffic jams and chaotic development patterns (e.g., the loss of open space in the Baltimore region; sprawling, commercial strip, gated community development and the loss of desert and the "sense of place" unique to Santa Fe; and a hostile pedestrian experience in Houston's Near Northside). The plans themselves are to guide future growth and development. Each project assumed growth, with pressure on the development of "greenfields" (typically undeveloped or agricultural land) in the cases of the Baltimore region and Santa Fe.

Each project is different in terms of scale, landscape, degree of anticipated change, governmental and regulatory structures, and the nature of public participation. As a result, each presents different challenges for simulation and visualization. The integration of simulation and visualization into the public participation process, facilitating informed discussion and decision-making, was guided by the number of people involved, propinquity, and localism. Regardless of the form the application of the technology took, it was consistently used to develop the plan in a variety of settings, from hands-on workshops to large "town hall" meetings, not merely to represent or visualize outcomes. Technology has typically been used toward the end of a planning process; used to sell the project or place but not to help the public develop the plan. Our objective has always been that simulations and visualizations are a means to the end of informing communities which are making decisions about their future.

The ESC has used simulations and real time interactive visualizations in planning and visioning since the early 1990's. To facilitate the process, the ESC developed its first 3D kit-of-parts (Fig. 1) in 1994 for a series of community planning and visioning prospects done in collaboration with the Regional Plan Association (Yaro and Hiss, 1996; Morgan, 1996). Advances in both hardware and software (e.g., ArcViewGIS and CommunityViz™ Planning and Design Decision Support System) made it possible to use real time 3D simulations and visualizations and GIS in the process of formulating a vision with citizens. The Santa Fe case study was among the first vision plans that fully integrates these digital tools in the process.

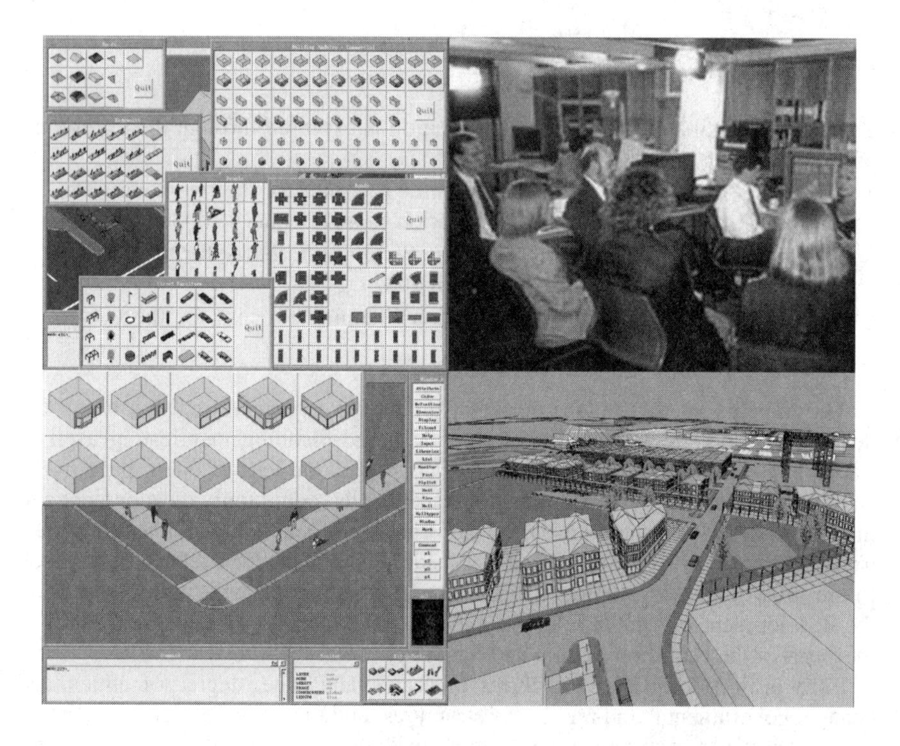

Fig. 1. 3D kit-of-parts developed by the ESC being used by the participants to design a transit village in "real-time".

While each of the planning projects is unique, all have commonalities that are useful for comparing why, how, and under what circumstances simulation and visualization are used to enhance public participation and, most importantly, decision-making. Conceptually, they all share a common base: all employ visioning processes based on the methodology first developed by Gianni Longo in his groundbreaking work with Chattanooga, Tennessee in the early 1980's, in which visioning is a citizen-driven process where all of the results are derived from public input. It is neither top-down nor bottom-up. It is inclusive of both the private-sector stakeholders (from business leaders to NGOs) and the public sector. Its purpose is to reach consensus on issues regarding values and group identity. It is less concerned with differences – what sets citizens apart – than with what a community, town, city or region share in common. It is about finding common ground in our pluralistic society.

"Place" plays a critical role in locating common ground. As Donald Appleyard (1979) observed,

> *(T)echnical planning and environmental decisions are not only value-based...but identity-based...(P)hysical planning decisions can, and frequently do, threaten the identity and status of certain groups while enlarging the power of others...The environment is divided into "ours" and "theirs;" the trees may be ours, the billboards theirs, the authentic is ours, the phony theirs, downtown may be ours or theirs, as may be the wilderness, the oil, or other natural resources. The city and the natural environment are areas of symbolic social conflicts and as such raise their own issues of social justice.*

From the perspective of the lay public, place -- their neighborhood, district, town, city or region -- is experienced as a whole in all of its glorious messiness, and not as a series of abstract planning categories (Casey, 1997). The "quality of place", the combination of its experiential and functional attributes and group values and identity are recognized as being synchronous in visioning. Visioning uses physical design as a form of inquiry, exploring the match and mismatch between words, numbers and images. Words and numbers are abstractions that have very specific real world implications. It is not unusual to hear, as has often been the case at the ESC, "That's not what I meant at all…." When the words and numbers in a standard master plan or zoning resolutions are simulated and visualized dynamically in three dimensions. The simulations and visualizations used in these three visioning processes and plans all play a similar role: grounding metaphor in reality.

Ironically, many of the proponents of "place" share an implicit tendency to be anti-technology and to regard the outcome of technology as "placelessness" (Relph, 1976; Lowenthal, 1985). In the three case studies we demonstrate that state-of-the-art technology can be used to help citizens and public officials *create place* and even to transform placelessness into place by using simulations and visualizations in a citizen driven process.

Regarding the digital technology, the projects were all done within a three year period, and made use of similar simulation and visualization tools and techniques (although the applications were obviously adapted to each project's needs and scale).

2. Baltimore Region

2.1 Vision 2030: Shaping our Region's Future Together

Over a 15-month period Vision 2030 explored six thematic areas that dealt with a broad range of issues that were, for the first time, brought together to form a comprehensive regional perspective. The areas were: Economic Development, Education, Environment, Government and Public Policy, Livable Communities, and Transportation.

The visioning process involved six interrelated and sequential steps:

- Step One: Understanding the Region – Perception and Reality
- Step Two: Involving Stakeholders
- Step Three: Prototypical Development Patterns and Scenarios
- Step Four: Gathering Ideas and Testing Results with the Public
- Step Five: Developing Vision Statements and Strategies
- Step Six: Testing the Vision Statements and Strategies with the Public

Within the overall context of the Vision 2030 process, simulation and visualization played a central role in helping the public and the project's Oversight Committee reach consensus on the "hot button" issue of where and how to accommodate growth in the region identified by the focus groups in Step One.

2.2 The regional workshop ("where to grow")

In response to the "hot button" growth issue, the Regional Workshop focused on "where to grow". Organized as a game, the purpose of the workshop was threefold:

1. To understand the complexity of thinking regionally,
2. To gain "intuitive" public input on future growth and land consumption considerations, and
3. To prepare for future subcommittee work (e.g., developing the vision statements, strategies, and principles that form the core of Vision 2030).

The participants consisted of 65 stakeholders, including elected officials, planners, educators, citizen activists, staff from NGOs, and business leaders. Participants were divided into eight groups, each with a facilitator.

As a first step, participants agreed on a percentage (average of all eight groups) of the region's total land they would like to protect over the next 30 years, in addition to land already protected. The next step was to agree on a common set of criteria, weighted differently by each of the eight groups, to help guide the choice of areas to protect for the future (e.g., the creation of contiguous natural environments, protecting forest and trail areas, etc.).

A three-foot by four-foot Geographic Information Systems (GIS) generated map of the region – which included layers delineating urbanized areas, areas already protected, agricultural areas, and unprotected land (e.g., forests, wetlands, etc.) – was overlayered (Fig. 2). Then each group was given green "chips", each representing one square mile of land, and asked to place them onto areas on the map that the group believed should be protected. During a brief break, the results of each group's approach to future land protection were hand-tabulated and a workshop average of protected land was calculated. The patterns of each group's placement of chips were compared and discussed by the workshop participants, revealing an underlying consistency in the choices made by each group: on average, 12 percent of the land was inside Priority Funding Areas (PFAs) (areas in the five county region that receive State of Maryland incentives for new development in urbanized areas) and 88 percent was outside the PFAs.

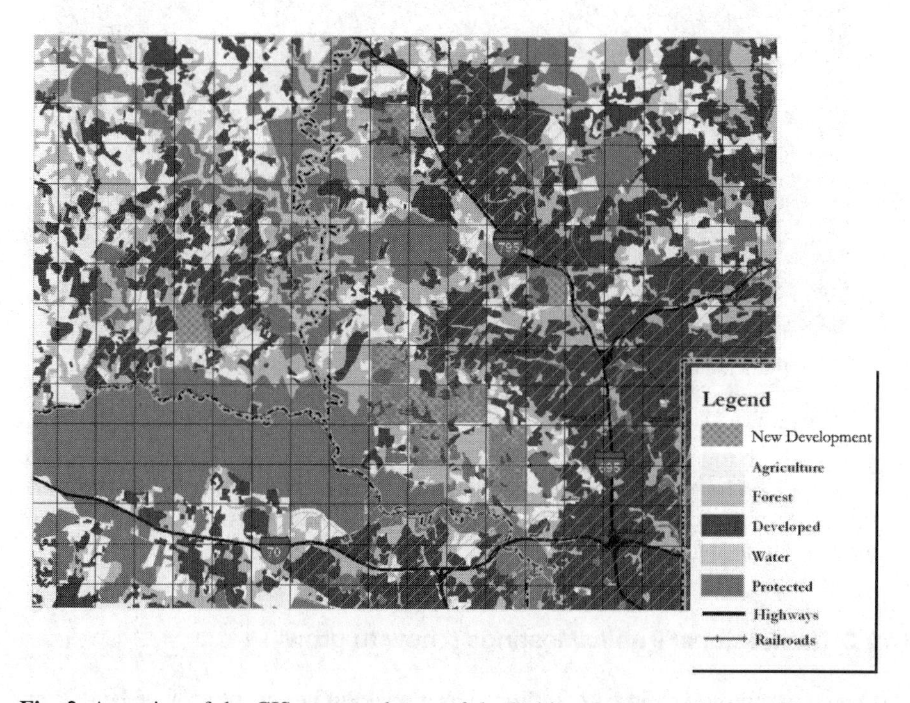

Fig. 2. A portion of the GIS generated map of the Baltimore region with one mile square grid overlay.

The next step was to determine where growth might be located. The groups were given brown chips that represented the amount of land that would be needed to accommodate the region's projected growth for the next 25 years (Fig. 3). In a similar way to the land protection exercise, the participants discussed and agreed on criteria that they could use to guide their decision-making regarding where growth should occur (e.g., along transit lines in already developed areas, in undeveloped areas, near employment centers, etc.) Again, the results were tabulated, averaged, and discussed, and revealed a consensus for redevelopment rather than "greenfields" development.

Fig. 3. Participants allocating the Baltimore region's projected growth on the GIS map.

The responses to the Regional Workshop "game" reflected consensus. Most groups chose to locate growth the region's developed areas and protect land in the outer areas. The groups placed an average of 70 percent of the growth within the PFAs, 6 percent of the growth in greenfields outside the PFAs, and 24 percent of the growth in Baltimore City.

2.3 The Regional Public Meetings ("how to grow")

Over a two-month period 17 facilitated Regional Public Meetings were held. Presentations were made of prototypical development patterns, region-wide development scenarios and the absolute and relative performance of each development scenario. Questionnaires were administered and small group idea sessions were conducted.

The Oversight Committee and its Thematic Subcommittee identified and tested three development patterns, four future regional development scenarios, and performance indicators. The development patterns became the building blocks for these future regional development scenarios. Each scenario showed how the region would develop depending upon the allocation of the development patterns/building blocks.

The three development patterns reflected trends that were occurring in the Baltimore region as well as those emerging nationwide. Each had different implications for land consumption, housing types mix, and proximity to jobs, shopping, and entertainment. They were:

1. *Type A: Conventional development pattern in undeveloped land.* This reflected a continuation of how the region had been growing with single-family detached houses, shopping entertainment, and employment in auto-centered malls (Fig. 4).
2. *Type B: Mixed-use walkable community on undeveloped land.* This assumed the creation of more compact neighborhoods with a mix of housing types and nearby shopping, entertainment, and employment (Fig. 5).
3. *Type C: Mixed-use walkable communities on redeveloped land.* This also assumed the creation of more walkable compact communities but on redeveloped land (Fig. 6).

Fig. 4. Type A: Conventional development pattern on undeveloped land.

Fig. 5. Type B: Mixed-use walkable community on undeveloped land.

Fig. 6. Type C: Mixed-use walkable community on redeveloped land.

Each building block or development pattern had the same goal: to accommodate 1,000 households with supporting commercial, schools, and open space. This allowed for "apples to apples" comparisons.

Unlike the Santa Fe and Houston vision plans, Vision 2030 was not focused on a particular place or places in the Baltimore region but rather on public policy, which required a focus on places characteristic of the region. In fact, given the region's emphasis on home rule (where each jurisdiction has land-use powers), it was critical that Vision 2030 not appear to be usurping local authority. Working collaboratively with the client, the ESC composited two characteristic places using the five-county GIS orthophotographs and database: one with large tracts of undeveloped land and existing suburban and rural development patterns with the possibility of "greenfields" development (Fig. 7) and another in an urbanized center with the possibility of infill/redevelopment (Fig. 8).

Fig. 7. Composited existing suburban and rural development pattern representative of the Baltimore region.

Fig. 8. Composited existing urbanized center development pattern representative of the Baltimore region.

Prototypical building types were created and introduced into the model development patterns that were included in the regional scenarios. Each development pattern was designed in the composite (layout of blocks, lots, streets, uses, open spaces, distribution of building types, etc.) and modeled in three dimensions to reflect the architectural character of historic and contemporary buildings in the region. The real-time 3D/GIS environment allowed the client to view the development patterns dynamically, to comment on the design of the development pattern, and to select views and real-time walk-through paths to be presented to the public at the 17 Regional Public Meetings. Because the development patterns were composed of a kit-of-parts that ranged from the building to the block (with each block having different combination of lot sizes), the ESC was able to quickly and efficiently respond in "real time" to subcommittee comments and suggestions in an iterative design process.

With the 3D simulations and visualizations of the three development patterns validated by the Thematic Subcommittees, the three development patterns (Types A, B, and C) were then used to assemble the four regional development scenarios identified by Vision 2030. The scenarios accommodated the forecasted population and employment growth for the region by using the development patterns in different combinations. The four prototypical regional scenarios were:

- Scenario 1: Current trends and plans
- Scenario 2: Emphasis on road capacity
- Scenario 3: Emphasis on mass transit
- Scenario 4: Emphasis on redevelopment

The compositing of development pattern types into scenarios is illustrated by comparing the mix of development types between Scenarios 2 and 3:

Scenario 2

Type A: Conventional development pattern on undeveloped land	75%
Type B: Mixed-use walkable communities on undeveloped land	20%
Type C: Mixed-use walkable communities on redeveloped land	5%

Scenario 3

Type A: Conventional development pattern on undeveloped land	25%
Type B: Mixed-use walkable communities on undeveloped land	37.5%
Type C: Mixed-use walkable communities on redeveloped land	37.5%

The comparison of the four scenarios was analyzed through an enhanced version of the Baltimore Metropolitan Council's travel demand model (enhancements by Smart Mobility Inc.). Model enhancements included improved sensitivity to transportation infrastructure and land-use, and improved fit with observed traffic data and travel survey data. The four scenarios were translated into transportation analysis zones ("TAZ") that were used to distribute the projected household and employment growth through the region. Distribution factors such as availability of land (for greenfield development only), vacancy rates for redevelopment, and proximity to public transit utilized GIS data and information derived from the

3D/GIS simulations and visualizations of the three development pattern types. Performance indicators (e.g. gasoline consumption, vehicle miles traveled, walking trips, transit trips, etc.) were used to quantify the performance of each of the four scenarios.

The analysis revealed that the *Current Trends* (Scenario 1) and *Emphasis on Road Capacity* (Scenario 2), would result in vehicle miles of travel and gasoline consumption that were three to four times greater than *Emphasis on Mass Transit* (Scenario 3) and *Emphasis on Redevelopment* (Scenario 4). The reason is that driving trips would be shorter as a result of a compact land-use pattern and the concomitant larger number of walk and transit trips. Further, the *Emphasis on Redevelopment* scenario performed particularly well because the region has a significant potential for redevelopment in areas close to the region's core.

The three development patterns (Types A, B, and C), the four regional development scenarios and the scenarios' performance as measured by the indicators were all presented at each of the 17 Regional Public Meetings, in conjunction with a questionnaire (entitled "Choices for the Future") that focused on the regional scenarios. The questionnaire presented a number of quality of life and transportation indicators related to the four prototypical regional development scenarios. Participants were first asked to select the scenario that they preferred according to the effect each had on each indicator. For example, the indicator "acres of new land consumed by development from the year 2000 to the year 2030" had a range of 41,242 acres to 138,316 acres, depending on the scenario.

The four scenarios and the performance indicators represented "what ifs"; hypothetical situations that were intentionally designed to offer a wide range of choices. Their abstraction, particularly when expanded to the five-county region was made palpable by the simulations and visualizations that employed 3D models and eye-level walk-throughs (Figs. 9a,b). They communicated to the public the question: "If you think or prefer this, then this is probably the kind of place that will result; is this acceptable?" The simulations and visualizations were compelling and, in conjunction with the performance indicators, provided the comfort level the respondents needed to complete the questionnaire. The results demonstrated overwhelming support for the *Emphasis on Redevelopment* and *Emphasis on Mass Transit* scenarios, both of which consumed less than half the amount of land in accommodating future growth as opposed to the other two scenarios.

a)

b)

Fig. 9. a)b) Two snapshots from the eye-level real time walkthrough simulations of the mixed-use walkable community on redeveloped land where a) shows existing conditions and b) redevelopment.

The ways in which the simulations and visualizations were used in Vision 2030 were a function of the scale of the community participation in the process in a five-county region. When working with the Oversight Committee's Thematic Subcommittee the participants had an opportunity to work in the real-time 3D/GIS environment Similarly, the Regional Workshop, with its groups of eight participants, allowed for direct participation in the workshop. The results of the "where to grow" and "how to grow" workshop and meetings needed to be presented to the region's constituency where they lived. By definition, the Regional Public Meetings could not be "hands on" in terms of determining or even changing the content, but rather were designed to take the pulse of the community, to register their response to a series of scenarios from which they were asked to choose. Given the possibility of checking "none of the above" it was clear that the public outreach and communications effort, the simulations, visualizations, and indicators of the development patterns and scenarios, and the iterative nature of the process (that at key points attempted to validate positions taken at that time in the visioning process) were effective at keeping people both informed and involved.

To insure the credibility of the Vision 2030 results, the core values and key strategies of the team were tested in a random regional phone survey. The survey focused on four primary goals: to understand which Vision 2030 issues were "hot button" issues for the region's residents, to test the degree to which the core values that came out of the visioning process resonated with the public, to compare regional attitudes to those nationally, and to derive a demographic and profile of residents across issue areas. The 1,200 random telephone interviews insured that there was a large enough number of interviews in each of the region's jurisdictions. The telephone survey validated the results of the visioning process, indicating that the region's citizens have a strong environmental ethic, are concerned about growth and sprawl, believe that there should be a balance between economic development and environmental protection, and have a heightened concern about traffic and congestion. Interestingly, notwithstanding the residents' support for public transportation, it would require a cultural shift of the respondents to actually use public transportation, assuming an adequate system was in place. Public participation in planning decisions was a mid-tier concern among those interviewed, although the interest in "encouraging public participation" was greater than "developing regional cooperation" and "coordination among the region's communities and counties". This revealed one of the problems facing regional planning: creating a regional identity that the residents could relate to.

3. Southwest Santa Fe

3.1 Southwest Santa Fe City/County Master Planning Initiative

The objective of this 25-week study was to develop a public participation process which would yield a policy framework for future development (a "vision plan"). The study was seen as the first step in the formulation of a master plan with corresponding development regulations that would implement the policy framework established by the vision plan.

Unlike Baltimore's Vision 2030, "the issue of "where to grow" was explicit, in that Southwest Santa Fe was, realistically, the only area left to accommodate growth within the City of Santa Fe. In fact, this area was already growing rapidly, if not chaotically, and all indications were that development pressure would continue. Southwest Santa Fe was at a crossroads. It could continue to develop as an entirely auto-dependent "suburb" of anonymous strip malls and "adobesque" gated communities or it could grow in a way that was sympathetic to Santa Fe's historic development pattern (Reps, 1979; Longo 1996), considered the values of the existing population, and conserved the high desert landscape. Hence, the public participation processes focus was on "how to grow".

3.2 Methodology/process

While the vision plan was to be a policy document, its focus was on a clearly delineated and hotly contested area in which existing residents had a clear, if not personal interest that resonated Appleyard's comment: "Planning is not only technical but value and identity based". Because the study area was in their backyard, passions ran high. If consensus was to be achieved to guide the area's future, the process needed to be inclusive, open, and informative. Most importantly, the vision plan needed to incorporate the values of all the participants and establish an area identity that all could support, relate to, and implement; that is, to find common ground.

The process that was designed to identify a vision for the Southwest Santa Fe area consisted of three sequential steps, each of which involved public participation, review, comment, and decision-making. The steps were:

1. Identify three prototypical areas where principles and possible development scenarios could be developed and applied to similar parts of the study area.
2. Develop land-use and urban design alternatives, including:
   - Identifying local patterns and conventions of development,
   - Translating those patterns and conventions into the development of 3D building blocks,
   - Extracting development principles from the building blocks, and
   - Applying the building blocks to each of the three prototypical areas.

3. Test the land-use and urban design alternatives and present them to the residents, the general public and the stakeholders in addition to the monthly meetings with the plan's Steering Group.

The following paragraphs explore the role of simulation and visualization throughout the visioning process, with a focus on creating the "building blocks", principles, and the planning area GIS for the three planning areas, and explains how they were used in the public review and decision-making process

The principle advocate for the stakeholders, the existing residents, and the NGOs was the Steering Group. This 28 member committee was continuously involved in all aspects of the process, meeting monthly with the consultants. They participated in the process of developing land-use and design alternatives for each prototypical area. The process included identifying recurring patterns of development, translating those patterns into development building blocks, and applying the building blocks to the prototypical areas of "New Development", "Corridor Development", and "Rural Protection".

Places to a great degree derive their character from conventions or patterns that are implicitly agreed upon by the community. Two examples are the more or less uniform setback of houses in New England, or the diversity of fences and walls that enclose the front yards of Santa Fe's houses. The patterns used by the ESC to design the building blocks derived from historic Santa Fe and from the planning area itself. Over fifteen large and small recurring patterns or urban design conventions in historic Santa Fe were identified.

The next step was to translate local patterns into digital 3D building blocks (blocks and lots consisting of houses, sidewalks, street widths, and on-street parking). The building blocks became the basic components used in developing the vision for each of the protypical areas. For example, for the New Development prototypical area building blocks included a variety of blocks and lots representative of Santa Fe, a hierarchy of streets, sidewalks and traffic calming devices, linear parks (based on the land subdivision of narrow long lots) and squares, mixed-use commercial buildings, and residential buildings. A draft of the 3D building blocks was presented to the Steering Group for its review and comment. Each block was reviewed individually and as aggregated patterns in the real-time 3D model.

The hierarchy of new streets and block sizes was reviewed in its entirety and as individual components, which determined each street's width, design speed, sidewalks and on-street parking patterns. Because safety of pedestrians (and particularly children) was one of the overarching themes that emerged from the stakeholder meetings and focus groups, the Steering Group carefully scrutinized the components prepared by the consultant team. Each street type, from alley to collector street, was modeled in 3D using the building blocks based on historic street widths in Santa Fe. The Santa Fe Department of Transportation standard issue street widths were also modeled for comparison.

The street patterns and widths were thoroughly discussed and voted on, and each street was assigned a grade based on comfort level (Fig. 10). With this input the Steering Group reduced the number of street types. Not all street patterns and building blocks survived the scrutiny of the Steering Group; in some cases they were modified to better match local conditions (as in the case of the commercial building whose height was reduced from three to two stories). In other cases, components were rejected as inappropriate for Santa Fe.

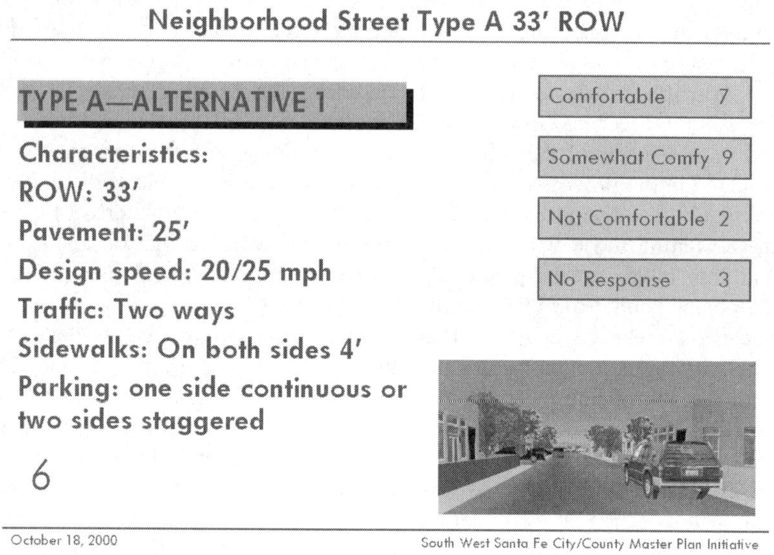

Neighborhood Street Type A 33′ ROW

TYPE A—ALTERNATIVE 1

Characteristics:
ROW: 33′
Pavement: 25′
Design speed: 20/25 mph
Traffic: Two ways
Sidewalks: On both sides 4′
Parking: one side continuous or two sides staggered

6

Comfortable 7
Somewhat Comfy 9
Not Comfortable 2
No Response 3

October 18, 2000
South West Santa Fe City/County Master Plan Initiative

Fig. 10. Example of neighborhood street type with characteristics, visual simulation using the Santa Fe "Building Blocks", and participant votes.

As the planning process proceeded, the individual building blocks were aggregated into development scenarios and reviewed by the Steering Group. The building blocks were applied to the prototypical areas using a GIS-based existing conditions 3D model of the three prototypical areas to show how they might go together to create a neighborhood, a mixed-use commercial area, or a development pattern in low-density rural protection areas.

Fitting the building blocks and land-use patterns to existing conditions demonstrated how the areas could develop incrementally and inclusively. The simulations and 3D visualizations were not meant to be finished, static designs rigidly applied, but rather used to illustrate how the building blocks might be organized to show a possible, but by no means singular end result (Fig. 11).

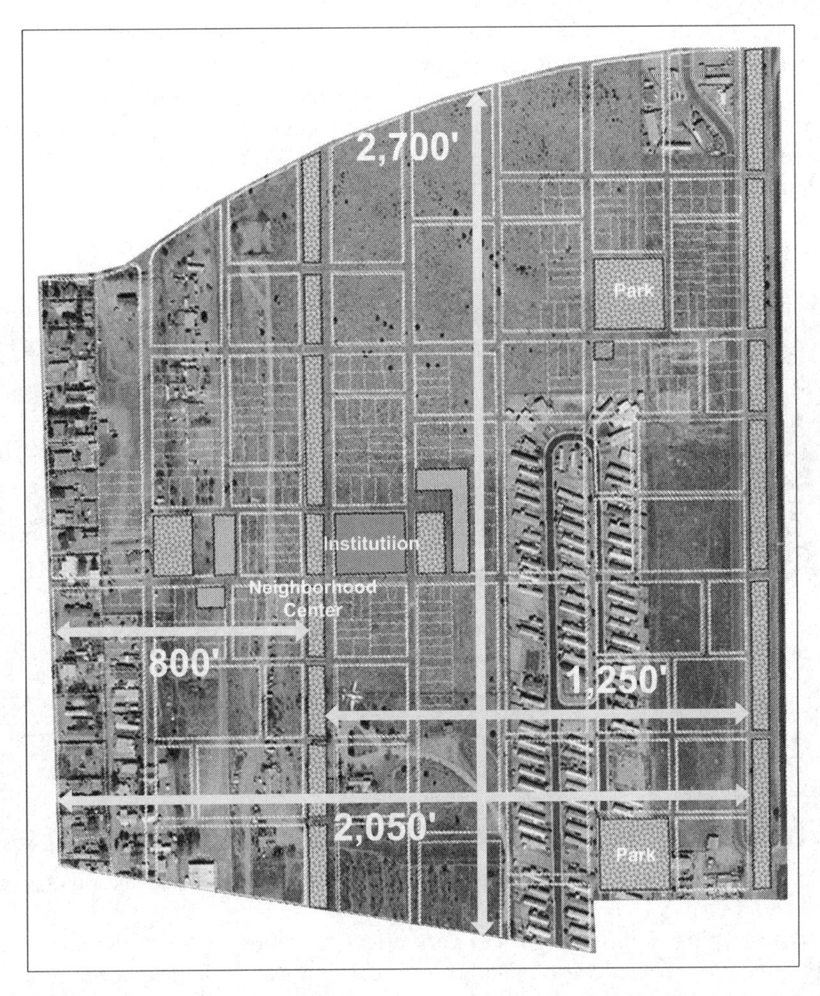

Fig. 11. Integrating the building blocks with existing land-uses and development. The measurements indicate walking distances.

The next step was to extract the development principles implicit in the development scenarios reviewed by the Steering Committee. Ten development principles were identified as representing the community core values. Each principle lists the action necessary, and the combination of building blocks required, to realize its intent. The principles generalized the results of the process, making the principles and the building blocks applicable throughout the entire planning area, rather than being limited in their application to a specific site or location within the prototypical area.

At the Public Forum each of the principles and corresponding combinations of building blocks were presented and discussed individually. To help the public better understand what kind of neighborhood would result if the principles were implemented in the aggregate, the ESC created a representative portion of a prototypical mixed-use neighborhood using the building blocks (a "cluster"). The cluster allowed the ESC to simulate – at eye level and in real-time within the context of the 3D model – what it would be like to leave your house in the morning, walk to the daycare center to drop off your child, pick-up a snack at the corner deli, and get on your bicycle and ride along the linear park to work at the commercial center (Fig. 12).

Fig. 12. An eye level visual simulation of a mixed use neighborhood and linear trail park.

The real-time walk-through was very effective in conveying to the public the future identity of Southwest Santa Fe. The real-time 3D model also made it possible to respond to requests to walk down other streets or what it would look like from someone's front yard. Being able to be "in the model" and to make choices was critical to engendering a spirited discussion of the pros and cons of the principles.

After considerable discussion, the Public Forum's participants were asked to rank the principles on a scale ranging from 1 (indicating the lowest level of support) to 5 (indicating the highest level of support), and an average score for each principle was tabulated. For example, Principle 5 ("Neighborhoods in Southwest Santa Fe should have a variety of lot sizes and building styles to allow for economic diversity, affordability and an inclusive community") received 4.13 out of 5 points. All ten principles received "strong" to "very strong" support at the Public Forum.

To increase public input, the designs and concept for Southwest Santa Fe were further reviewed and scrutinized through seven small scale workshops with neighborhood associations and professional groups and through a survey widely distributed to residents, business owners, students and others throughout the planning area.

The second of the Community Choices Workshops presented the emerging community consensus around the principles for the three prototypical areas. Up to this point density had not been mentioned. As in Baltimore it was critical for the methodology to separate "what we want" from "how we get there". This is critical to the visioning process, so that the vision is not compromised and consensus is built to support implementation. For example, Principle 5 ("Neighborhoods in Southwest Santa Fe should have a variety of lot sizes and building styles to allow for economic diversity, affordability and an inclusive community") is extremely difficult to achieve in new developments and communities due to bank lending preferences (where economic stratification is preferred), zoning regulations (that set minimum lot and house sizes), and difficulty marketing the development to a public that has an expectation of uniformity. Simulations and visualizations were done that showed "apples to apples" comparisons of a neighborhood where each block had the identical lot and house size, and another where the same number of lots and houses were randomly mixed on each block. The latter was closer to a typical Santa Fe neighborhood that has developed over time (not "cookie cutter", regimented or stratified economically). This principle was one of the highest ranked, notwithstanding its difficulty of being achieved, only because implementation was not to be considered in the voting. The same was true with density.

During the visioning process density was not discussed because density per se is not a principle but rather a means, and not an end in and of itself. Instead of the abstraction of density numbers (e.g., dwelling units per hectare), building blocks that encapsulated the public's value system were used to design the new neighborhood the public wanted. The translation of the building blocks into principles was the process of extracting policy from the design of the neighborhood place. Used this way, the simulations and visualizations supported design "as inquiry" rather than "as a product".

With consensus on the principles achieved, the next step was to explain how density would help achieve the vision plan. The visualized and simulated neighborhoods could be developed at different densities, all meeting the same amount of anticipated future growth. To help the community understand how much of undeveloped Southwest Santa Fe would be needed to accommodate future growth, the consultants used Santa Fe's GIS. To get a general "feel" for density, a figure/ground map was created of the historic core of Santa Fe and superimposed on Southwest Santa Fe adjacent to a mobile house park that occupied the same amount of land. This demonstration was an "eye-opener". One was a complete world and the other...just a mobile home park. In addition, by utilizing the GIS building footprint areas and parcel maps, existing densities of representative neighborhoods, well known to the participants, were calculated, visualized, and used to inform the discussion of density.

The effects of density (or the lack thereof) on the consumption of land was demonstrated by the use of GIS-based dynamic maps and charts that graphically delineated the amount of land needed to meet the area's projected housing needs at existing zoning density (3 dwelling units per acre), and 5 and 8 dwelling units per acre (8 dwelling units per acre being the density of Santa Fe's historic core). These maps and charts, in combination with the 3D model that represented the proposed prototypical development types, effectively conveyed both the quantitative and qualitative effects of development at "sprawl" and "urban" densities (Figs. 13 a,b,c).

a)

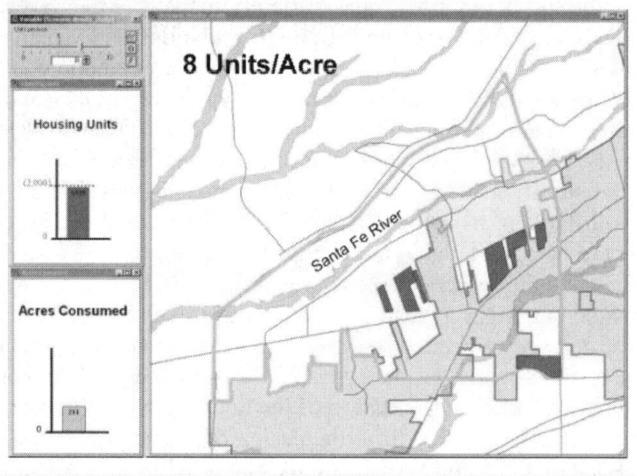

b)

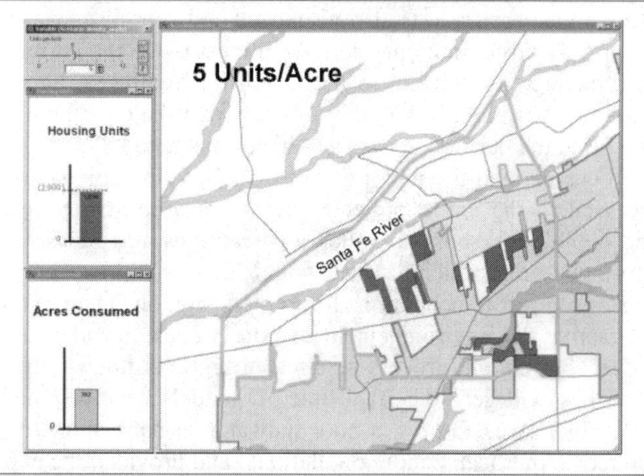

c)

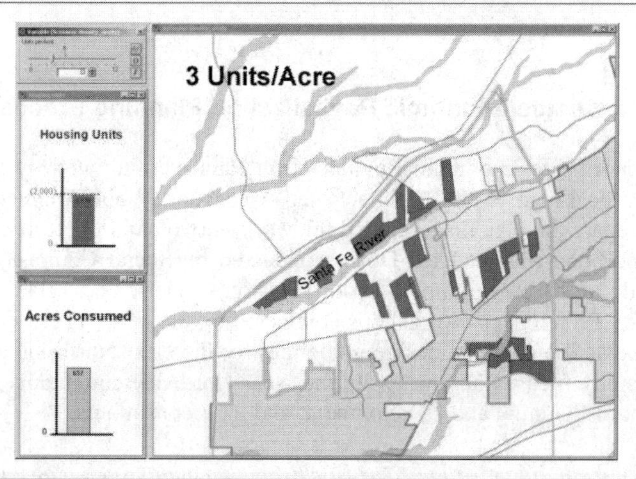

Fig. 13. Land consumption to meet projected housing needs at a) 8 dwelling units per acre, b) 5 dwelling units per acre, and c) 3 dwelling units per acre.

What became instantly clear was that at current zoned densities all available land, including land fronting the Santa Fe River, would be developed in contradiction to the agreed upon principles. The amount of land consumed at the levels of 5 and 8 dwelling units per acre was also visualized and consensus was reached on a density of over 5 dwelling units per acre (or almost twice the current zoned density). Without having first developed the vision for Southwest Santa Fe through simulation and visualization, it is highly unlikely that the community would have supported doubling the density, and no less have considered it.

Given the focus on land-use and urban design, GIS simulations and 3D visualizations were vital to the entire project. They were used at Steering Committee meetings and public workshops to validate perceptions, test ideas, understand alternatives, formulate the principles that would become the backbone of the area plan, and to develop recommendations for implementation. The use of a real-time, data-rich interactive 3D environment to explore values and ideas as an integral part of the decision-making process resulted in high confidence that "what they meant is what they will get". The real-time 3D models went beyond abstractions such as FAR (Floor Area Ratio – a poor indicator of density and/or intensity of use, found in most American zoning regulations) and provided the experiential basis to discuss in concrete terms the kind of place the residents and stakeholders wanted.

4. Near Northside, Houston

4.1 Near Northside Economic Revitalization Planning Process

The Near Northside Economic Revitalization Plan was initiated by the City of Houston. A local firm, Webb Architects and Associates (Webb Architects) was selected to prepare the plan that included the work of two parallel efforts: The Houston Neighborhood Market Drill Down conducted by Social Compact Inc. (retail market study of the bargaining power of Houston's inner-city neighborhoods) and "Community Preferences" workshops conducted by the ESC. These efforts led to the implementation of urban design guidelines for the Near Northside community. This case study focuses on work with the Near Northside community, Houston's Department of Planning and Development, and their consultants.

> *The following plan for the Near Northside, like many such plans, has many facts, figures, and maps. But behind these lies a vision for how a neighborhood might grow and develop. A vision not developed by city planners or outside consultants though such individuals played a vital role in development. Rather, it is a vision for a neighborhood developed by that very neighborhood. Through countless steering committee meet-*

ings and public sessions over a year's time, community members developed this plan. It is their plan for their neighborhood.
Gabriel Vasquez, Council Member, District H (*Northside Village Economic Revitalization Plan,* June 2002)

The Near Northside, located adjacent to Houston's Downtown, is a predominantly Latino, yet diverse neighborhood, and is one of two neighborhoods in Houston selected for HUD's Community Technology Initiative in 2001. One of the primary objectives of the Initiative was demonstrating the role simulation and visualization could play in helping communities collaboratively plan and reach consensus on their future. An essential part of making the initiative sustainable was technology transfer. Houston's Planning and Development Department staff was trained in the use of visualization software and most importantly, its application in future community-based visioning and planning workshops (Illus. 20).

4.2 Organizational structure

A Steering Committee of community stakeholders was established to provide community input, act as a sounding board, share their intimate knowledge of the community with the consultants, and review the work of the consultants. The Steering Committee met frequently over a nine month period; members included neighborhood civic associations, neighborhood service organizations, the school district, business associations, local development corporations, and property owners. A separate Advisory Committee provided input from governmental agencies. In addition, three community-wide workshops were held. All meetings and workshops took place in the Near Northside neighborhood.

4.3 Methodology

The overall plan's methodology/ process was somewhat compromised by a series of factors that made coordination and, at critical moments, collaboration between the three consultants difficult. Because each consultant was funded by a different source, whose objectives and timing were not always consistent with the general goals of the revitalization effort, coordination suffered. For example, the ESC's work focused primarily on the commercial corridors, and in the best of worlds should have reflected the market potential of the neighborhood to attract and support a broader range of services and retail than currently existed. Unfortunately, the market work done by Social Compact, Inc. came too late in the process, focused only on neighborhood buying capacity (to the exclusion of neighborhood demand for goods and services), and was not designed to provide information such as how much additional commercial space the neighborhood could support. As a result it was not particularly useful in formulating the development scenarios, simulations, and visualizations. To complicate matters, the consultants did not begin their work in unison, making coordination, the sharing of information, and the development of an integrated process difficult. Notwithstanding the hurdles, the collective efforts of the consultants, the Steering Committee, the Advisory Committee, Houston's Planning and Development Department staff, and the Near Northside community led to the adoption and implementation of the vision plan.

Interviews were held by Webb Architects with property owners and other stakeholders rather than groups. The difference between interviews and a group discussion is that with interviews the participants are not engaged in a dialogue. This had the unfortunate result of sustaining often contradictory, mutually exclusive expectations rather than achieving consensus, both in terms of content and process.

Three community-wide workshops were held. Initially, there were to be two, but for reasons to be explained below, an additional workshop was required. The first workshop was held about a month before the ESC's contract with the City of Houston took effect and, as a result, was without input from the ESC. Over 100 people participated in a facilitated process that identified issues, problems, and assets, and a question-and-answer session.

The second workshop, held about two months later, was keyed to ensuring that the consultant team was on-target regarding community issues, transportation, strategies for change, and concepts for new development. The ensuing discussion clearly revealed deep flaws in the Webb Architects-led process. Expectations were high and frustrations deep, because the workshop was virtually all presentation with little time for discussion. The ESC was asked to simulate and visualize the development implications of a commercial corridor, with and without the proposed light rail system. The real-time simulations and visualizations were mislabeled and introduced by Webb Architects at the end of a three-hour meeting as the "virtual reality tour" of future corridor development. When confronted with possible development scenarios in real photorealistic 3D, which they had no role in formulating, the participants were outraged. They were frustrated because they felt that they were being asked to choose a development alternative, notwithstanding that the simulations and visualizations raised substantial questions about the values, future, character, and identity of the Near Northside neighborhood.

What became clear in the ensuing confrontation was that the community assumed that the ESC's visual simulations were being used to sell a conclusion rather than inform and engage the citizens in developing *their* vision for the neighborhood. Moreover, while the ESC envisioned the simulation and isualizations as the beginning of a PROCESS to stimulate discussion, the community REGARDED them as product of a process in which they had not role. The Webb Architects process/ methodology had misinformed the community about the way simulations and visualizations were to be used to help the Near Northside community develop its vision. Once the mismatch was identified, the Steering Committee and Near Northside community decided to start over and hold another workshop. It is this last workshop that is discussed in more detail. One lesson learned is that people take visual simulations literally. The second is how threatening simulations and visualizations can be when presented out-of-context and without a clear explanation of how they will be used in the planning process.

4.4 Focus on simulation / visualization

The neighborhood's second public workshop focused on presenting the polices and programs formulated in response to the first workshop, but failed to address the issues of values and identity – the character, look and feel, and sense of place. The Webb Architects process was a traditional planning process of identifying what needed fixing, a litany of policies and programs, and a grab-bag list of everything from a department store to repairing sidewalks. Many of the community-suggested policies and programs were often mutually exclusive or contradicted other community objectives. Moreover, and this is where the community's outrage emerged, there was no attempt to reconcile the inconsistencies and contradictions, nor to give form and reality to these policies by translating them into a 3D representation of the neighborhood using visual simulations.

This is exactly what the ESC did. Taking the policies, words, numbers, and wish-lists, the ESC gave them physical form, illustrating their implications, and asked, "This is what you said, is this what you mean?" For example, in the case of the light rail alternative for the Irvington Commercial Corridor, the recommendations included:

- an intermodal garage for commuters working downtown,
- densities needed to support light rail,
- mixed-use development that included the introduction of new building types into the neighborhood, and
- attracting a large chain store typically found in shopping malls.

The ESC's real-time visual simulations of three light rail development scenarios from moderate to high density – by Houston neighborhood standards) showed radical change. The development scenarios were based on discussions with Webb Architects regarding the results of the first workshop and the level of consensus reached on both program and community identity. Given the response to the simulations and visualizations of the three development scenarios, it was clear there was a considerable gap in perception and expectations between the community and Webb Architects. The jump from the programs and policies to physical design appeared to be a disconnect in the process. First, the simulations and visualizations, by necessity, involved making design decisions, many of which were fundamental, that assumed a value system and (possibly implicit) identity that had not been explicitly discussed with the Near Northside community. Second, poor management of time and lack of clarity of purpose by Webb Architects at the workshop resulted in a limited time to discuss the issues related to each development scenario, and led to the high level of community frustration.

As a result, a Community Preferences Workshop was added. Its purpose was five-fold:

- Reach agreement on the community values and sense of its identity (where it is and where it wants to be);
- Help the community understand the rules of the game such as the City of Houston development regulations;

- Translate the policy and programs presented at the proposed second workshop into a series of principles that would profoundly affect the way the Near Northside neighborhood developed;
- Understand the community's perception of itself as it currently exists; and
- Comprehend the implications of light rail on future development.

In order to both reveal and test the community's perceptions of itself, its common values, and sense of identity, the ESC focused the Near Northside community's workshop on combining a cognitive mapping narrative exercise and 3D simulations and visualizations. The mapping narrative approach was employed rather than a questionnaire, so as not to restrict or channel the residents' responses. Members of the Steering Committee were given a "brief" that was distributed to their constituency. The brief asked each participant to describe through words and images his or her daily experience (Fig. 14). The approach was purposely open-ended, leaving the length, format, and organization to the individual. Themes that had emerged in prior workshops (e.g., walkability, pedestrian-friendly, shopping, housing, landscaping, etc.) were mentioned in the brief, but were not required to be referred to in the narrative (Portugali, 1996).

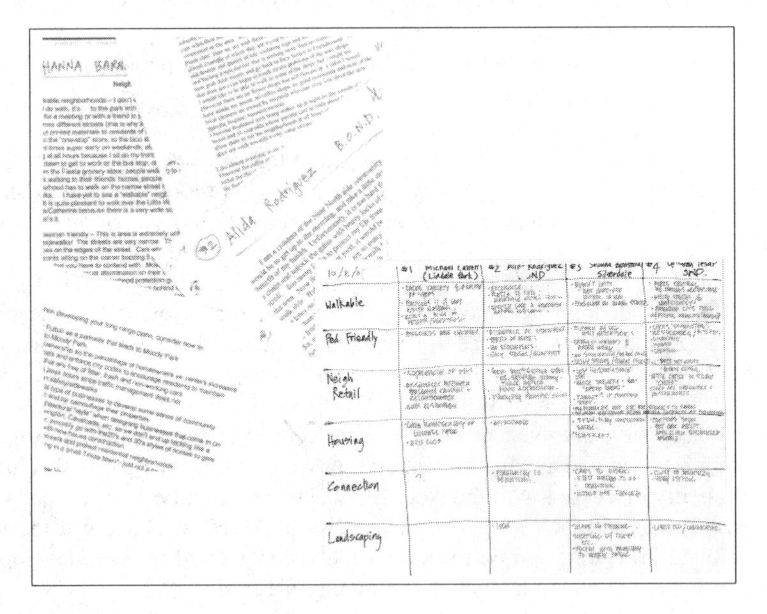

Fig. 14. "Narratives" describing a participant's daily experience and their interpretation.

The narratives were then interpreted for environmental images. Not surprisingly, the environmental images of the residential areas were generally positive while those of the daily shopping experience were uniformly negative. The narratives revealed that shopping was generally done on foot rather than by automobile, in part because of the Latino culture and economics (where most families had one car that was used to get to work in Houston's decentralized environment). As a result, in the Near Northside neighborhood, walking to reach destinations within the community is a way of life. While the narratives identified many strong service and retail destinations, they are dispersed, making them difficult to access on foot. Furthermore, the character of existing development is suburban and based on the automobile. The narratives noted the unpleasantness of negotiating curb cuts, parking lots, and discontinuous sidewalks. They also discussed the lack of shade (e.g., from trees), security (e.g., from lighting), and a place to sit (e.g., from street furniture). Implicit in the narratives was a desire to concentrate and connect compatible activities. How these objectives would be achieved, given form, and agreed upon was the goal of the Near Northside Community Preferences Workshop.

In addition to interpreting the narratives, the ESC produced a set of graphics, to explain the relationship between Houston's parking requirements for specified land-uses, and a photo-realistic real-time 3D digital model to simulate and visualize potential development in the commercial/residential areas, with and without the introduction of the light rail (Figs. 15 a,b). The 3D model was disaggregated into its constituent components (buildings, parking, streetscape, etc.) creating a set of building blocks that could easily be assembled and re-assembled to represent alternatives. The photo-realistic 3D models of retail, office, restaurants, and housing used typical Houston building types, for purposes of familiarity and local context to the participants. The buildings used represented an array of types, sizes, uses, and densities. Photographs of the buildings were texture mapped on the 3D massing model of the building. The use of familiar buildings avoided the issue of architectural design, since the workshop was focused on urban design and the siting and configuration of open/ public spaces and the location of parking. The workshop was concerned with principles leading to the formulation of urban design guidelines, rather than a particular design solution (which in any case is virtually impossible in Houston's regulatory and development culture). As a result, a representative section of blocks, including residential blocks, were selected for simulation and visualization.

The choice of a real-time environment allowed the ESC to simulate a pedestrian's experience of walking from a house to the commercial corridor. It also served the additional purpose of building confidence in the openness of the workshop process, since a participant could locate themselves anywhere in the 3D model rather than be limited to either static images or animations, where the viewers' path and focus are predetermined. It was important that the participants saw the 3D digital model as support for decision-making rather than as a means to manipulate decision-making.

At the workshop, the ESC presented its interpretation of the narratives and how they informed the design of the workshop, and the issues to be discussed and acted upon. The validation of the ESC's interpretation interpretation of the narratives, and their translation into a series of issues dealing with connectivity, continuity, and compatibility was critical to the workshop progress. Without validation by the community, the workshop would have been redirected to better understand the meaning of the narratives. As a methodology for understanding environmental images, the narratives proved to be, according to the participants, a non-threatening and non-manipulative means of inquiry.

The visualization of Houston's parking requirements proved to be an eye-opener for the participants. By visualizing the area of a typical block in the Near Northside neighborhood, consumed by a single land-use and that use's required parking, it became instantly clear what the implications of the parking requirements meant in terms of compact, walkable commercial district. While the development rules were done in the abstract, a single use at a time, the rules were used in the simulations and visualizations to accurately reflect the reality of the regulations.

An additional "reality test" was the simulation and visualization of what are called "big box" uses such as mall-style department stores. While many residents articulated a desire to have one or two in the neighborhood, their sheer size, the amount of parking, and the servicing requirements raised issues regarding pedestrian access, traffic and conflicts with the bordering residential use. Moreover, there were only one or two blocks within the neighborhood commercial areas that could accommodated a big box store. The simulations and visualizations added clarity and substance to the discussion.

Each of the objectives were simulated and visualized using the same set of building blocks, allowing the residents to better understand that, all things being equal, the location of buildings and parking lots on a block greatly affect the pedestrian's experience. In addition, the possible introduction of light rail in the area also raised substantive issues, such as whether or not to "hold the corner" with a pedestrian-oriented building activity, or allowing parking lots to be located at the corners. Typically, a business-as-usual scheme (i.e., automobile-oriented, with parking lots in front of buildings) was contrasted with one that was pedestrian-oriented (i.e., parking lots behind or next to buildings), all other things being equal (Figs. 15 a,b).

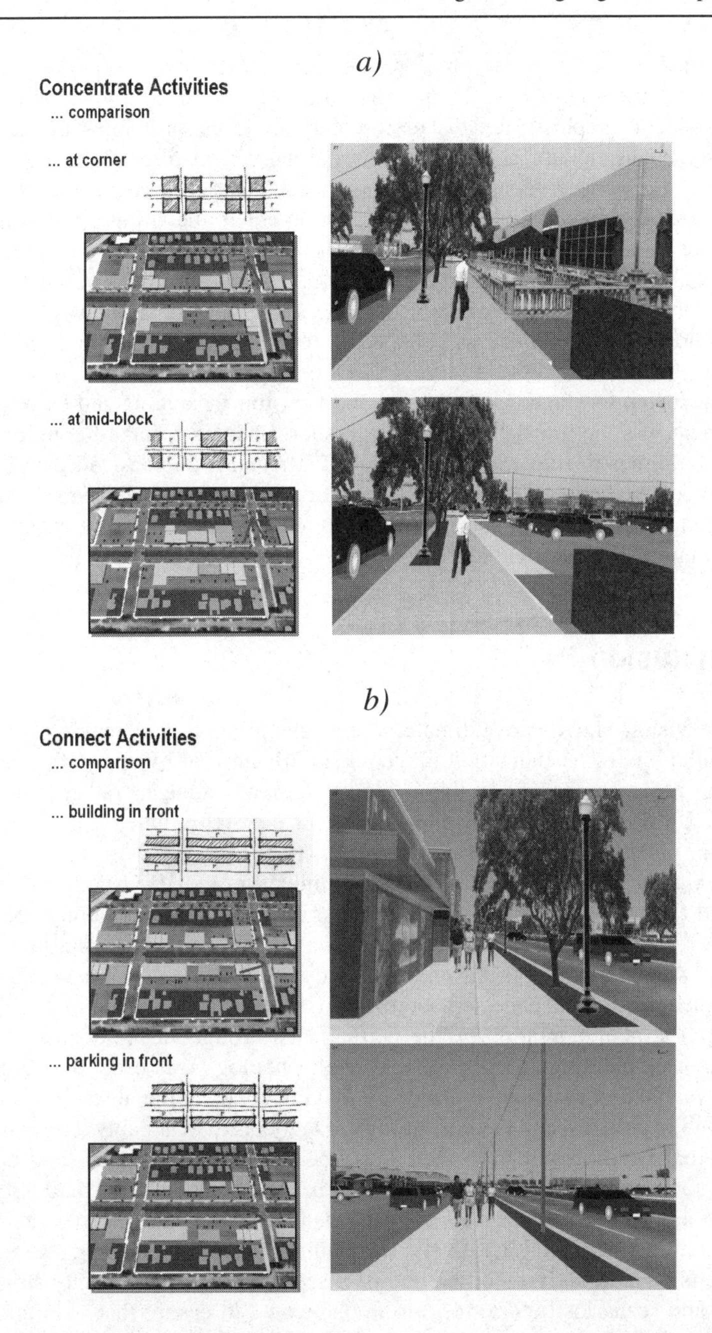

Fig. 15. Snapshots of 'apples to apples" 3D visual simulation of equivalent development along the Irvington Corridor indicating pedestrian friendly versus automobile friendly development patterns as experienced by a pedestrian.

Each of the objectives was presented in side-by-side comparisons as well as in synthesized virtual reality walk-throughs. Interestingly, the side-by-side comparison allowed for a sophisticated discussion of urban design principles and aesthetics based on the participants' analysis of their alternatives relative to whether and why they were auto-centric versus pedestrian-centric. What became clear as the workshop progressed was that, all other things being equal, the siting of buildings and parking lots either favored the driver or the pedestrian. This became a fundamental choice that participants were asked to vote on. The question was posed as an "either-or" or "both-and" set of choices. The final vote unanimously favored a neighborhood whose identity was shaped by the pedestrian experience and the potential of light rail.

The next step for the consultant team and Houston's Planning and Development Department was the translation of the workshop findings/points of consensus into design guidelines to implement the vision. Participants emphasized the power of three dimensional computer images of the place, both static and dynamic, to effectively guide the discussion and lead to future development that is consistent with the communities' values and identity.

5. Conclusion

All three vision plans were ultimately about people and placemaking. It is here that simulation and visualization played a significant role by supporting physical design as a form of inquiry. In this modality, design is a means rather than an end in itself. Unlike the traditional planning linear hierarchy, where physical design follows or illustrates policy, physical design *is* policy.

In the three case studies the use of digital simulation and 3D visualization greatly enhanced the public decision-making process and building community consensus when they were an integral part of the planning and visioning process in two fundamental ways. First, the visual simulations were not used to "sell the plan/vision" but as a means to enhance the democratization of planning, by using the technology to inform the plan's creation by the participants throughout the decision-making and consensus-building process. Second, the technology was used to help citizens *create* liveable and sustainable places by making concrete the abstractions of scenarios, public policies, and the like through 3D models that palpably represented the place(s) that would result from their implementation. Unlike maps and physical models, participants could place themselves in the 3D models and randomly walk through them at eye level, as well as query the underlying data. Moreover, the visual simulations were used iteratively, responding to participant's suggestions. The 3D models were quickly modified and new scenarios created, visually simulating policies and scenarios before they are implemented to ensure that -- to quote Dr. Seuss's Horton the elephant -- "I meant what I said and I said what I meant."

The Baltimore Vision 2030 demonstrated that digital simulation and 3D visualization could be applied to help a regional citizenry, living in cities, towns and in rural areas, to determine the future of their region, by evaluating the implications of a range of agreed-upon growth patterns and scenarios. The Master Planning Initiative in Santa Fe was also faced with a question of "how to grow", but in the much more limited geography of a new district in Santa Fe. There, the visual simulations played a critical role in formulating the development principles that would guide future growth, achieve consensus, and provide the foundation for the design of zoning regulations consonant with the community's vision. The experience of the Near Northside Economic Revitalization Planning Process demonstrated the role simulation played in creating a new neighborhood identity based on enhancing the pedestrian's walking experience and the support of light rail in a city reliant on the automobile. It also showed how the process can go awry when the use of visual simulation has not been carefully integrated in the public participation and decision making process. In each case study physical design, mediated through visual simulation, was employed to assist citizens better understand their own values and sense of individual and group identity, stimulate informed discussion about design and placemaking in concrete terms rather abstractions, and lead to community consensus.

References

Appleyard, D. (1979). The Environment as a Social Symbol: Within a Theory of Environmental Action and Perception, *Journal of the American Planning Association* 45 (2):143-145.

Casey, E. S. (1997). *The fate of place- A philosophical history*. University of California Press, Berkeley.

Lowenthal, D. (1985). *The Past as a Foreign Country*. Cambridge University Press, London.

Portugali, J. (ed.)(1996). *The Construction of Cognitive Maps*. The GeoJournal Library. Kluwer Academic Publishers, Dordrecht.

Relph, E (1976). *Place and placelessness*. Pion, London.

Reps, John W. (1979). Cities of the American West: A History of Frontier Urban Planning. Princeton, New Jersey.

Longo, G. (1996). *A Guide to Great American Public Places*. Urban Initiatives, New York.

Morgan (1996) New York – New Jersey Highlands Demonstration Planning Prospect.

Yaro, R. D., and T. Hiss. (1996). A region at risk: The third regional plan for the New York-New Jersey-Connecticut metropolitan area. Washington, DC: Island Press.

Notes

The core technology employed by the ESC was ArcViewGIS, a spatial database whose output is typically expressed in maps, charts, and tables, none of which by themselves either resemble the world of everyday experience nor are easily accessible to lay people. It is axiomatic that people experience the world in three-dimensions, in time, and in motion. In response to this need the ESC has developed software and applications that seamlessly integrate words, numbers, maps, and images in a real-time 3D environment (e.g., CommunityViz™). Further, the tools have been designed to support both deductive reasoning (analysis) and inductive reasoning (design, what ifs).

Vision 2030: Shaping our Region's Future Together background

Vision 2030: Shaping our Region's Future Together was an initiative of the Baltimore (Maryland) Regional Transportation Board of the Metropolitan Planning Organization for the Baltimore region. Additional support was provided by the Baltimore Metropolitan Council, a private non-profit regional planning agency and the Baltimore Regional Partnership, an alliance of civic and environmental groups that share a common agenda of enhancing quality of life through community revitalization and environmental protection.

ACP –Visioning and Planning (ACP) with its expertise in conducting regional visions was the lead consultant. The ESC was a member of the ACP team of five subconsultants.

Southwest Santa Fe: City/County Master Plan Initiative background

The study was initiated by the Santa Fe City Council. Subsequently, after discussions with the County and members of the public, the planning area was expanded to coincide with the area undergoing development pressure. The Southwest Santa Fe Planning Area falls within both City and County jurisdictions and, as a result, two entirely different sets of land development regulations. A Steering Group, representative of the Southwest Santa Fe community, provided project oversight and the Planning Department of Santa Fe provided logistical and technical support.

The prime consultants were ACP and the ESC. Local consultants contributed their knowledge of the Santa Fe region and assisted in the implementation of the public outreach and participation program.

Near Northside Economic Revitalization Planning Process background

The Near Northside Economic Revitalization Planning Process was an initiative of the City of Houston. It was supported by grants from the Federal Highway Administration (FHWA), U.S. Department of Housing and Urban Development (HUD), and the Main Street Revitalization Project, with contributions from Avenue CDC, a local community development corporation. The prime consultant, Webb Architects Associates was selected to prepare the plan. The work also encompassed two other parallel efforts: 1) the "Houston Neighborhood Market Drill Down" by Social Compact Inc, and 2) the Community Preferences Workshops, conducted by the ESC and funded by HUD's Community Technology Initiative grant.

Planning Support Systems Evolving: When the Rubber Hits the Road

Richard K. Brail

I study nuclear science
I love my classes
I got a crazy teacher
He wears dark glasses
Things are goin' great
And they're only gettin' better
I'm doin' alright
Gettin' good grades
The future's so bright
I gotta wear shades
Pat MacDonald, 1986

Abstract. Planning support systems (PSS) have moved from concept to application. One of the core assumptions of PSS is that these computer-based systems can be applied in actual planning situations and found useful as decision support tools. Based on previous experiences with applied computing efforts, we need to think carefully about how best to support successful implementations of PSS. Lessons can be learned from both experiences with large-scale urban models and with the four-step urban transportation planning process. We examine the potential of PSS across four dimensions – data availability, acceptance and support, ease of use, and appropriate and useful output.

1. Introduction

This is an exciting time for simulation modeling and visualization tools in planning and public policy. Planning support systems (PSS) have moved from concept to application. Is this future so bright that we need to buy those shades? This paper will explore the potential for PSS, assessing the need to visit the nearest kiosk.

Described by Klosterman as a "fully integrated, flexible, and user-friendly system," PSS would assist the planner in selecting and applying appropriate projection and impact models, where outputs would be displayed graphically as well as numerically (Klosterman, 1997:52). Ideally, such a PSS would operate in real-time and support group decision processes, contributing to the goal of collective design. The PSS concept has been implemented in a number of commercially available packages (Ospina and Stephens, 2004). Scenario 360 from CommunityViz is the new ESRI ArcGIS version of perhaps the fullest implementation of PSS in planning (www.communityviz.com). The package integrates scenario development, impact analysis, and 3d visualization. The original path-breaking CommunityViz implementation also contained an innovative agent-based projection model, Policy Simulator (Kwartler and Bernard, 2001). There are other PSS packages commercially available. *What If?* projects alternate future land use scenarios based on the integration of market forces and policy choices and offers a sophisticated ranking and allocation mechanism (Klosterman, 2001). INDEX is an ArcGIS based PSS that nicely combines mapping with graphic and numeric indicators and contains a documented set of impact equations (Allen, 2001; www.crit.com). Already established models such as DRAM/EMPAL have been reconstituted as PSS. (Putman and Chan, 2001)

One of the core assumptions of PSS is that these computer-based systems would be applicable to actual planning situations and "useful" as decision support tools. Based on previous experiences, caution flags abound. The history of large-scale operational urban models (LSUMs)[i] shows the problems inherent in offering computer-based guidance to government. While Wegener was optimistic in conclusion about the viability of urban models, he also indicated that: "Nowhere in the world have large-scale urban models become a routine ingredient of metropolitan plan-making." (Wegener, 1994:17) Whether one agrees or disagrees with Lee's requiem paper (1973), the arguments presented were more than a warning shot across the bow. They were directed at the mainsail, and sent more than a few ships to seek a quiet port.

Both model development and theory have evolved in the three decades since Lee's paper (Batty, 1994; Timmermans, 2003). There are a number of healthy large-scale modeling efforts underway. Focusing on the United States we can point to UrbanSim (Waddell, 2002), the California Urban Futures suite (Landis, 2001) and Putman and Chan (2001). However, questions remain about the prospects for the full integration of planning decision support into public sector organizations. As we will discuss, a major proposal for a statewide PSS in the United States met an unhappy fate.

Our question then: under what conditions could a decision support system contribute to operational planning on a continuing and widespread basis? To explore this question we need to better describe the term PSS. These planning decision support systems have as their purpose either projection to some point in the future or estimation of impacts from some form of development. The ideal system would have an explicit visualization component. Existing packages contain a variety of visual representation alternatives, ranging from detailed 3d "flying" landscapes to 2d GIS-based maps to static images. The systems also would have the capacity to operate in real time and to support group decision processes.[ii]

There are issues of scale in PSS development. In the broadest sense the regionally focused LSUMs are planning decision support tools. There is also a second class of systems that have emerged, focused on localized impact modeling and visualization

---

[i]. From Lee (1994).

[ii] The recent San Francisco conference, Tools for Community Design and Decision Making: Working Session V, held in December 2003 and led by Ken Snyder from PlaceMatters.com, was an important effort at using a variety of PSSs in a "digital charette" process. Teams using different packages were given realistic planning situations and asked to do an intense problem solving exercise in a three-day period.

at the sub-regional scale, but not projection to future years. Examples of these visualization and impact models (VIMs) are INDEX and Scenario 360 by CommunityViz. These VIMs operate on smaller urban environments – sites, neighborhoods, and small cities. These distinctions are always messier than we might like. INDEX literature promotes its use at a variety of scales, and the earlier CommunityViz originally contained Policy Simulator, an agent-based simulation model. We will focus our comments on existing and emerging systems that claim usefulness to actual planning situations, do either projection or impact analysis, contain a visualization component, and have connections to interest group processes as a goal.

While PSS is still finding its way, there are other model systems that have achieved recognition and success. We will look at one of these successful implementations, the four-step urban transportation modeling system (UTMS), to develop a framework for exploring the necessary evolution of PSS. Currently under serious scrutiny for revision, UTMS has a story to tell.

2. UTMS: a historical perspective

The traditional four-step urban transportation modeling system (UTMS) of trip generation, distribution, model split and traffic assignment has a long history in modeling terms, dating back over half a century. Currently available commercial models offer default parameters, relative ease of use, guaranteed calibration, and wide scale acceptance. These models have been used in virtually every major metropolitan area in North America, as well as around the world. UTMS has also been widely critiqued with proposals for improvements, extensions and alternatives.[iii] The new agent-based competitor is TRANSIMS. Yet the traditional structure is widely used to this day, and represents a core methodology in metropolitan transportation planning.

The UTMS success story is rooted in four elements – data availability, governmental support and acceptance, appropriate and useful output, and relative ease of use for calibration and prediction purposes. UTMS is quite data-hungry, and it is rather amazing that the very expensive home interview surveys that provide so much essential information are still widely performed. Virtually, every major metropolitan area has performed a household survey since 1990 (Cambridge Systematics, 1996). These surveys cost upwards of a million dollars for sample sizes between two and twelve thousand households. The home interview study was not viewed favorably in the early years. An analysis of area transportation studies performed in the 1960s in the US (Levin and Abend, 1971) confirm this:

> *The cooperating federal and state highway agencies entertained more limited but clearer expectations ... They were interested in ... the development of a low-cost transportation model that would permit future studies to be conducted with considerably less data collection – in particular, fewer expensive road and home interviews. (Levin and Abend, 1971:4)*

[iii] See the Travel Model Improvement Program web site (http://tmip.fhwa.dot.gov/).

In spite of these early expectations, the home interview surveys have continued. Data availability for UTMS is required on both the travel demand and network supply sides. Accurate highway and transit network coding is essential for metropolitan areas. In spite of the early concerns about costs, household and corollary studies supporting UTMS continue. While funding agencies might wish for cheaper solutions in gathering transportation data, the evidence clearly points to continuing substantial data collection expenditures supporting UTMS.

There has also been strong continuing governmental support for UTMS, both the traditional models and proposed revisions. The healthy research budgets of the Bureau of Public Roads and Federal Highway Administration in the US have fed these transportation models. The early mainframe version, Urban Transportation Planning System (UTPS), was developed with federal dollars. The workstation versions followed and devolved into commercial products offered by multiple vendors. Federal and state funding supported excellent training tools, including software, training materials, and workshops.[iv]

UTMS also produced useful outputs. The development of robust traffic assignment models meant that engineers and planners could project traffic volumes with some degree of confidence, providing a base for infrastructure decisions. The expenditures were not trivial. Billions of dollars were allocated to the construction of new roadways. If UTMS has not been developed to provide traffic volumes and associated congestion measures, highway engineers and planners would have had to invent a similar system as a base to design decisions. In the broadest sense, UTMS provided a useful set of transportation system indicators. These indicators fed investment decisions.[v]

Finally, UTMS worked. The craftiness, and self-serving side, of the calibration and prediction process in UTMS is well known. The singly-constrained gravity model typically used requires a balancing routine for trip productions and attractions, as well as providing the capacity to modify specific zone-to-zone socioeconomic factors to ensure better calibration results. The commercial versions of UTMS generated reasonable numbers for planning and engineering purposes.[vi] The four-step process was disseminated in the literature, in universities, and in short courses. Professors and consultants taught it[vii], and planners and engineers used it. By all accounts UTMS was a success in spite of methodological flaws. Its history provides the basis for a useful framework in evaluating current developments in planning support systems.

[iv] See the training materials developed by the National Highway Institute (2003) and the National Transit Institute.

[v] It is interesting that the recent broadening of transportation policy objectives in the United States, based on ISTEA and subsequent legislation, appears to be diminishing the influence of modeling in the transportation planning process (Kramer and Mierzejewski, 2003).

[vi] An example of this continuing interest in the validity of UTMS packages is the paper by Yu, Yue and Teng (2003) comparing EMME/2 and QRS II.

[vii] Meyer and Miller (2001) is perhaps the best introductory text.

3. PSS evaluation framework

We will look at PSS in the context of the four criteria outlined above:

- Data
- Acceptance and support
- Ease of use
- Appropriate and useful output

3.1. Data

The first dimension is data availability. It is here that our current generation of planning support systems, both LSUMs and VIMs, is often quite comfortable. The advent of GIS and its wide availability has provided a firm foundation for display and analysis. Typical layers such as land cover and land use, roadway networks, water and wetlands, and environmentally sensitive land use are widely available. While parcel layers are less likely, they are slowly coming. The move to agent-based models has complicated the situation. Policy Simulator, the agent model in CommunityViz, required a wide variety of local databases in its early iterations. Successive iterations of the model required less data as input.

The PSS community could take a clue from transportation planning, which appears to have little fear of requiring extensive date inputs and of offering automatic data generation tools. TRANSIMS, the agent-based transportation model, is very data-hungry. The roadway network at the core of TRANSIMS includes information on all local streets, land use on each block face, all lane connections at intersections, intersection signalization, and transit system schedules, routes, and stops (Los Alamos National Laboratory, 2002). There has been clear recognition of these data needs with the development of automatic generators for various data inputs (IBM, 2003).

There are data issues in PSS at two levels -- data availability for smaller, poorer or more rural areas, and 3d visualization. Sprawl in the United States is continuing at a rapid pace. It has been estimated that from 2000 to 2025 18 million acres will be converted to residential and non-residential uses. Under an "uncontrolled" scenario fully 14 million of these acres will be either agricultural or environmentally fragile land (Burchell, et al., 2002:194-195). Ironically, it appears that exactly in those areas where development will occur and a PSS might help, there may be fewer planners and less collected data.

The 3d visualization data issue centers on the quality of the artificial environment that can be created. Based on my experiences with graduate classes doing 3d landscapes, there is a need for rapid prototyping of realistic and site-specific building objects. The landscapes that students created took significant time, and the learning curve was steep. How a local planner could take the time or commit the resources is a major question. Student experience does show that reasonably realistic building objects can be constructed in a few hours after training: the problem is that a small landscape can contain a large number of objects.

3.2. Acceptance and support

UTMS has achieved wide acceptance. Considerable resources are spent by the public sector on both developing data inputs and running the various models. When we look at planning support systems the picture is much more mixed. There are three levels on which we can look at the acceptance and support issue.

- Feasibility
- Mandate
- Peer involvement

Feasibility- Can we do the model with the resources available? This feasibility question sits very much at the root of local acceptance. The early transportation modeling efforts in the 1950s and 1960s worked well enough (Creighton, 1970; Meyer and Miller, 2001). Resources were available and the required outputs produced. The early land use models were another story. Large-scale land use modeling efforts of the period were not well received (Levin and Abend, 1971; Lee, 1973). The reasons why reflect their relative position in the hierarchy. Modeling complex land use decisions proved to be difficult and the early models were not always up to the job. The early LSUMs were laboratories where experiments were tried, sometimes discarded and sometimes failed. Not a happy situation for a client waiting at the door for a finished product.

There also is a scale and locational issue. PSS and embedded simulation models can be quite resource intensive. Are they only the purview of larger and richer places? As we have suggested, urban growth in the US will invade the more rural communities with poor infrastructure and questionable planning resources. These communities have limited abilities to use these more complex tools in their local decision-making. Local governments with limited resources need simple and workable tools to assist the decision process. It is still unclear to me if it is even reasonable to assume that locally focused VIMs will be useable in places where significant growth is expected. The research community is quite comfortable with the more complex and data-intensive modeling and simulation efforts. Timmermans (2003) argues that complex phenomena require complex models. TRANSIMS is one extreme example. Do we need to think more about simpler models in terms of accuracy, predictability and usability, where we do a marginal analysis of the expected benefits and costs of complexity? The original design of UTMS was flawed, but offered useful outputs. Was it so flawed that it should not have been used, or were the outputs reasonable enough for planning purposes?

Mandate - The Intermodal Surface Transportation Efficiency Act of 1991 (ISTEA) created a mandate that metropolitan areas test alternative development scenarios, fostering the use of LSUMs. Air quality considerations are central to transportation planning in the United States. Congress passed the Clean Air Act Amendments of 1990, which strengthened the requirements that urban areas respond to air quality standard violations. The 1991 ISTEA law followed by establishing a 1.5 percent set-aside from a variety of federal transportation funding categories to support data and modeling efforts in metropolitan areas. In the earlier period of the 1950s and 1960s there was the mandate to develop land use and transportation models feeding highway construction programs. The federal mandate in the 1990s focused on models to support alternative metropolitan development scenarios. Where is the current federal mandate for integrated planning support systems? Pieces of government interest in model development exist, particularly at the state and local level (Waddell, 2002), but it is scattered. We can hope that a more extensive commitment to planning decision support will emerge, based on a broader national interest in sustainable development and smart growth.

Peer involvement - The modeling fraternity is not monolithic and thrives as much on competition as cooperation. Urban transportation researchers created a set of accepted models that then were developed and modified as needed. At one time we could choose among MINUTP, TRANPLAN, TMODEL2 and other commercial packages. While different in design and implementation, these packages used a common language and conceptual framework. While recent consolidation has occurred among vendors, the basic UTMS framework continues. TRANSIMS is the new federally supported regime, and we might expect a substantial future effort to push this model to the forefront.

The commonality that permitted teaching and workshops of UTMS concepts and methods has not existed for LSUMs or VIMs. Various developers have offered workshops and training, but the broad commitment to a common conceptual framework is still in flux. Wegener (1994) suggests that the majority of models reviewed employ some variant of random utility theory, yet vary widely in their treatment of markets. Timmermans, in his excellent review of integrated land use and transport models, argues:

> *... operational models are still largely based on traditional location theories and models that may have been adequate to describe traditional cities .... but that seems inadequate to describe the evolution of modern cities, dominated by service industries and information technology. (Timmermans, 2003)*

The theory is inadequate and the models vary widely in design. The regional planning organization faced with a desire to develop projections of population, jobs and land use to a future year faces an interesting task. In another venue, the choice of a GIS vendor to develop an enterprise system for a local government was often done through a "shootout." The GIS vendors competed across a set of defined tasks. There have been some recent research studies focusing on the comparison of LSUMs (US Environmental Protection Agency, 2000; Hunt et al., 2002). These incipient shootout efforts are only the beginning, and along with others will begin to push the modeling envelope.

3.3. Ease of use

How likely is it that local and regional planners will be able to run any form of PSS? The idealized PSS proposed by Klosterman talks of real-time interactive modeling. Obviously the need to do real-time analysis is keyed back to the nature of the problem. While LSUMs look to distant years at a regional scale, the more locally focused VIMs are designed to respond more immediately to issues. What are the impacts of that new shopping center? How will that new residential development look? The promise of PSS as a real-time decision tool is that local planners and analysts will be able to use these packages. It is unclear how often a local planner will ever run LSUMs. In transportation planning the UTMS consultant firms under contract often run commercial packages, used at both metropolitan and local scales. The case may be different for more locally focused PSS.

The smaller scale VIMs are promoted and sold to local public sector planners and consultants, and are designed to be easy to use. Both Scenario 360 by CommunityViz and INDEX are available for purchase. My own experiences with the earlier version CommunityViz are a reminder that technology challenges. I have taught two graduate seminars using CommunityViz. In each the students were exposed to the different components of the package. The students quickly gravitated to the 3d component and spent goodly amounts of time developing building objects. They were able to get their hands around the concrete creation of a 3d world. The agent-based modeling and impact analysis components were of less immediate interest. Students were willing to work with impact equations and agent models, but found formula building and model manipulation challenging. While a wonderful teaching device, developing one's own impact equations is not a priority for the practicing planner or policy analyst. There is an important need to develop and document sets of impact equations useful in VIMs that are both generally accepted in the professional community and widely disseminated. As illustration, INDEX contains a broad range of impact models along with documentation.

3.4. Appropriate and useful output

It is difficult for an operating agency to know how to deal with models that claim that they are designed for understanding, not prediction, or that the results are an approximation. While the model builder may desire to put cautionary notes on any model outputs, the planner needs numbers. The UTMS computer packages produce useful outputs at the appropriate scale. At both regional and local scales the UTMS models generate the traffic volume and transit utilization data required.
Ideally there would be a corresponding match between the scale of LSUM output and decision-making. LSUMS performed for a regional agency need direct connection to local governments within the region. In the United States these connections vary widely. While there are wide variations across states, land use decisions are predominantly local and made at the municipal level. This mismatch creates problems because regional and local planners each live in different worlds. While many planners would agree that we should all take a regional perspective, the reality is much different.

There can be serious ramifications of a mismatch between vision and acceptance of planning support systems for local government. The sad case of PSS in New Jersey points to the issues that occur over implementing a potentially important decision tool. In the late 1990s staff at the New Jersey Office of State Planning (NJOSP) proposed a planning support system for local governments in the state. There are 566 municipalities in New Jersey, and NJOSP proposed to develop a VIM for every single one, from large cities to rural hamlets. The proposed PSS would help towns deal with the immediate and cumulative impacts of development proposals. The statewide system would contain GIS-enabled state regulations and maps for wetlands, runoff, etc. as well as a host of local impact models. Initial funding was found, based on presentations to high-level state officials. However, the proposed PSS program never developed broad support among decision-makers. NJOSP was also the developer of the State Plan for Development and Redevelopment, and the lead agency for state-level planning. Regardless, in 2001 the entire office was disbanded and the staff fired. There is some speculation that its expansive (and perhaps unrealistic) view of the potential role of planning support systems contributed to the demise, but this will never be known. Selected elements and staff have been reconstituted as the Office of Smart Growth, but the statewide PSS effort is moribund. There are current selected efforts in some communities to implement PSS, but the grand plan is no more.

4. Conclusion – do we need those shades?

We can summarize this discussion with the following points. First, planning support systems have not reached their full potential in assisting public sector decision-making. The reasons vary for the two varieties of PSS, LSUM and VIM. The LSUM issues center on moving to a common theoretical framework and application, which engenders broad government and peer support. The VIM packages are simply too early in the development process to offer planners satisfying and useful outputs that they themselves can generate.

Second, the story of the transportation demand modeling and the successful application of UTMS computer-based packages offer lessons for PSS. The broad support of a common methodology, the availability of extensive governmental resources, and user acceptance were key factors. UTMS offered identifiable and needed outputs for infrastructure investments. The story with LSUMs is much more muddy – many competing models, uneven levels of support, and disconnect between outputs and planning decision-making. We are only at the beginning of the story of VIM use and acceptance. Planning in the trenches will require further developments in model consolidation and extension for both LSUMs and VIMs.

Third, data issues in supporting PSS appear to be manageable. As we suggest, there appears to be continuing willingness to fund expensive travel surveys for transportation. GIS applications will continue to expand across urban and rural areas. Developers of agent-based models, including TRANSIMS and the Policy Simulator module in CommunityViz, have been cognizant of the severe data requirements in generating synthetic households and travel behavior, and have offered automated alternatives.

Fourth, we face the educational issue. Planners are becoming increasingly sophisticated in the use of GIS, and planning students have taken to GIS with abandon. Simulation models and analytic tools are less interesting – they require mathematical sophistication and offer less concrete application. Visualization tools do capture attention, and sometimes provide a way into PSS.

Fifth, the web will offer promise and potential, but only if planners learn to use these tools. Full credit must be given to Paul Waddell for a downloadable UrbanSim (www.urbansim.org). But then who will run these models locally? We are back to the education – graduation education, professional workshops and instructional software.

The full potential of planning support systems has not been realized. Yet, research continues, technology evolves, and theory evolves. We may yet need those shades.

References

Allen, E. (2001). INDEX: Software for Community Indicators in *Planning Support Systems: Integrating Geographic Information Systems, Models, and Visualization Tools*, (R Brail and R Klosterman, Eds.), ESRI, Redlands, 229-261

Batty, M. (1994). A Chronicle of Scientific Planning: The Anglo American Modeling Experience, *Journal of the American Planning Association* 60(1) 7-16

Burchell, R, et al. (2002). *Costs of Sprawl – 2000,* Transit Cooperative Research Program Report 74 (National Academy of Sciences, Washington).

Cambridge Systematics, (1996). *Scan of Recent Travel Surveys* (Cambridge, Cambridge Systematics

Creighton, R. (1970). *Urban Transportation Planning* (University of Illinois, Urbana)

Hunt, J.D., Johnston, R., Abraham, J.E., Rodier, C.J., Garry, R., Putman, S.H. and de la Barra, T. (2002). Comparisons from the Sacramento Test Bed, *Transportation Research Record* 1780 53-63.

IBM,(2003), *TRANSIMS-DOT* TRB Briefing Document (http://www.transims.net)

Klosterman, R. (1997). Planning Support Systems: A New Perspective on Computer-Aided Planning, *Journal of Planning Education and Research* 17(1) 45-54

Klosterman, R. (2001). The *What If?* Planning Support System, in *Planning Support Systems: Integrating Geographic Information Systems, Models, and Visualization Tools* (R Brail and R Klosterman Eds.), ESRI, Redlands, 263-284

Kramer, J., and Mierzejewski E. (2003). Innovations in Long-Range Transportation Planning: Observations and Suggestions, *Transportation Research Record* 1858 1-8

Kwartler, M. and Bernard R. (2001). CommunityViz: An Integrated Planning Support System in *Planning Support Systems: Integrating Geographic Information Systems, Models, and Visualization Tools* (R Brail and R Klosterman Eds.), ESRI, Redlands, 285-308

Landis, J. (2001). CUF, CUF II, and CURBA: A Family of Spatially Explicit Urban Growth and Land-Use Policy Simulation Models, in *Planning Support Systems: Integrating Geographic Information Systems, Models, and Visualization Tools* (R Brail and R Klosterman Eds.), ESRI, Redlands, 157-200

Lee, D. (1973). Requiem for large scale urban models, *Journal of the American Planning Association* 39 163-178

Lee, D. (1994). Retrospective on Large-Scale Urban Models, *Journal of the American Planning Association* 60(1) 35-40

Levin, M and Abend, N. (1971). Bureaucrats in Collision: Case Studies in Area Transportation Planning (MIT Press, Cambridge)

Los Alamos National Laboratory,. (2002). *TRansportation ANalysis SIMulation System (TRANSIMS)* Portland Study Reports, Volumes 1-7

Meyer, M. and Miller, E. (2001). *Urban Transportation Planning* (McGraw-Hill, New York)

National Highway Institute, (2003). *Introduction to Urban Travel Demand Forecasting, Participant's Notebook* NIH Course No. 152054 (Federal Highway Administration, US Department of Transportation, Washington)

Ospina, M. and Stephens, R. (2004). World Beaters, *Planning* 70 (7) 28-31

Putman, S. and Chan S. (2001). The METROPILUS Planning Support System: Urban Models and GIS, in *Planning Support Systems: Integrating Geographic Information Systems, Models, and Visualization Tools* (R Brail and R Klosterman Eds.), ESRI, Redlands, 99-128

Timmermans, H. (2003). The Saga of Integrated Land Use-Transport Modeling: How Many More Dreams Before We Wake Up?, Presented at the 10th International Conference on Travel Behavior Research, Lucerne, August

US Environmental Protection Agency, (2000). *Projecting Land-Use Change: A Summary of Models for Assessing the Effects of Community Growth and Change on Land-Use Patterns* EPA/600/R-00/098, prepared by Science Applications International Corporation (Office of Research and Development, Washington)

Waddell, P. (2002). UrbanSim, Modeling Urban Development for Land Use, Transportation, and Environmental Planning, *Journal of the American Planning Association* 68 297-314

Wegener, M. (1994). Operational Urban Models: State of the Art, *Journal of the American Planning Association* 60(1) 17-29

Yu, L., Yue, P. and Teng, H. (2003). Comparative Study of EMME/2 and QRS II for Modeling a Small Community, *Transportation Research Record* 1858 103-111.

Index

O

P

Q

R

S

T

U

V

W

Printing and Binding: Strauss GmbH, Mörlenbach